POLYMERS

The Origins and Growth of a Science

Emperor Wilhelm, (in the white coat) with his generals, watching his car being fitted with the first synthetic rubber tires, in 1912. (Courtesy Bayer A.G.)

POLYMERS

The Origins and Growth of a Science

HERBERT MORAWETZ
Polytechnic Institute of New York
Brooklyn, New York

A Wiley-Interscience Publication
JOHN WILEY & SONS
New York / Chichester / Brisbane / Toronto / Singapore

Allen County Public Library
Ft. Wayne, Indiana

Copyright © 1985 by John Wiley & Sons, Inc.

All rights reserved. Published simultaneously in Canada.

Reproduction or translation of any part of this work beyond that permitted by Section 107 or 108 of the 1978 United States Copyright Act without the permission of the copyright owner is unlawful. Requests for permission or further information should be addressed to the Permissions Department, John Wiley & Sons, Inc.

Library of Congress Cataloging in Publication Data:

Morawetz, Herbert.
  Polymers: the origins and growth of a science.

  "A Wiley-Interscience publication."
  Bibliography: p.
  Includes index.
  1. Polymers and polymerization—History. I. Title.
QD381.M663   1985   547.7′09   84-26996
ISBN 0-471-89638-1

Printed in the United States of America

10 9 8 7 6 5 4 3 2 1

To Herman Mark

for the inspiration of his example
for the warmth of his friendship

Preface

In August 1980 I received a letter from Dr. Magda Staudinger, who wrote that in view of a widespread interest in the history of macromolecular chemistry it would be desirable to have a survey of its early developments published for the hundredth anniversary of Hermann Staudinger's birth. She asked me whether I would be willing to undertake such a task since I "had experienced the whole development." In my response I pointed out that far from having been a witness to the "whole development" of polymer science, I had my first contact with it at a time when most of the essential foundations had been laid, but I admitted that I have occasionally toyed with the idea of writing such a history, although I would be unable to do much about it for Hermann Staudinger's centenary in 1981.

My eventual decision to embark on this undertaking was stimulated by various considerations. I was greatly influenced by the example of J. T. Edsall and J. S. Fruton, two eminent biochemists, who had become equally eminent historians of their discipline. After reading with great pleasure R. Olby's *The Path to the Double Helix*, I became increasingly troubled by the contrast between the detailed documentation of the origins of molecular biology and the lack of a corresponding description of the emergence of polymer science. I was also concerned about solving my very personal problem: How would I keep intellectually alive at the time when I would have to give up or restrict my career as a teacher and investigator? Maybe the history of polymer science would provide a continuing fascination.

Dr.Magda Staudinger helped me greatly by sending me ten volumes of her late husband's collected works—this saved me many trips to the library. One of the unexpected benefits of this project turned out to be the numerous sessions with Herman Mark; throughout our association of more than thirty years I had never seen so much of him. Listening to his recollections was frequently a revelation; his memory of events lying back sixty years or more never ceased to amaze me. Beyond doubt I was helped by my fluency in reading German and French and by the circumstance that I was working in New York with its outstanding library facilities. I found out quickly that there is no substitute for reading every reference cited—second-hand citations are incredibly unreliable. Unfortunately, the insistence on citing only publications I had read has led undoubtedly to some neglect of the Russian and Japanese contributions; for this I apologize but it was a price I had to pay if I was to cite only what I knew in detail.

I owe a great debt to H. Mark, J. A. Moore, and M. Schlecht for having read all of the manuscript and for having pointed out errors or passages which needed clarification. A major part of the manuscript was read by R. C. Schulz, J. J. Hermans, J. T. Edsall, M. Gordon, R. Ullman, and B. H. Zimm and I am grateful for their comments. Many other colleagues and friends gave me valuable suggestions—often providing an obscure but revealing piece of information which I would have been unlikely to come across. I was thus aided by G. Allen, N. Bikales, J. Blackwell, R. Brill, W. Burchard, V. Crescenzi, J. V. Crivello, C. M. Dannenberg, H. Eisenberg, F. R. Eirich, J. S. Fruton, J. Hengstenberg, J. H. Hill, E. Katchalski, S. Krimm, H. Kuhn, J. B. Lando, H. Logemann, R. Lontie, S. Lifson, H. Markovitz, C. S. Marvel, P. H. Marchessault, J. Meissner, H. W. Melville, K. M. Mislow, L. Monnerie, F. Montanari, P. W. Morgan, S. Narang, P. Pino, E. Trommsdorff, O. Vogl, B. Vollmert, and W. T. Winter.

In the summer of 1983 I visited Germany to inspect some of the sources in that country. I was most graciously received at the Deutsches Museum in Munich by H. Grätz, who allowed me to study for a full day the Staudinger Archive. Dr. C. Priesner, who is associated with this archive, has been most kind in responding at various times to my inquiries. Later I visited BASF, Höchst, and Bayer and received the most cordial cooperation and a wealth of material at all three companies. The archives of Höchst are most remarkable—they certainly constitute the most valuable source of the history of the polymer industry. I am also greatly indebted to H. Martin of the Max Planck Gesellschaft für Kohlenforschung for having made available to me material pertaining to the discovery of high-density polyethylene by K. Ziegler and his students. In the United States I received important assistance from R. Pariser of DuPont, who obtained for me a copy of a historical sketch written by Carothers on the development of polyester and polyamides.

Eric S. Proskauer suggested to me that a book on the history of polymer science would be enlivened by pictures of some of its pioneers. Most of them I obtained from H. Mark; P. J. Flory's picture I owe to his secretary. J. M. Lehn sent me a picture of M. Berthelot, H. Kuhn a picture of W. Kuhn and J. C. Bevington a picture of H. W. Melville. The photograph of Emperor Wilhelm looking on as his car is fitted with the first synthetic tires I owe to J. Witte of Bayer AG.

Finally I want to acknowledge my great debt to Mrs. Dorothy Luyster, who typed the manuscript with infinite patience from my nearly illegible handwriting. It is no figure of speech that without her attention to detail, spotting of inconsistencies, and the patience with which she put up with frequent changes in the text this book could not have been accomplished.

New York, New York
February 1985

HERBERT MORAWETZ

Contents

Introduction	xiii
PART 1 FROM EARLY BEGINNINGS TO 1914	**1**
1. Berzelius Coins a Name	3
2. Toward Concepts of Molecular Structure	7
3. Early Observations of Addition Polymerization	16
4. Molecular Weights	26
5. Natural Macromolecules	32
Hevea Rubber 32	
Starch 36	
Cellulose 38	
Proteins 41	
6. Colloids	47
7. Beginnings of a Polymer Industry	53
Rubber 53	
Cellulose Nitrate and Regenerated Cellulose 56	
Bakelite 58	
PART 2 1914–1942	**61**
8. The Beginnings of a Synthetic Rubber Industry	63
9. The Impact of X-Ray Crystallography	70
10. Staudinger's Struggle for Macromolecules	86

x   Contents

11. **New Methods for the Determination of the Molecular Weight of Macromolecules**   99
    The Ultracentrifuge   99
    Solution Viscosity   103
    Light Scattering   110

12. **Polycondensation**   113

13. **Addition Polymerization**   124

14. **Advances in the Understanding of Proteins**   139

15. **The Elasticity of Flexible Chain Polymers**   148

16. **The Nonideality of Polymer Solutions**   153

17. **Polyelectrolytes**   157

18. **Scientists in a Time of Crisis**   160

PART 3   1942–1960   167

19. **Free Radical Polymerizaiton**   169

20. **Ionic and Coordination Polymerization**   182

21. **Advances in Polycondensation and Ring Opening Polymerization**   193

22. **The Rise of Molecular Biology**   197
    Proteins   198
    Nucleic Acids   203

23. **The Impact of Spectroscopy**   214

24. **Polymer Solutions**   221
    The Shape of Chain Molecules   222
    Colligative Properties   225
    Light Scattering   226
    Intrinsic Viscosity   229
    Solutions of Helical Polypeptides   231
    Polyelectrolytes   234
    Characterization of the Molecular Weight Distributions   239

25. **Polymers in Bulk**   241
    The Glass Transition   241
    Polymer Crystallinity   242
    Rheology   248

Epilogue   252

References	**254**
Name Index	**287**
Subject Index	**301**

Introduction

The emergence of synthetic polymers with their manifold uses has affected modern life so profoundly that our world is hardly imaginable without them. In a way we may liken this expansion of materials available for human use to the discovery of the smelting of ores or the firing of clay, but while we shall never know much about that magical moment when molten bronze was first seen by our ancestors, the growth of understanding of polymeric materials can be documented in detail. Yet until now no effort seems to have been made to write a comprehensive history of polymer science. This is surprising in view of the large number of scientists who are engaged in the study of polymers. General histories of chemistry hardly mention synthetic polymers (although they give good accounts of the developing understanding of natural macromolecules). Typically, Partington's monumental *History of Chemistry*,[1] published in 1964 (eleven years after Staudinger received the Nobel Prize for his pioneering work on polymers), refers only to Staudinger's study of diazomethane and ketene. (There is no entry for "polymers" or "polymerization" in the index.) Ihde[2] did somewhat better in devoting eight of the 823 pages of his *Development of Modern Chemistry* to polymers, but even that seems hardly commensurate with the importance of the field.

Chemistry and physics are sciences in which problems of current interest could not even have been conceived fifty years ago. (In this, these sciences differ strikingly from mathematics, whose classical problems formulated in some cases centuries ago are still considered challenging.) It is, then, problematical whether a scientist fully occupied with today's researches will have the time and inclination to read about the ideas of a vanished era, the false starts and controversies which have long been resolved. Obviously, the attitude will vary from person to person: Conant is quoted[3] as feeling that a knowledge of its historical development is essential to a true understanding of a scientific field, whereas Luria[4] has been outspoken about his lack of interest in the history of his specialty. I can testify only for myself that the study required for writing this book gave me in many ways a new outlook on a science that had occupied me for forty years.

In organizing the material, I decided to divide the book into three parts. The first, which carries the story up to the outbreak of the first world war, deals with speculations regarding the nature of biological macromolecules—rubber, starch, cellulose—and early observations of polymerizations. The latter part of this period saw the rise of the colloid concept which was to have such a profound influence on the views then current concerning the nature of "high molecular" substances. That time also witnessed the establishment of the first industries to produce artificial polymeric materials, cellulose nitrate, viscose, and phenol-formaldehyde resins. The second era, which extended from 1914 to 1942, may be regarded as the classical period of polymer science. During those years, dominated by the leadership of Staudinger, Mark, Meyer, Carothers, Schulz, and Kuhn and encompassing the beginnings of Flory's scientific career, the existence of long-chain molecules was firmly established, crystallography provided detailed structural information for some macromolecules, the mechanism and kinetics of polymerizations were elucidated, and the concept of a flexible "molecular coil" was advanced with an attempt to analyze the physical consequences of that molecular structure. The third period, from 1942 to 1960, may be regarded as the time when polymer science reached full maturity. The decision to cut off the story at that particular date was, of course, arbitrary. Another author at some future time will have to try to cover more recent developments. During that time Flory made crucial contributions to almost all aspects of polymer science and a large number of other brilliant scientists advanced more restricted areas of the field.

A difficult decision concerned the extent to which the developing understanding of proteins and nucleic acids should be included in this account. On the one hand, the story of the emergence of "molecular biology" is too vast to be treated adequately in the context of this book. On the other hand, this constitutes the part of macromolecular science that has profoundly affected our view of life, being the part of modern science which is most exciting to our imagination. It seemed absurd to leave it out, particularly since it has influenced in some subtle ways many of those working on synthetic polymers. I can only hope that the compromise solution that I adopted will be considered reasonable by the readers for whom the book was intended.

Part of the history of a scientific discipline is the organization of symposia, the publication of monographs and specialized journals, and the foundation of academic centers and specialized scientific societies dedicated to its research. The first scientific meeting at which the question of the nature of "high molecular" substances was to be debated took place in Düsseldorf in 1926; the lectures delivered on that occasion are contained in Vol. 59 of *Berichte*. On that occasion Staudinger was alone in championing uncompromisingly the existence of long-chain molecules; the organic chemists present were committed to the concept of "association colloids" while the crystallographic evidence was ambiguous. Two international meetings of the Faraday Society provided an opportunity for advances to be recorded and controversies to be aired. The first took place in September 1932 under the title "The Colloid Aspects of Textile Materials and Related Topics." At that time the existence of long-chain molecules was generally accepted, but the meeting was highlighted by the controversy between Staudinger's insistence on rigid rod-like chains and opposing views that stressed chain flexibility. The second meeting, sponsored by the Faraday Society at Cambridge in 1936 under the title "Phenomena of Polymerization and Polycondensation," dealt with the kinetics and mechanisms of polymer formation. After the second world war it was largely as a result of Mark's influence that the Union of Pure and Applied Chemistry (IUPAC) decided to organize a Commission on High Polymers within the Division of Physical Chemistry. This Commission sponsored a first International Polymer Symposium in Liège in 1947 and followed it with similar symposia in Amsterdam (1949), Strasbourg (1952), Uppsala (1953), Torino (1954), Rehovot (1956), Prague (1957), Nottingham (1958), Wiesbaden (1959), and Moscow (1960). These IUPAC symposia soon developed into central events of the international community of polymer scientists.

The first monograph on polymers was restricted to a discussion of rubber, cellulose, starch, and proteins. Published by Meyer and Mark in 1930,[5] it was based largely on the authors' crystallographic studies introduced only two years earlier. Two years later Staudinger[6] published his influential book on natural and synthetic macromolecules. It is perhaps indicative of the limited interest in polymers at that time that neither of these books was translated into English. The two-volume *Hochpolymere Chemie* by Meyer and Mark[7] was published in 1940, a significant wartime date because both Meyer and Mark had been forced to leave Germany well before their work appeared in print. This time an English translation was provided in 1942. As interest in polymers grew a number of books that deal with the subject was published but none approached the impact of Flory's *Principles of Polymer Chemistry*,[8] published in 1953. In spite of all the advances of the intervening years it is still the most authoritative statement of what polymer science is all about.

Staudinger had long been anxious to dedicate a journal to macromolecules but for many years his efforts were frustrated. Eventually, in 1940, he was made editor of the *Journal für praktische Chemie*, a title that was supplemented, beginning with Volume 155, by *unter Berücksichtigung der makromolekularen Chemie*.* In August 1943 the title was changed to *Journal für makromolekulare Chemie*, but because of the wartime paper shortage the journal was discontinued in the following year.

After the war the *Journal of Polymer Science* started publication in 1946, *Die Makromolekulare Chemie* in 1947. Although most of the crucial papers in the polymer field continued to appear in journals with broader specialization, the existence of journals devoted to polymers was undoubtedly important to the development of polymer science as a distinct discipline. An "Institute of

---

*"with notice of macromolecular chemistry."

Polymer Research" was created by Herman Mark at the Polytechnic Institute of Brooklyn in 1946 and in 1947 the "Centre de Recherches sur les Macromolécules" was organized in Strasbourg by Charles Sadron. Finally, the award of the Nobel Prize to Hermann Staudinger in 1953 may be regarded as belated recognition by the scientific community not only of the laureate's contribution but also of the importance of the field which he had fought so hard to establish.

References

1. J. R. Partington, *A History of Chemistry*, Macmillan, London, 1964.
2. A. J. Ihde, *The Development of Modern Chemistry*, Harper and Row, New York, 1964.
3. J. B. Conant, as paraphrased in E. Mayr, *The Growth of Biological Thought*, Harvard University Press, Cambridge, 1982, p. 20.
4. S. E. Luria, *A Slot Machine, a Broken Test Tube*, Harper and Row, New York, 1984, p. 123.
5. K. H. Meyer and H. Mark, *Der Aufbau der Hochpolymeren Organischen Naturstoffe*, Akadem. Verlagsges, Leipzig, 1930.
6. H. Staudinger, *Die Hochmolekularen Organischen Verbindungen, Kautschuk und Cellulose*, Springer, Berlin, 1932.
7. K. H. Meyer and H. Mark, *Hochpolymere Chemie*, Akadem. Verlagsges, Leipzig, 1940; *Natural and Synthetic High Polymers*, Interscience, New York, 1942.
8. P. J. Flory, *Principles of Polymer Chemistry*, Cornell University Press, Ithaca, New York, 1953.

POLYMERS

The Origins and Growth of a Science

Part 1. From Early Beginnings to 1914

1. Berzelius Coins a Name

For any of us practicing chemistry in the last quarter of the twentieth century it is hard to imagine the difficulties a chemist faced one hundred and fifty years ago. To read papers written at that time is a sobering experience and we may well question whether today's scientists, accustomed to their sophisticated instrumentation and equipment, would be able to duplicate the achievements of the pioneers of that early era under their working conditions.

We may then wonder, when reading references to Berzelius' use of the term "polymeric" in 1832, how that concept could have arisen at that time. The structure of even the simplest molecules was still a matter of controversy. Dalton's "relative weights of ultimate particles" were based only on elemental analysis—none too accurate.[1] Because the ratio of hydrogen to carbon was twice as high in carburetted hydrogen (methane) as in olefiant gas (ethylene), he concluded that methane molecules contained one carbon and two hydrogens, CH_2, ethylene one carbon and one hydrogen, CH. Since he knew of only one compound of hydrogen and oxygen he assumed that water contains one atom of each of these elements and used for it the formula HO. Dalton disbelieved Gay-Lussac's demonstration that gases combine in simple volume ratios and its implication that equal volumes of gases contain equal numbers of molecules.[2] How would it then be possible for nitric oxide, NO, to have the same volume as the sum of the volumes of nitrogen, N, and oxygen, O, from which it originated? How could we explain that the density of water vapor, HO, was lower than that of oxygen, which was only part of the water molecule?[3] This contradiction should have been resolved by Avogadro's 1811 paper[4] which suggested that "constituent molecules" of hydrogen, oxygen, or nitrogen could divide into "elementary molecules" when they participated in a reaction. Avogadro, however, made little impact on his contemporaries.

Three years after Avogadro's paper Ampère published a "letter to Count Berthollet", in which he outlined his thoughts on molecular structure.[5] Whereas Avogadro was motivated by a desire to explain Gay Lussac's chemical results, Ampère started from a physicist's reflection on the nature of a gas. He wrote:

> One has to assume that a molecule is an assembly of a fixed number of atoms* in a defined arrangement with the space between them incomparably larger than the atomic volumes and so that this space should have comparable dimensions in the three directions the molecule must contain at least four atoms.

Further on he reasoned:

> I start with the assumption that when a substance becomes gaseous, its molecules are separated from each other by the expansive force of caloric to distances much larger than those at which the cohesive forces are appreciable so that these distances depend only on the temperature and the pressure exerted on the gas and that at equal temperatures and pressures, the molecules of all gases, simple or compound, are at the same distance from each other.

It may be noted that these two passages express quite independent principles and that it is hard to think of a logical argument that would have led Ampère to insist on particles with "comparable dimensions in the three directions" or which would have, even if this principle were granted, excluded the existence of a gas consisting of monatomic particles. Ampère's paper suggests that this model of a gas was postulated *before* it was found that it provided an interpretation for Gay-Lussac's chemical data, but one may be skeptical in this regard. Anyway, his picture of hydrogen, oxygen, and nitrogen molecules, each consisting of four atoms, does not seem to have been adopted by any chemist.

In 1825 Faraday examined an oil that separated under a pressure of thirty atmospheres from a gas produced by heating whale oil sold in cylinders for home illumination.[6] On fractional distillation of this material he discovered benzene (which he called "bi-carburet of hydrogen" since he took a carbon atom to be six times as heavy as an atom of hydrogen). However, he seems to have been much more intrigued by the most volatile fraction of his fluid. It was liquid at 0°F, boiled below 32°F, and its gas had a density 27–28 times that of hydrogen. One volume of the gas required six volumes of oxygen for its combustion and yielded four volumes of carbon dioxide. Sulfuric acid condensed one hundred volumes of the gas in a process similar to that observed earlier with "olefiant gas" (ethylene). In short, Faraday had discovered butene. He was particularly struck by the comparison of butene and ethylene.

> It is a remarkable circumstance . . . that though the elements are the same and in the same proportion as in olefiant gas, they are in a very different state of combination. . . . 4 volumes or proportionals of hydrogen = 4 are combined with 4 proportionals of carbon = 24 to form one volume of the vapour, the specific gravity† of which would therefore be 28.

---

*For clarity we use here the modern meaning of molecule (which Ampère called particle) and atom (which Ampère called molecule).
†Defined as the ratio of the density of the gas to that of hydrogen.

Faraday's "remarkable circumstance" did not fail to astound Berzelius,* who reported it in his "Annual Report on Advances in the Physical Sciences."[7a] He also noted the "most peculiar finding" that two substances which were clearly different should have the same elemental composition.† A few years later he returned to the subject and wrote:

> It has been assumed in physical chemistry that substances of the same composition . . . must necessarily have the same chemical properties. Faraday seems to have found an exception. . . . I suggest that substances of equal composition but different properties be called isomers.[7b]

In fact, this discussion confused several phenomena of very different nature. On the one hand, there was ethylene and butene with the same elemental composition but different gas densities. Then there were examples of what would today be described as isomerism; for example, cyanates and fulminates. Finally, there were different crystalline modifications of the same compound, such as tin oxide. A year later Berzelius drew attention to these distinctions and wrote:[8]

> It is necessary to define precisely the concept of isomerism. I understand this to refer to substances containing the same absolute and relative number of atoms. . . . We should not confuse this with the case where the relative number of atoms is the same but the absolute number is different. For instance, the relative number of carbon and hydrogen atoms in olefiant gas and in oil of wine is exactly equal . . . but an atom of the gas contains only one atom of carbon and two of hydrogen, CH^2, while oil of wine contains four carbons and eight hydrogens, C^4H^8.‡ To describe this equality of composition coupled with a difference of properties I suggest that such substances be called polymeric.

It is ironic that this first reference to a "polymeric state" should contain two obvious errors. First, it is surprising that Berzelius should have used the formula CH^2 for ethylene after Faraday had noted that ethylene is half as dense as butene. Second, it seems strange that Berzelius should not have chosen as an example of an "ethylene polymer" the butene whose identity had been so well established by Faraday, but "oil of wine," which cannot be regarded as an ethylene polymer by Berzelius' definition. Concerning this substance he quotes Hennell,[9] who regarded it as "a compound of sulfuric acid and carbon and hydrogen, the latter in the same proportions as in olefiant gas."

---

*In citing Berzelius, we refer to the German translation of his Annual Reports published under the editorship of F. Wöhler. The translations are dated one year after the Swedish original.
†Berzelius need not have been surprised since Gay-Lussac had written as early as 1914 (*Ann. Chim. Phys.*, **91**, 149) that "the composition of acetic acid does not differ significantly from that of the ligneous matter. Here are two substances composed of carbon, oxygen and hydrogen in the same proportions which have quite different properties. This is a new proof that the atomic arrangement in a compound has the greatest influence on its character. . . . Sugar and starch lead to the same conclusion, since they have very different properties in spite of an identical elemental composition."
‡Berzelius used 12.25 for the ratio of the atomic weights of carbon and hydrogen, whereas most chemists of his era used 6.

According to Beilstein,[10] "oil of wine" was by no means a well defined substance but a mixture of diethyl sulfate with olefins. This is also indicated by Hennell:

> ... we can only infer the composition of the hydrocarbon which is combined and neutralizes the sulfuric acid; for in all the specimens of oil of wine ... I have found a variable quantity of hydrocarbon held in solution part of which separates in a crystalline form. ... "

Hennell's elemental analysis in fact points to a mixture of diethyl sulfate with olefins, although the sulfur and carbon analyses are not consistent with each other. To define the nature of the hydrocarbon component Hennell converted his "oil of wine" into potassium ethyl sulfate and noted that this product was identical with the potassium salt obtained from the product obtained on treating sulfuric acid with ethylene or alcohol. He analyzed this material obtaining 28.84% potash, 48.84% sulfuric acid, 13.98% carbon, 2.34% hydrogen, and 7% water. From this he concluded that the salt contained one "proportional" of potash, two of sulfuric acid, and four each of carbon and hydrogen. How he arrived at this conclusion is obscure, but it led him to state that "it would appear that in these salts ... four proportionals of carbon with four of hydrogen are saturating one of sulfuric acid." It was on such evidence, flawed even in the context of the time, that Berzelius based his suggestion of the term "polymeric." In any case, the expression took root. It was used in a similar context but with sounder experimental basis when Berthelot found 34 years later[11] that acetylene could be converted on heating to its "polymers" benzene and styrene. Although in contemporary usage "polymer" refers to a chain with a large number of monomeric units, thus being distinguished from "oligomer" in which relatively few units are joined to a chain molecule, this distinction was not made for a long time.[†] As late as 1930 and 1932 papers entitled "polymerization of ethylene"[11a] and "polymerization of butadiene"[11b] described dimerizations. In fact, occasionally, "polymer" was used to refer to a substance with a higher molecular weight and the same elemental composition without any regard for molecular structure. Thus Arnold's "Repetitorium der Chemie," published in 1896, cited (p. 265) the series formaldehyde, CH_2O, acetic acid, $C_2H_4O_2$, lactic acid, $C_3H_6O_3$, and glucose, $C_6H_{12}O_6$, as an example of "polymerism."

---

*Berzelius used 12.25 for the ratio of the atomic weights of carbon and hydrogen, whereas most chemists of his era used 6.
†The term "oligomer" seems to have been suggested first by L. V. Larsen for a laboratory manual published by G. F. D'Alelio in 1943 (Larsen, *Chem. Eng. News*, January 23, 1984, p. 58). Earlier, B. Helferich used the terms "oligosaccharide" and "oligopeptide" [*Ber.*, **63**, 989 (1930); *Ann.*, **545**, 178 (1940)].

2. Toward Concepts of Molecular Structure

Dalton's belief that elemental analysis was sufficient for the characterization of a chemical compound shows that little thought was given in his time to the arrangement of atoms in a molecule. The discovery of isomerism, however, made it obvious that elemental composition did not define a compound and that the question of atomic arrangement would have to be dealt with. A related problem which then presented itself concerned the nature of the forces acting between the atoms of a molecule. Berzelius was convinced that these forces were electrical, establishing bonds between an electropositive and electronegative atom. This concept seemed inconsistent with Dumas' finding[12] that hydrogen in acetic acid may be replaced by chlorine, that is, that the electronegative chlorine can take the place of the electropositive hydrogen. Dumas' discovery of trichloracetic acid therefore led to a heated exchange.* When Berzelius insisted[13] that "an element as electronegative as chlorine can never enter an organic radical; this idea is contrary to the first principles of chemistry," Dumas responded passionately:[14]

> Obviously I have . . . basing myself on fact . . . ignored electrochemical theories. . . . But are these concepts . . . based on such clear evidence that they should become articles of faith? Or do they at least have the ability to adjust to facts? . . . One must admit that it is not so.

An investigation of compounds obtained from benzaldehyde, published by Wöhler and Liebig in 1832,[15] proved to be most influential in the development

---

*Liebig felt free, as editor of the *Annalen*, to add to the published papers his comments at times endorsing, at other times condemning, the authors. In the case of Dumas' report on trichloroacetic acid he decided to ridicule the paper by adding a communication under the pseudonym "S. C. H. Windler" (swindler), who "reported" replacing not only the hydrogen but also the oxygen and carbon of acetic acid with chlorine. A footnote reads: "I have just learned that shops in London carry fabrics made of woven chlorine; these are highly prized in hospitals for night caps, etc."

of concepts of chemical structure. According to their interpretation, the benzoyl radical $C_{14}H_{10}O_2$ could combine with hydrogen to form benzaldehyde, with oxygen to form "anhydrous benzoic acid" (which could be hydrated to benzoic acid), or with chlorine to yield benzoyl chloride. They also prepared ethyl benzoate, benzamide, and benzoin.* Evidence that such a series of compounds all contain the complex benzoyl grouping was, at the time, so unexpected and exciting that it induced the authors to start their report with this rhapsodic introduction:

> When one succeeds to chance on a flash of light in the dark area of organic chemistry which may provide a true understanding and exploration of this realm, one has good reason to wish for good fortune. . . ."

Their enthusiasm was fully shared by Berzelius,[16] who commented:

> The circumstance that a substance consisting of carbon, hydrogen and oxygen . . . combines with other substances . . . like a single element proves that there exist ternary compound atoms. . . . Your facts . . . may be viewed as a new day in vegetable chemistry.

Another factor that pointed the way toward the organization of atoms in molecules was the gradual discovery of various series of similar "homologous" compounds with a regular change in properties. Thus the boiling points of "oxyhydrates" (alcohols) of a series of hydrocarbon radicals seemed to differ by 18°C for any two neighboring members.[17] In a more daring suggestion it was proposed[18] that with a knowledge of the properties of the series X_iY and XY_j, in which X_i and Y_j are homologous organic radicals, the properties of any compound X_iY_j could be predicted.

For a long time different values were used by different investigators for the relative atomic weights of hydrogen, carbon, and oxygen. Berzelius suggested the ratio 1:12.25:16, but most chemists based their formulas on the ratio 1:6.12:8. This meant that formulas could be understood only when the convention used by the author was known. The ratio of the weights of carbon and oxygen, based on the relative densities of oxygen and carbon dioxide (ignoring deviations from ideal gas behavior), led to difficulties in the interpretation of the elemental composition of organic compounds,[19] a problem that was resolved only when Dumas and Stas[20] established this ratio as 0.75 in experiments in which they burned diamonds and graphite.

To express the manner in which organic compounds reacted it became

---

*It is interesting to note that Wöhler and Liebig named the benzoyl radical after benzoic acid (which had been described by de Vigenère as early as the sixteenth century) without suspecting its relation to benzene. This was demonstrated two years later by Mitscherlich [*Ann.*, **9**, 39 (1834)] who obtained benzene by distilling calcium benzoate. The comments appended by Liebig to this report make it clear, however, that the relation between benzene and benzoic acid was still not understood.

customary to describe them by "rational formulas." These unwieldy structures often varied for the same compound from investigator to investigator and much energy was expended in arguing their merits; for instance, the following rational formulas were championed by three highly respected scientists for lactic acid:

$$\text{HO}\left(\text{C}_4\begin{vmatrix}\text{H}_4\\ \text{HO}_2\end{vmatrix}\right)\text{C}_2\text{O}_2,\text{O} \qquad \left.\begin{array}{c}\text{C}_3\text{H}_4\text{O}\\ \text{H}_2\end{array}\right\}\text{O}_2 \qquad \left.\left(\begin{array}{c}\text{CO}\\ \text{C}_2\text{H}_4\\ \text{H}\end{array}\right\}\text{O}\right\}\text{O}$$

Kolbe[21] \qquad\qquad Wurtz[22] \qquad\qquad Wislicenus[23]

In 1858 Kekulé published his celebrated paper in which he stated[24a] that carbon atoms are tetravalent and that "carbon atoms attach themselves to one another, thereby part of the affinity of one is naturally engaged with an equal part of the affinity of the other." He concluded that "the number of hydrogen atoms joined to n atoms of carbon which are attached in this manner to one another is $2n + 2$." Yet, in spite of this insight, rational formulas, like those depicted above, were in widespread use for a number of years. Structural formulas resembling those in modern use were first proposed by Crum Brown in 1865[25] and strongly recommended in a paper by Frankland and Duppa[26] in which lactic and paralactic acid were represented by

Lactic acid

Paralactic acid

It is important, however, to note that Frankland and Duppa did not try to convey the impression that their formulas represented the manner in which atoms are linked to one another. They state most emphatically that "it is hardly necessary to repeat Cum Brown's remark that such formulae are meant to express only the chemical and not the physical atomic positions." This point is made more explicitly by Kekulé, who continued, three years after his 1858 paper, to champion the use of rational formulas:[27] "which of the various rational formulas should be used depends on the purpose. . . . [They] are only transformation and not constitutional formulae . . . and in no way express the position of atoms in these compounds." He pointed out that two-dimensional representations cannot in principle convey the steric relationship of atoms in three-dimensional space. But his criticism went further: "The way atoms leave a changing or decomposing compound cannot indicate how they are arranged in the existing and unaltered compound." In Kekulé's view only physical measurements might in time clarify atomic positions in organic compounds.

This statement is interesting also because it is characteristic of Kekulé's speculations concerning the mechanism of reactions. There is a passage in his 1858 paper [24b] that deals with this subject. In a discussion of "double decompositions" he draws a diagram

$$\begin{array}{c|c} a & b \\ a_1 & b_1 \end{array} \qquad \begin{array}{cc} a & b \\ a_1 & b_1 \end{array} \qquad \begin{array}{cc} a & b \\ \hline a_1 & b_1 \end{array}$$

Before   During   After

which he describes as showing that "while the connection between atoms a and a_1 and b and b_1 is continuously loosened, that of atoms a and b and a_1 and b_1 continuously increases." This may be regarded as an early precursor of modern transition state theory.

There is no evidence in Kekulé's writing to suggest that he thought of optical activity when he commented on the possible role of physical measurements in clarifying molecular structure. Yet at that time Pasteur had already studied optically active organic compounds for 10 years and his results were most suggestive in that regard. He summarized his investigations in a series of beautiful lectures before the Chemical Society of Paris in 1860.[28] He started by recounting how he had been troubled as a student by a report that salts of tartaric acid and "paratartaric" acid, identical in all other respects, differed from each other in that solutions of tartrates were optically active, whereas paratartrate solutions (racemic tartrates) were not. How could two chemically identical compounds have different physical properties? Did this not invalidate, he wondered, the very definition of a chemical species? This led him to the unexpected discovery that paratartrates formed equal numbers of hemihedral crystals that were mirror images of one another.* He separated them and found that their solutions, while exhibiting an equal magnitude of optical activity, were characterized by opposite directions in which the plane of polarized light was rotated. He describes the emotional scene which took place

---

*Actually only the sodium ammonium and the sodium potassium salts of racemic tartaric acid yielded hemihedral crystals that allowed Pasteur to separate the optical isomers. Many other salts of racemic tartaric acid were prepared by Pasteur [*Ann. Chim. Phys.*, [3], **28**, 56 (1850)] but their crystals exhibited no hemihedry. Pasteur wondered about it but did not understand that in these cases the *crystals* were racemic.

when he called on Biot, the long-time champion of optical activity research, at the Collège de France. Biot made Pasteur repeat the experiments under his supervision. When he was satisfied that a salt of the *levotartaric* acid, which had never been seen before, could, in fact, be isolated by a separation of right-handed and left-handed crystals from the inactive paratartrate, Pasteur recounts, "the illustrious old man visibly moved seized my hand and said 'My dear child,[†] I have loved science so much throughout my life that this stirs my heart.'"

Pasteur clearly understood that the dextrorotary and levorotary tartrates must consist of molecules that are mirror images of each other and speculated: "Are the atoms of the right acid grouped on the spiral of a dextrogyrate helix[‡] or placed at the summit of an irregular tetrahedron?" He also found that optically active compounds may be racemized on heating and that optically active bases may be used to resolve racemic acids. He investigated mesotartaric acid which is inactive and cannot be resolved into optically active components and discovered that fermentation may attack only one of a pair of optical isomers. He was most intrigued by the fact that optical activity is characteristic of compounds produced by living organisms and engaged in a great deal of speculation about the forces that might be responsible.

In spite of the admiration which we owe Pasteur for this imaginative and monumental work, it is only if we define its limitation that we can understand why it should have taken another fourteen years before the relation between optical activity and molecular structure was clarified. Pasteur believed that optical activity was induced by twisting a molecule in one direction or the other; in this view mesotartaric acid was simply the "untwisted" tartaric acid molecule. In fact, he believed that all organic compounds can exist as dextrorotary, levorotary, racemic, or inactive species and urged that "all our efforts should be bent to produce them for *each particular species*." Because Pasteur did not consider the molecular structure of tartaric acid, he could not arrive at the conclusion that the meso compound has molecules with a plane of symmetry. When he found that a great deal of heat was liberated when solutions of dextro- and levotartaric acid were mixed and that this mixing led to crystallization, he assumed that the mixing of antipodes led to a chemical reaction in which the dextro and levo acid were combined.[*]

One may wonder why Pasteur's provocative work had so little influence on the chemical studies of succeeding years. As late as 1873 Wislicenus[29] wrote about the difference between the levorotatory lactic acid isolated from muscle and the optically inactive acid obtained by fermentation suggesting, without mentioning Pasteur, that it might be due to some "geometric isomerism."

[†]Pasteur was 26 years old at that time.
[‡]The suggestion that helical structures of molecules would lead to optical activity anticipated by more than a hundred years work on proteins and synthetic polypeptides in which optical activity characteristics were interpreted as indicative of helical conformations.
[*]The racemic tartaric acid crystal is less soluble than crystals of the optically active species. Thus mixing saturated solutions of the D- and L-acid led to crystallization with evolution of heat.

The relation between optical activity and molecular structure was finally clarified in the brilliant paper published in 1874 by the twenty-two year old van't Hoff.[30]* He first remarked that a planar structure of a quadrivalent carbon compound would imply two isomers for doubly substituted methane (cis and trans) and three isomers for methane substituted by three different groups (with R,R', or R" trans to H). In fact, no isomerism has been observed for doubly substituted methane, whereas methane substituted by three different groups generally yields two isomers that appear, on the basis of their optical activity, to be mirror images of each other. Van't Hoff therefore suggested that the quadrivalent carbon be represented by a tetrahedron with the substituents at the apices. Thus the optical activity of a molecule obtained by attaching different substituents at the four apices would correspond to the two arrangements of the substituents leading to structures which are mirror images of each other. He cited a number of instances where a chemical transformation in which an "asymmetric carbon" is eliminated leads to a loss of optical activity. Instances in which no optical activity was observed in spite of the presence of asymmetric carbons could be due to three causes:

1. The sample may be a racemic mixture.
2. A low compound solubility may render the observation of optical activity difficult.
3. The asymmetry of a carbon may not be a sufficient condition for optical activity, which should also depend on the nature of the substituents.

It should be noted that the tetrahedral representation of the carbon valences was suggested five years before van't Hoff by E. Paterno.† His suggestion resulted from an erroneous report by Butlerov that there exist three isomers of dibromoethane. Butlerov tried to explain this by assuming that the valences of carbon are not equivalent; Paterno suggested that this assumption is not necessary if one isomer is 1,1-dibromoethane and if 1,2-dibromoethane may exist as two rotational isomers. The drawing in his paper, representing these structures in the eclipsed form with the internal angle of rotation 0 and 120°, respectively, is astonishingly similar to modern molecular models. Paterno did not understand that rotation around C—C single bonds is normally too rapid to allow separtation of rotational isomers. In 1889 van't Hoff wrote to Paterno to thank him for a copy of his 1869 paper,‡ remarking "It is impossible that I could have seen this work previously (you say that you sent it to me) since the drawings would have left an unforgettable impression." In his book *"The*

---

*Van't Hoff's life has been recounted in loving detail by his devoted student E. Cohen,[30d] who became his successor at the University of Utrecht. In 1943 Cohen was deported to the notorious Auschwitz extermination camp where he died in a gas chamber one day before his seventy-fifth birthday (H. R. Kruyt, *Jaarbok Kon. Nederl. Akad. Wettensch.*, 1949–1950).
†*Giornale di Scienze Naturali ed Economiche di Palermo*, **5**, 117–122 (1869).
‡*Gazz. Chim. Ital.*, **43**, II, 503 (1913).

Arrangement of Atoms in Space" (second revised and enlarged edition, Longmans Green, London, 1898) van't Hoff considered compounds of type $R_1R_2R_3C$—$CR_4R_5R_6$ in which two asymmetric carbons are joined. He wrote (pp. 55–56):

> Free rotation being admitted by the fundamental conception, the mutual action of groups R_1,R_2,R_3 on the one hand and R_4,R_5,R_6 on the other will lead to a single "favored configuration." It is for the present indifferent which we call the "favored configuration" and we may take as such . . . where R_1 is above R_4, R_2 above R_5, R_3 above R_6.

Thus van't Hoff did not use "free rotation" in the modern sense but applied that expression to indicate merely the possibility of a rotation around the C—C bond *until the most stable conformation was obtained*. In fact, the existence of energy barriers in the rotation around C—C bonds was postulated by C. A. Bischoff* with experimental data that supported this concept well before the publication of the book cited above, but because van't Hoff makes no reference to this work it may be assumed that he did not consider it to be in conflict with this theory.

The confidence that van't Hoff felt in his theory is proved best by his willingness to make predictions on its basis; for instance, he predicted that optically active amyl alcohol would have the structure $CH_3CH_2CH(CH_3)CH_2OH$ and that the hydroxyl in citric acid, which is optically inactive, must be attached to the β- rather than the α-carbon.

The tetrahedral representation of the carbon atom also led van't Hoff to important inferences that concern the structure of compounds with carbon-carbon double bonds. In these compounds the two tetrahedra could be represented as being joined by a common edge so that molecules of the type RHC=CHR could have the R substituents either cis or trans to one another. Again, van't Hoff did not hesitate to make predictions. He proposed that maleic and fumaric acid are isomers of this kind, whereas Kekulé[31] had interpreted their difference as due to the fact that the double bond of fumaric acid is replaced in maleic acid by two unsatisfied valences. Another example he cited is the relation between crotonic and isocrotonic acid. Wislicenus[32] had assumed that the two compounds could differ only by the position of the double bond; that is, $CH_3CH=CHCOOH$ and $CH_2=CHCH_2COOH$, but van't Hoff pointed out that they yield the same oxidation products and are therefore cis→trans isomers. This assumption was later confirmed.[33] Somewhat later[34] van't Hoff elaborated his theory in several respects:

1. Stereoisomerism will also exist in disubstituted ring compounds.
2. Maleic acid must be the cis compound because of its easy conversion to the anhydride.

*$Ber.$, **23**, 623 (1890); **24**, 1074, 1085 (1891).

3. Tartaric acid exists in three stereoisomers, the meso form lacking optical activity because of a mutual cancellation of the effects due to the two asymmetric carbons.*

He went even further to predict the number of stereoisomers in compounds with a larger number of asymmetric carbons.

Van't Hoff's paper produced violent reactions from various quarters. Berthelot[35] claimed that styrene had "definitely" been shown to be optically active, although its molecules contain no asymmetric carbon. Van't Hoff disproved this contention in an elegant experiment.[36] He dissolved Berthelot's styrene, derived from storax, the resin of the *Liquidambar orientalis* tree, in alcohol and measured the optical activity of the solution. He then heated the solution to convert the styrene to "metastyrene" (polystyrene) which precipitated from the solution and was filtered off. After diluting the filtrate to the original volume of the solution van't Hoff found that the optical activity was the same as in the original solution. This showed that this activity was not due to styrene but to an impurity identified as an alcohol. Incidentally, van't Hoff also found, in the first optical activity measurement on a synthetic polymer, that solutions of the polystyrene had no optical activity.†

Berthelot's dogmatic statement that "the rotatory power of styrene being established, all theory incompatible with this property is hereby shown up as inexact"[35] was far from being the worst of the reactions van't Hoff had to face. Kolbe's scurrilous blast established an all-time low in the scientific literature. He wrote:[37]

> ... the domination of weeds which try to appear profound but are nothing but a trivial philosophy of nature‡ ... which like a tart is putting on finery and makeup and is being smuggled into good society. ... A certain Dr. J. H. van't Hoff, employee of a veterinary college has no taste for exact chemical science. He finds

---

*J. A. LeBel published a theory of optical activity in the same year as van't Hoff [*Bull. Soc. Chim. Fr.*, [2], **22**, 337 (1874)]. His paper is written in a more abstract style and therefore made less of an impact, although the arguments were essentially the same as van't Hoff's. He added the point that molecules with a plane of symmetry would have no optical activity and he gave mesotartaric acid as an example. He was wrong in proposing, on the basis of a mistaken benzene structure, that disubstituted benzenes should exhibit optical activity. This is surprising eight years after Kekulé's paper on the hexagonal benzene molecule. In the introduction to his book *The Arrangement of Atoms in Space* van't Hoff wrote that it was "purely fortuitous" that he and LeBel had both worked in Wurtz's laboratory; they "never exchanged a word about the tetrahedron there." He credits Wislicenus' speculation about lactic acid[29] with the stimulation that led him to his result.

†Berthelot was in no mood to admit that he had been mistaken, even after van't Hoff demonstrated that the optical activity of styrene derived from storax resin was due to a impurity. Two years later [*Ann. Chim. Phys.*, [5], **15**, 145 (1878)] he reported that a slab of "metastyrol" 23 mm thick turned the plane of polarized light to the left by $0.42°$ and considered himself justified to make some rather contemptuous comments about van't Hoff and Le Bel. Because he mentioned the biological origin of storax, it seems that he was at least partly influenced by Pasteur's conviction that optical activity is somehow related to life processes.

‡The expression "philosophy of nature" does not convey the flavor of "Naturphilosophie" used by Kolbe, which referred to a romantic school tinged with mysticism whose major protagonists,

it easier to mount the Pegasus (borrowed from the veterinary college) to announce . . . how atoms appeared to him in his flight . . . belief in witchcraft and ghosts. . . .

Why should van't Hoff's theory have excited such passion? It must have seemed incredible to an experimentalist like Kolbe that pure reflection by a young man with no previous record of accomplishment should be able to solve a central problem of chemistry. Also, as we have seen, there was a deep prejudice against the notion that chemical properties can be interpreted in terms of the relative spatial positioning of atoms in a molecule. It was probably particularly irritating that van't Hoff should have presumed to make *predictions* on the basis of his speculations. Finally, Kolbe was clearly incensed that this new theory should have received the backing of the highly respected Wislicenus, who went so far as to induce his assistant, F. Herrmann, to prepare a German book based on van't Hoff's *Les atoms dans l'éspace* (Rotterdam, 1875); this was published in Braunschweig in 1877 under the title *Die Lagerung der Atome im Raume*, with Wislicenus' introduction.

F. W. J. Schelling and L. Oken, had a substantial following in Germany at the beginning of the nineteenth century. (See H. Hörz, R. Löther, and S. Wollgast, *Naturphilosophie-von der Spekulation zur Wissenschaft*, Akademie-Verlag, Berlin, 1969.) Liebig is quoted as having written: "Even I lived through this period, so rich in words and ideas, so poor in true knowledge; it cost me two precious years of my life; I cannot describe my horror as I awoke from this intoxication." (W. Heisenberg, *Das Naturbild der heutigen Physik*, Rowahlt, Hamburg, 1955, p. 98.) It was, of course, ridiculous for Kolbe to characterize van't Hoff's closely reasoned arguments as "Naturphilosophie."

3. Early Observations of Addition Polymerization

The first recorded observation of a polymerization reaction seems to be due to Simon, "an apothecary in Berlin," who in 1839 distilled storax resin with a sodium carbonate solution,[38] and obtained an oil which he analyzed as containing 89.25% carbon, 10.24% hydrogen, and 0.029% oxygen. He named the substance styrol and noted that "with old oil the residue which cannot be vaporized without decomposition is greater than with fresh oil, undoubtedly due to a steady conversion of the oil by air, light and heat to a rubberlike substance." He assumed that this product was styrene oxide. A few years later Blyth and Hofmann, in a reinvestigation of styrene,[39] confirmed Simon's ratio of carbon to hydrogen and noted that this ratio was the same as in benzene and "cinnamol," an oil obtained by Gerhardt and Cahours[40] by the distillation of cinnamic acid with barium hydroxide. In trying to determine the number of carbon and hydrogen atoms in the styrene molecule they found that "the determination of the specific gravity of its vapour was no longer to be thought of after we had observed the peculiar metamorphosis produced by heat" and therefore decided to use styrene derivatives for molecular weight determination. With this approach they interpreted the N/C ratio in nitrostyrene as indicative of a formula C_8H_8.*

Blyth and Hofmann then proceeded to a study of the "peculiar metamorphosis," writing as follows:

> On first observing this phenomenon we thought it arose from the conversion into resin by the action of air on a portion of the raw oil which had been standing in a bottle only half-filled. . . . We soon convinced ourselves that this was erroneous by distilling a second time a portion of the newly rectified oil. A very considerable

---

*Here and in what follows we "translate" formulas given by authors of that period to conform to atomic weights of 12 for carbon and 16 for oxygen.

residue. . . remained in the retort. The rapidity with which styrole changes into the solid substance indicates sufficiently that it could not be from a combination with oxygen. . . . A portion of oil was confined for several months over mercury in a tube filled with oxygen without the least diminution of the volume of the latter.[39]

After the rubbery mass produced by heating styrene had been extracted with ether and dried, it was subjected to elemental analysis. This proved to yield the same results as unaltered styrene. Blyth and Hofmann then suggested that it be called metastyrol,* and to establish its formula resorted once again to nitration. This time they found a ratio of 1:7 for N/C of the nitrated product "from which it is seen that by the metamorphosis of the body C_8H_8 the number of equivalents of carbon and hydrogen in the new compound has diminished by one eighth." They also found that metastyrol may be largely reconverted to styrol by careful heating, which seems inconsistent with the assumption that styrol, C_8H_8, had been converted to metastyrol, C_7H_7. (Later investigators used the term metastyrol but generally assumed that its formation involved a polymerization.) Their paper contains an intriguing footnote: "There may be no organic solid which has a refractive power equal to that of metastyrol. It is not unthinkable that it could be used for optical applications."[39]

Blyth and Hofmann raised the possibility that "cinnamol" and styrene might be the same substance and this was proved by Kopp to be the case.[41] Yet this conclusion was not universally accepted and Berthelot claimed as late as 1866 that although the two substances seemed to behave similarly in many of their reactions, "cinnamol" exhibited a much smaller tendency to polymerize.[42] This seemed plausible because Gerhardt and Cahours had been able to measure the vapor density of the substance at its normal boiling point.[40] Berthelot also insisted that styrene is optically active, whereas "cinnamol" is not. As we have seen, it took van't Hoff's elegant experiment to show that the supposed optical activity of styrene is due to an impurity,[36] and following this demonstration no more was heard of "cinnamol." Yet it is unclear at this time why styrene obtained from cinnamic acid should have been relatively slow to polymerize. Styrene conversion to "metastyrol" was found to be catalyzed by sulfuric acid[43] and by alkalis[44] and to be inhibited by iodine and sulfur.[44]

During the time of all these investigations the structure of styrene was unknown. Its description as vinyl benzene was established by Erlenmeyer in 1866.[45]†

---

*Blyth and Hofmann also observed that this "metastyrol" was formed when styrene was exposed to sunlight, while it remained unchanged in the dark. This is probably the first report of photopolymerization.
†Berthelot showed that styrene could be obtained by the pyrolysis of ethyl benzene [*Ann. Chim. Phys.* [4], **16**, 156 (1869)] and he also claimed its preparation from benzene and ethylene [*Ann. Chim. Phys.* [4], **12**, 137 (1867)]. He wrote its formula as $C_6H_4(C_2H_4)$.

On April 27, 1863, Berthelot† delivered a lecture to the Chemical Society of Paris which includes the first general discussion of polymerization in the scientific literature.[46] In this lecture polymers were treated as a special case of isomers; thus Berthelot adopted the term "polymer" proposed by Berzelius without restricting the concept of isomerism according to his suggestion.[8] The lecture referred only briefly to styrene. Pinene polymerization by heat or catalytic action had been studied by Berthelot ten years earlier;[47] pentene and its oligomers were prepared by heating amyl alcohol with zinc chloride, followed by fractional distillation of the products.[48]‡

Berthelot stated that the ability of pentene to polymerize "could have been foreseen from its general affinity, that is, from its ability to add chlorine, hydrogen chloride, water, ammonia, etc., etc." He felt free to proclaim a sweeping generalization that "all compounds able to add hydrogen, chlorine, water, must be able to add molecules identical to themselves." He predicted that "incomplete" (i.e., unsaturated) alcohols or acids—of which he knew no examples—should also polymerize. As for aldehydes, they are "by definition incomplete compounds since they are formed from alcohols by loss of hydrogen" so that their polymerization is not surprising.

According to Berthelot there is, in principle, no limit to the degree of polymerization because the addition of two "incomplete" compounds leads to an "incomplete" product.* However,

---

†P. E. M. Berthelot (1827–1907) was an unusually creative scientist who published papers on inorganic, organic, physical, analytical, agricultural, and physiological chemistry. His writings also include *The Origins of Alchemy*, *Science and Philosophy*, and *The Chemical Revolution, Lavoisier*. He was president of the "Scientific Committee for the Defense of Paris" during the Franco-Prussian war, thus embarking on a career in politics. He served in two French governments as Minister of Foreign Affairs (1895) and Minister of Education (1896–1897). (*Dictionary of Scientific Biography*, Scribner, New York, 1970.)

‡Berthelot gives 35° as the boiling point of pentene, which is close to the boiling point of 2-pentene, and we must assume that this isomerized to 1-pentene before polymerization. Bauer[48] gives the boiling points of the dimer, trimer, and tetramer as 165, 245, and 390°, respectively; Berthelot found 160, 250, and 320°.

*Here it should be pointed out that at that time a great deal of confusion existed concerning the structural basis of unsaturation. A. M. Butlerov suggested [*Ann.*, **144**, 1 (1867)] that the dehydration of normal butanol, which involved the loss of the hydroxyl and a hydrogen attached to any one of the four carbon atoms, may lead to four different butenes:

$$\begin{cases} CH_2' \\ CH_2 \\ CH_2 \\ CH_2' \end{cases} \quad \begin{cases} CH_3 \\ CH' \\ CH_2 \\ CH_2' \end{cases} \quad \begin{cases} CH_3 \\ CH_2 \\ CH' \\ CH_2' \end{cases} \quad \begin{cases} CH_3 \\ CH_2 \\ CH_2 \\ CH'' \end{cases}$$

where the primes (apostrophes) stand for unsatisfied valences. As late as 1877 (*Ann.*, **188**, 104) Fittig considered the following formula for methacrylic acid

$$CH_3-CH-CH= \\ | \\ COOH$$

while theory predicts the possible existence of such compounds, it does not indicate that they necessarily exist. This depends on numerous conditions. . . . No compound can exist without satisfying them . . . but they do not ensure equilibrium of molecular groupings.

An extensive section of Berthelot's presentation was devoted to the thermochemistry of polymerization. From the heat of combustion of pentene, its dimer and tetramer, he estimated the heats of dimerization and tetramerization as 26 kcal/mol and 156 kcal/mol, respectively.[49] Heats of combustion of homologous alkanes led to an increment of 157 kcal/mol for each added CH_2 residue; by subtracting this quantity from the heat of combustion of ethylene he arrived at a heat of formation of "methylene" endothermic to the extent of 14 kcal/mol. "Compounds formed from the elements with heat absorption are intrinsically unstable," he wrote, and concluded that "this, no doubt, is the reason why methylene has not been prepared." Processes which should yield methylene, such as the dry distillation of acetic acid salts, lead instead to olefins, that is, methylene polymers.[50]

Berthelot cited three methods by which unsaturated compounds can be polymerized: (1) the direct application of heat, (2) the action of a simultaneous reaction, and (3) the influence of the "nascent state." He found that a very small amount of boron trifluoride was sufficient to polymerize pentene or pinene. He believed that the acid first reacted with the unsaturated compound to release heat, which induced neighboring molecules to polymerize with further heat evolution so that "once the process is started it spreads through all the mass like a fire." To exemplify the action of the nascent state Berthelot cites an experiment in which propylene was absorbed in sulfuric acid and, after some aging, was released from its combination with the acid by a sudden addition of a large amount of water. The oily layer that separated contained fractions boiling between 250 and 300°, "suggesting that they contain 24 to 30 equivalents of carbon. These condensed hydrocarbons were formed under the influence of the nascent state at the instant when the compound of propylene with sulfuric acid was destroyed."* Here, also, Berthelot suspected that the evolution of heat was the fundamental cause of polymerization, but high temperatures were known to lead to decompositions and this seemed to present Berthelot with a dilemma. He discussed it in the following terms:

> It is obvious that the heat liberated by the formation of a compound can never be sufficient for its complete decomposition. . . . This would produce a vicious circle, leading to perpetual motion. If the final state could be the same as the initial

---

*The concept of the "nascent state" goes back to Priestley, who applied it to an enhanced reactivity of hydrogen at the time of its release. B. C. Brodie [*Phil. Trans.*, **140**, 759 (1850)] assumed that the high reactivity of "nascent oxygen" is due to its transitory monatomic state. Berthelot's suggestion that the nascent state of a relatively complex molecule may lead to its polymerization was similar to that used half a century later by Harries in his speculation concerning the biosynthesis of rubber (see p. 34).

state . . . after having led to an evolution of heat during compound formation, the result would be the production of heat without expenditure of work. . . . This is contrary to the laws of mechanics.

It is surprising that a man of Berthelot's sophistication did not realize that any heat evolved in polymerization would be absorbed in depolymerization. More than eighty years earlier Lavoisier and Laplace had stated[51] "if in the combination or change of state there is an increase in free heat, this new heat will disappear in the return of the substances to their initial state".

It is striking that in the course of this long lecture on "polymerism" Berthelot never hesitated to assign what we should call today "degrees of polymerization," although no means were available to estimate molecular weights. He used two criteria: boiling points in the case of propylene and pentene oligomers and densities in the case of pinene and its presumed dimer. He assumed that "metastyrene" was the styrene dimer without giving any justification for his choice.

In the second part of his lecture Berthelot speculated about the "unity of nature" which would extend the phenomenon of polymerization to the elements. He wonders whether ozone is not a polymer of oxygen (although he remarks in a footnote that Soret† had recently established the ratio of ozone and oxygen densities as 3/2). He suggests correctly that the transition of liquid sulfur to a viscous state at around 160°C is due to a polymerization, but he draws from this the incorrect conclusion that the density of sulfur vapor should undergo an analogous change in this temperature region, which should be observable as a change in the gas density.* His final extension of the unity-of-nature theme concerns the build-up of the elements. Citing various series in which "equivalent weights" appear to be multiples of a base unit (e.g., $O = 8$, $S = 16$, $Se = 40$, $Te = 64$), he wonders—while stressing the speculative nature of such thoughts—whether these relations have a deeper significance.

Six years after delivering the lecture to the Chemical Society of Paris in which he outlined his ideas on polymers and polymerization, Berthelot published a paper that contained the details of his experiments on the polymerization of ethylene, propylene, pentene, and pinene.[52] He obtained high-boiling olefins by exposing ethylene to boiling alkali and he described the fraction that boiled in the range of 280–300°, which he estimated to be hexadecene, as a "polyethylene," surely the first use of that word. Propylene was polymerized by sulfuric acid—he no longer claimed that the nascent state was involved in the process—with fractions boiling at 200–220° and 260–280° being assumed to be the tetramer and pentamer, respectively.

Attempts by Goryainov and Butlerov[52a] to apply sulfuric acid catalysis,

---

*Meyer and Go[51a] first proved that sulfur may form long chain molecules. Gee[51b] studied the conversion of the cyclic S_8 to the chain polymer in the melt, whereas in the vapor the cyclic S_6 and S_8 were shown to decompose to S_2 at high temperatures.

†J. L. Soret, *Compt. Rend.*, **61**, 941 (1865).

effective for the polymerization of olefins, also to the polymerization of ethylene, ended in complete failure, eliciting the comment: "This peculiar stability of ethylene under conditions leading to the polymerization of its homologs points again to its symmetrical structure in contrast to propylene, isobutene, etc." The photopolymerization of ethylene to "a liquid with an odor of heat degraded wax" was reported in 1910 by Berthelot and Gaudechon,[52b] who were also the first to polymerize acetylene to a solid product. Another route to polyethylene was established in 1900 by Bamberger and Tschirner[52c] in a study of diazomethane reactions which led them to conclude that free methylene is a reaction intermediate. They identified the precipitate that formed spontaneously in ether solutions of diazomethane (in the presence of unglazed porcelain but not in its absence) as polymethylene. (The same "white flocculant material which could be crystallized from chloroform" had been described one year earlier by Pechmann,[52d] who had obtained too little of it for characterization.)

In his 1863 lecture Berthelot made no claim to have presented a comprehensive review of all addition polymerizations known at that time. In fact, Regnault had prepared vinylidene chloride as early as 1838, noting that "the liquid is unstable, when left in a sealed tube it soon becomes turbid and deposits a white non-crystalline substance which is an isomeric modification."[53] A similar observation made some years later concerned vinylidene bromide and here the suspicion was expressed that exposure to air may favor the transformation.[54] An analogous behavior of vinyl bromide was described in 1860 by Hofmann[55] as "a simple molecular transformation" fifteen years after his work on "metastyrene." How little he understood the nature of this process emerges from his comment that vinyl bromide "may be considered an ether of allyl alcohol and hydrogen bromide but . . . the peculiar molecular transformation seems to indicate a relationship to the aldehyde which is isomeric with allyl alcohol." Hofmann noted candidly that he "tried without success to produce the transformation at will." Twelve years later Baumann reported "the transformation of vinyl bromide and vinyl chloride into isomeric bodies"[56] on their exposure to sunlight, but the author seems to have had no clear idea of the nature of the reaction. This is rather surprising and one wonders whether Baumann was unaware of Berthelot's discussion of the polymerization of unsaturated compounds. Much later, when many chemists assumed that polymers were produced from cyclic oligomers by secondary valence forces, Ostromislensky[57] wrote that the vinyl bromide oligomer, from which the vinyl bromide polymer was formed, could not be dibromocyclobutane or tetrabromocyclooctane (which were known at that time) and that its basic cyclic unit had to contain at least 12 carbon atoms. He also claimed that the polymer produced by ultraviolet irradiation was different from that formed in visible light.

For many years a number of investigators had studied the destructive distillation of natural rubber, hoping to obtain from the fragments clues to the structure of the material. Working along that line, Williams in 1860 isolated isoprene and its dimer (called "caoutchine"). In discussing his results, he wrote[58]:

> I am anxious to call attention to the fact that the atomic constitution of caoutchouc appears to bear a simple relation to the hydrocarbon resulting from its decomposition by heat. The composition of caoutchouc coincides with that of isoprene . . . to a degree which is remarkable when we consider that caoutchouc, in addition to being non-crystalline, is scarcely able to be purified by chemical means.*

Williams made the point even more explicity by stating,[59]

> The author considers the action of heat on caoutchouc to be merely the disruption of a polymeric body into substances having a simple relation to the parent hydrocarbon. He deduces this view from the similarity in composition between pure caoutchouc, isoprene and caoutchine.

Not surprisingly, these statements inspired many attempts to reverse the breakup of rubber into isoprene. Bouchardat's first experiments[60] in which isoprene was heated for 10 hours to 280° resulted in the production of a viscous oil from which a dimer boiling at 176–181° was isolated. No attempt was made to study the residue that remained after the distillation. In 1879 Bouchardat obtained a more spectacular result and the exultant title "Action of hydracids on isoprene: reproduction of rubber"[61] testifies to his excitement. By exposing isoprene to concentrated aqueous hydrochloric acid he obtained, apart from HCl addition products, a solid that "has the composition of isoprene. . . . Moreover, it possesses the elasticity and the other characteristics of rubber. . . . This product subjected to dry distillation yields the same products as rubber." A few years later, Wallach discovered[62] that, although it is difficult to polymerize isoprene by heating, the reaction proceeds readily on illumination at room temperature. He ends his report on this observation with an exclamation mark, clearly unaware of the photopolymerizations reported by Blyth and Hofmann in 1845 and by Bauman in 1872.

The possibility of polymerizing isoprene to a material that seemed to be extremely similar to the rubber produced from the sap of the *Hevea* tree overshadowed for a long time other polymerization reactions, particularly when the birth and growth of the automobile industry increased at dramatic rate the consumption of this material. An important milestone in this development was the discovery that alkali metals act as polymerization catalysts, found independently within three days of one another by Matthews and Strange in

---

*We are meeting here for the first time a prejudice common among organic chemists that the inability to purify polymers by crystallization makes their identity suspect. This was to prove a major handicap in the development of polymer science, being used as late as the 1920s as an argument against Staudinger's macromolecular hypothesis.

England[63] and Harries in Germany.[64]*† The polymerization of isoprene led to another concept that was to play a crucial role in the development of polymer science and technology. It seemed probable that conjugated dienes other than isoprene would behave chemically in an analogous manner. While Couturier[65] considered the polymerization of 2,3-dimethylbutadiene by sulfuric acid to be an annoying impediment in his attempt to effect the hydration of the hydrocarbon to the glycol, others were clearly motivated by a desire to produce varieties of rubber. Thus Kondakov[66] demonstrated the polymerization of 2,3-dimethylbutadiene and Lebedev[67,68] reported the same process for unsubstituted butadiene. The rubbers produced by these processes were found to have different properties; in fact, Harries observed that polymers of the same monomer were different when obtained at a high temperature without a catalyst and when sodium was used to catalyze the polymerization.[64] (Metal-organic compounds such as diethyl zinc or ethyl sodium were other effective catalysts for the polymerization of butadiene and its homologs.[68a]) This suggested that it might not necessarily be desirable to imitate the natural product because modifications of it could lead to properties advantageous for specific applications.

A fascinating passage in a paper published by Kondakov[68b] as early as 1901 which dealt with the polymerization of dienes contains for the first time a reference to the role of the polarity of the double bond, a concept which was to play an important role more than forty years later. He wrote:

> The essential cause of the polymerization of divinyl compounds is the dependence of the properties of unsaturated carbon compounds on their electronegativity. . . . Ethylene derivatives in which hydrogen is substituted by electronegative atoms or groups are apt to polymerize by light, heat, or catalysts. . . . Since a double bond has electronegative properties, ethylene substituted by such a residue should also polymerize.

Other polymerizations of vinyl derivatives were also described during this early period. Redtenbacher[68c] discovered acrolein in 1843 and described in a most amusing paper its conversion to a resin only four years after Simon's report on the polymerization of styrene. He called the resin "disacryl" (a designation that was to be used for eighty years) and suggested for it an

---

*The strange coincidence that two laboratories should make such an important discovery virtually at the same time led to a lengthy controversy, with strong nationalistic overtones, concerning the "first rubber synthesis." (See Ref. 93, pp. 48–58.) The bad temper of these arguments was probably due to a considerable degree to Harries' vituperative style; for instance, in 1913 (*Ann.*, **395**, 211) he attacked C. B. Lebedev for using ozonolysis in his study of rubbers and made the outrageous claim that "it is customary to leave a method to its discoverer as long as he is engaged in its use."
†Because the discovery of the polymerization of isoprene at elevated temperatures by F. Hofmann is intimately related to synthetic rubber production in Germany during the first world war, I discuss it in Part 2 of this book.

empirical formula corresponding to a dimer of acrolein. (Redtenbacher also described the conversion of acrolein to an acid he named acrylic acid.) In 1878 Wislicenus[68d] synthesized vinyl methyl ether and attempted to add iodine to the double bond of this compound. He was surprised that instead of this expected reaction "the mixing produces a violent reaction . . . the mobile liquid thickened to the consistency of beet oil . . . one part of iodine for 200 parts of vinyl methyl ether was sufficient to complete the transformation." At about the same time Fittig[69] found, on distilling methacrylic acid, that white flakes separated from the liquid. In attempting to characterize this product, he noted that "a large number of experiments aimed at clarifying the structure by splitting the substance failed because of its great stability." He also noted that it was better not to attempt a purification of the ethyl methacrylate, which was an intermediate in the synthesis of the acid, because part of the ester was lost by polymerization on each redistillation. Both Fittig and Kolbe[70] reported that polymerization of methacrylic acid takes place at room temperature in the presence of hydrochloric acid. Fittig found that methacrylic acid polymerizes slowly even below its freezing point, and Kolbe observed that calcium methacrylate crystals gradually lose their water solubility, which he ascribed to polymerization. These were undoubtedly the first reports of polymerizations in the crystalline state. Moureu[71] found that melted acrylamide heated to 150° was suddenly converted to a "horn-like amorphous substance." It was insoluble in water, and because it had lost only 0.4% of its weight by ammonia evolution, Moureu concluded that a polymer of very high molecular weight must have been formed. Many years would have to pass before the concept of crosslinking would be properly understood leading to the realization how few units in a long polymer chain have to react to produce insolubility.

Finally, we may mention Butlerov's studies[72] in which catalysis by sulfuric acid led to the dimer and higher polymers of isobutene. He noted that propylene may be polymerized by BF_3 but was troubled by his inability to isolate the dimer and other low oligomers. The attitude of his time is well characterized by his comment that "obviously, one can hope to clarify the mechanism of polymerization only if one starts with the study of the simplest products."

While the investigations listed were concerned with the polymerization of specific monomers, a paper published by Engler[72a] in 1897 "On the origin of petroleum" dealt with the general problem of polymerization. He wrote that

> even if we assume that only in a small fraction of the molecules the double bonds are partially dissolved, their slow addition to complex structures is easily understood. Moreover, one need not assume that only similar molecules assemble.

This passage is of interest for a variety of reasons. Whereas "assemble" *(sich zusammenlagern)* suggests a loose association, the reference to the "partial dissolution of double bonds" would seem to imply covalent bond formation between the monomers. Thus the passage is typical of the lack of precision with which the term polymerization was used. Ten years after publication of this paper Staudinger joined Engler's laboratory as "Extraordinärer Professor," and it seems most probable that Engler's interest in polymerizations planted a seed in Staudinger's mind, particularly because Staudinger's memoirs express great admiration for Engler's personality. It should also be noted that Engler understood that different olefins can copolymerize, a concept that was surprisingly difficult for researchers thirty years later.

Engler refers in his paper to a study of polymerization undertaken by his assistant Kronstein. This study was published in 1902[72b] and seems to be the first attempt to follow the course of a polymerization. Kronstein believed that the glassy polystyrene is insoluble in the styrene monomer and concluded that the absence of a phase separation during the polymerization of styrene points to the existence of a reaction intermediate. He believed that his "discovery" of this intermediate explained the inconsistencies in older reports on styrene polymerization.

4. Molecular Weights

In the first half of the nineteenth century gas densities could have provided the only means of determining the relative weights of molecules. However, this would have required an acceptance of the hypothesis of Avogadro and Ampère that a given volume of gas contains the same number of molecules of different substances and few chemists were willing to concede the general validity of this principle. The acceptance of the theory was aided substantially by the publication, in 1858, of Cannizzaro's "Outline of a course of chemical philosophy given at the Royal University of Genoa".[73] This paper contains an informative history of the controversies generated by "Avogadro's principle"; for instance, the prestigious Berzelius was willing to accept it only for the elements, not for compounds. Cannizzaro's many examples demonstrated the impressive consistency of experimental data with Avogadro's theory and his suggestion that a "molecular weight" be defined as twice the ratio of the density of a given gas to the density of hydrogen was gradually adopted by the chemical community.

However, it should not be assumed that Canizzaro's arguments were generally accepted.* The nature and violence of the controversy surrounding the subject are well illustrated by the account of the discussion that followed a lecture given by Williamson as late as 1869 on the "atomic theory".[74] Williamson contended that

> the whole course of investigation had been one grand confirmation of the assumption that compounds must have molecular weights corresponding to at least the smallest atomic proportion which would represent the actual numbers of analysis. The perfectly independent observations which had been made with reference to the boiling points of homologous liquids, the phenomenon of diffusion and the equality of volumes of masses containing an equal number of these molecules under similar conditions in the gaseous state: all concurred to corroborate the conclusions necessitated by the atomic theory.†

---

*As late as 1863 Friedel and Crafts wrote that "whatever importance one may attach to the Avogadro and Ampère law . . . it still remains a physical hypothesis which must be submitted to the largest number of chemical verifications".[415b]

†A full statement of Williamson's arguments is contained in *J. Chem. Soc.*, 22, 328 (1869). In spite of his complete commitment to the reality of atoms and molecules, Williamson wrote in a passage which was far ahead of the thinking of his time: "the question whether our elementary atoms are built up of smaller particles is one upon which I, as a chemist, have no hold . . . in chemistry the question is not raised".

Following this closely reasoned argument, Frankland, the same man who two years earlier had championed the use of structural formulas[26] is quoted as saying

> it is impossible to get at the truth as to whether matter was composed of indivisible particles or whether it was continuous—the question belonged to what metaphysicians termed the unknowable; but he acknowledged the fullest use of the theory as a kind of ladder to assist the chemist.

This, at least, conceded a pragmatic use of the molecular concept, but other participants in the discussion were much more negative, with one quoted that "as a chemist he is not particularly interested in whether matter is infinitely divisible" while another asserted that "no one has ever seen an atom and in this respect the atomic theory has an exact parallel in the theory of phlogiston." It can hardly be said that Sir Benjamin Collins Brodie, professor of chemistry at Oxford, presided at the discussion in an impartial manner. In his concluding remarks

> He found that works of Kekulé and Naquet scribbled over with pictures of molecules and atoms arranged in all imaginable ways for which no adequate reason was given . . . it was a mischievous thing . . . mixing up fiction with facts.

Eight years later Kekulé presented his views in a lecture given as he was assuming the rectorship of the University of Bonn.* The published text[75] contains these passages:

> We must certainly conclude that *at present* the observed facts can be deduced as necessary consequences of the atomic theory only . . .we may doubtless use the atomistic theory *for the time being* as a basis for further reflections in the domain of natural science. . . . (emphasis by Kekulé)

The tentativeness of these statements is in surprising contrast to a later passage in this lecture:

> After Avogadro's hypothesis . . . had gained general recognition and the relative weights of gas particles could be deduced from the specific gravity of gases;

---

*At the beginning of his lecture Kekulé complained that "one-sided representatives of so-called humanistic studies, who also confuse the applications of chemistry with its scientific task, tend towards the unjustifiable view that chemistry ought to be taught at polytechnic schools and not at universities". This seems strangely similar to the patronizing attitude of much of today's chemical "establishment" toward polymer chemistry. The bias of nineteenth-century German universities against any "practical" activity is also exemplified by a passage in a letter written on December 19, 1844, by J. v. Liebig to M. Faraday: "What struck me most in England was . . . that only works which have a practical tendency . . . command respect; while the purely scientific which possess far greater merit are almost unknown. . . . Practice alone can never lead to the discovery of truth. . . . In Germany it is quite the contrary . . . no value or at least but a trifling one is placed on practical results. . . . for both nations the *golden mean* would certainly be a real good fortune." (Bence Jones, *The Life and Letters of Faraday*, Vol. II, Longmans, Green, London, 1870, p. 184.)

after, on the other hand, we had learned to determine the relative weights of chemical molecules by chemical considerations, it appeared that the two values were identical, we arrived at the conception . . . not necessary previously, that gas particles and chemical molecules are identical, that heat is able to subdivide matter to chemical molecules.

For the polymer chemist a further passage in Kekulé's lecture is of particular interest:

> The hypothesis of chemical valence further leads us to the supposition that a considerable number of single molecules may, through polyvalent atoms, combine to *net-like*, and, if we like to say so, to *sponge-like* masses . . . which resist diffusion and which, according to Graham's proposal are called colloidal.

This is certainly the first reference to the concept that later came to be described as the crosslinking of macromolecules. On the other hand, we may note that in proposing crosslinking as a possible cause of slow diffusion Kekulé diverted attention from the possibility that a low mobility could result from a great length of linear molecules. He also endorsed—somewhat hesitantly—an idea advanced by the biologist Pflüger, whom he described as "our genius" that "these mass molecules . . . through constant change of position of polyvalent atoms show a constant change in the connection between the individual molecules so that the whole . . . is in a sort of living state."

Surveying all these controversies from our vantage point, it may seem strange that chemists of the third quarter of the nineteenth century seem to have paid no attention to the progress accomplished by physicists in the understanding of the gaseous state, which placed Avogadro's law on a firm theoretical foundation. In 1857 Clausius published his remarkable paper[76] in which he formulated the kinetic gas theory that extended (and in part corrected) suggestions made earlier by Krönig[77] and Joule.[78] He pointed out that the energy of a gas molecule contains contributions from translation, rotation, and vibration and that the fraction of the energy contained in the translational kinetic energy therefore tends to decrease as the number of atoms joined in a molecule increases. Citing Regnault,[79] who pioneered studies of the deviations from Boyle's and Gay-Lussac's law (and who proposed the concept of an "ideal gas" as applying to gas behavior in the limit of low pressures), Clausius proposed that gas nonideality is due to the volumes and the mutual attraction of the molecules. A quantitative theory of this deviation from gas ideality was then formulated in van der Waals' celebrated doctoral thesis, published in 1873[80] at a time when much more extensive experimental data on gas behavior were available. It should be noted that van der Waals fully acknowledged the fact that the origin of the attractive forces between the molecules was unknown. (In fact, comprehension of these forces had advanced very little when Jeans[81] celebrated the fiftieth anniversary of van der Waals' thesis.) I have noted these developments not only because they made the concept of molecules, to which a variety of physical properties could be assigned, much more real, but also because the deviation from gas ideality was to find later a parallel in deviations from "ideal solution" behavior.

Whereas the observation of the freezing point depression of a solvent by solutes goes back to Watson's[82] and Blagden's[83] experiments, published as early as 1771 and 1788, de Coppet was first to note a relation to the molecular weight of the solute when he found that aqueous alkali halide solutions of equal molality exhibit the same freezing-point depression.[84] An extensive program of study of the freezing points of solutions in water and organic solvents was later undertaken by Raoult.[85] He took great care to avoid supercooling, but his temperature could be measured only to within 0.1°C, so that data could not be collected for very dilute solutions. An early paper that dealt with aqueous ethanol solutions[85a] shows clearly the influence of preconceived ideas. Raoult reported that the ratio of the freezing point depression ΔT_f to the concentration of ethanol remains constant (to *three* significant figures, 0.377) for ethanol concentrations up to 10.5 g/100 g of water and exhibits another constant value (0.528) in the concentration range between 29.7 and 51 g ethanol in 100 g water. Accepting Blagden's rule that ΔT_f should, in principle, be proportional to the solute concentration, Raoult interpreted his data as indicating the formation of an ethanol monohydrate that would reduce the amount of "available" solvent. (In this he followed earlier suggestions that ΔT_f of aqueous salt solutions reflects the formation of salt hydrates. The concept that ΔT_f is affected by compound formation between solute and solvent was not used in Raoult's subsequent papers because he considered it too ambiguous.) Later, Raoult measured the freezing points of 29 organic compounds in water[85b] and found to his surprise that the ratio of ΔT_f to the molality m varied only within narrow limits (1.55 for phenol and 2.29 for oxalic acid, with an average $\Delta T_f/m = 1.85$), although the molecular weights of the solutes varied by a factor of more than 10. He tried to account for the variation of the "molar depression" $\Delta T_f/m$ by assuming that it depends on the average of parameters characteristic of the atoms from which the molecule is constituted; this parameter was given as 1.5 for carbon and hydrogen and 3 for oxygen and nitrogen. On extending these studies to organic solvents, Raoult found[85c] that each solvent was characterized by a "molecular depression" $\Delta T_f/m$, and suggested that freezing-point depressions could be used to obtain molecular weights of solutes.* He also noted that in some cases (e.g., solutions of alcohol or carboxylic acids in benzene) the freezing-point depressions had only half the value expected from the "normal" molecular depression constant and concluded that the solute molecules were joined in pairs. He reasoned[85b] that "vaporization does not always separate molecules completely and it is natural to think that neither does dissolution." Because organic compounds exhibited a molar freezing-point depression in water only half as large as salts such as

---

*The melting points of the solvents given by Raoult differ in some cases substantially from currently accepted values. For benzene and ethylene bromide he gives melting points of 4.96° and 7.92°, whereas the modern values are 5.49° and 9.97°.

alkali halides, Raoult was led to the conclusion that organic compounds associate to pairs in the aqueous medium.[85c]

In considering the implication of his findings, Raoult recognized that "their most important application will be in the molecular weight determination in the numerous cases where it is impossible to measure gas densities."[85b] On the other hand, he did not understand how the cryoscopic constants were related to the properties of the solvent. His data seemed to indicate that these constants are approximately proportional to the molecular weight of the solvent, and since aqueous solutions did not conform to this pattern he proposed that water is trimeric in the neighborhood of its freezing point. It is surprising, in view of his sustained interest in cryoscopy, which extended over many years, that Raoult was unaware of Guldberg's demonstration in 1870[86] that freezing-point depressions are related to the decrease in solvent vapor pressure by the heat of melting of the solvent. Raoult's measurements of the relation between the vapor pressures p of a solution and $p°$ of the solvent led him to conclude that $(p° - p)/p°$ is proportional to the mole fraction of the solute,[87] with the proportionality constants varying between 0.95 and 1.10, but although he suspected that this constant would tend to unity at high dilutions he did not prove the point. It remained then for van't Hoff[88] and Planck[89] to derive rigorously the relation between the cryoscopic constant and the latent heat of melting from the second law of thermodynamics.*

In his classical 1887 paper[88a] van't Hoff also formulated the theory of osmotic pressure (Π) by demonstrating its analogy to the gas laws.† He took special pains to stress that the simple formulation $\Pi V = nRT$ should be applied only "to ideal solutions, that is, to solutions which are diluted to such an extent that they are comparable to 'ideal gases,' mutual action of the dissolved molecules and also their volume in comparison with the volume of the solution being negligible." (In an earlier Swedish publication of this work van't Hoff specified that the ideal solution laws are applicable "to solutions so dilute that the heat of dilution is negligible.")[88b]

The requirement of semipermeable membranes presented great practical difficulties to the exploitation of osmometry for molecular weight determinations. On the other hand, cryoscopy was substantially advanced by Beckmann's development of a differential thermometry method,[90] which allowed the use of thermometers graduated to 0.01°C and reasonable estimates of fractions of this temperature interval. Beckmann exploited the improved sensitivity of the method to collect data over a broad range of concentration up to high dilutions. He plotted "apparent molecular weights" calculated from $\Delta T_f/m$ (assuming that a single cryoscopic constant applies to all concentra-

---

*Planck wrote that according to Raoult's data *either* organic compounds were dimerized *or* salts were dissociated in water. The thermodynamic derivation of $\Delta T_f/m$ from the heat of melting of ice proved that the second interpretation applies.

†For a description of the circumstances that led van't Hoff to this result see G. Wald, *Science*, **217**, 1084 (1982).

tions) and extrapolated the curves to $\Delta T_f = 0$. (Raoult had suggested extrapolation of $\Delta T_f/m$ to $\Delta T_f = 0$.)[85d] In acetic acid he found good agreement with known molecular weights. In benzene solution acetic and benzoic acids behaved as dimers with no indication of dissociation at the highest experimentally accessible dilutions, whereas oximes exhibited a behavior that suggested a progressive dimerization as the concentration increased. With ethanol the "apparent molecular weight" showed an extremely steep increase with an increasing freezing-point depression (which was interpreted as gradual association to large aggregates) but the data extrapolated to the correct molecular weight. Influenced by van't Hoff's paper, Beckmann wrote that the concentration dependence of the apparent molecular weight should depend on the relative magnitude of forces between solute and solute and between solute and solvent molecules "a higher molecular weight and greater chemical affinity leading to steeper curves." This may be regarded as a first attempt to interpret colligative properties in terms of solvent-solute interaction. On the other hand, Beckmann was mistaken in implying (although he did not state this explicitly) that in ideal solutions the freezing-point depression should be proportional to the solute molality over an extended concentration range. Twenty years had to pass before Porter[91] and Trevor[92] derived the relation between osmotic pressure and solvent activity for solutions of any concentration, from which it follows that the freezing-point depression of an ideal solution is proportional to the logarithm of the solvent mole fraction.

5. *Natural Macromolecules*

Long before chemists aimed at the preparation of synthetic polymers they were faced with the problem of understanding the chemistry of substances such as cellulose, starch, rubber, and proteins produced in living organisms. The full history of this search is beyond the scope of this book and I have had to restrict myself to some of the highlights, referring the reader to more specialized treatments of these various materials.

Hevea Rubber

From the biological point of view rubber is undoubtedly the least important of these substances because it is produced by only a small number of plants and cannot be said to play a crucial role in the life of the organism. Yet from the point of view of the history of polymer science its study proved in several respects to be a greater stimulus than that of the natural polysaccharides and proteins. Among these materials rubber alone could be broken down to simple molecules with known structure and reformed from these "building blocks." This feature not only raised the question whether the material could be synthesized but led to the important insight that there is no need to duplicate in detail the structure of the natural substance to arrive at materials with its useful properties. Also, the elastic properties of rubber were so strikingly different from those of other solids with which nineteenth-century chemists had to deal that they excited their curiosity.

A beautiful account of the early history of rubber was written by Törnqvist.[93] Columbus reported that he had learned of a game played by natives of Haiti with balls of an elastic resin exuded from a tree but the first detailed account of the origin and use of rubber was written by de la Condamine, who explored Ecuador from 1735 to 1743.[94] He described the collection of rubber latex from incisions in the "Hhevé tree" (from which the scientific name Hevea originates) and reported the use of the hardened sediment to waterproof fabrics and shoes and to fashion elastic water bottles. He wrote about rubber balls which "when allowed to fall on a horizontal surface keep rebounding" and even suggested that rubber might be used to manufacture diver suits. De la Condamine reported great efforts to locate the sites at which rubber trees could be found but cautioned that "all one wants to make from this resin has to be made on the spot where the trees stand since the milky juice dries and thickens rapidly once drawn from the tree." Because rubber trees had been found in French Guiana he added hopefully that "this will be an exclusive article of commerce from the colony where this treasure is found."

It is significant that Faraday worked as early as 1826 on the analysis of rubber,[95] writing that "much interest attaches to this substance in consequence of its many peculiar and useful properties." He washed a latex sample obtained from Mexico, coagulated the rubber, dried it, and carried out an elemental analysis. His C/H weight ratio was 6.812 (as against the correct value of 7.5), but because he used the atomic weight of 6 for carbon he concluded that rubber contains eight carbon atoms to seven of hydrogen. Faraday also noted that the material precipitating from the aqueous phase of the latex when it was boiled, yielded on heating a great deal of ammonia. He concluded that "it resembles albumen more than any other substance."

We have already noted the attempts to learn more about the chemistry of rubber by subjecting it to destructive distillation to yield products that could be more easily characterized. By this approach Himly[96] in 1836 obtained a fraction boiling between 33 and 44°C, which he called faradyine in honor of Faraday's study. The isoprene isolated in the more careful work by Williams in 1860[58,59] boiled at 37–38°C (as against the boiling point of 34°C for pure isoprene); apparently it was contaminated by 2-methyl-2-butene. The structure of isoprene remained long uncertain; in 1884 Tilden[97] proposed five possible formulas for C_5H_8. The correct one was proved in 1897.[98,99]

However, in spite of the degradation of rubber to isoprene (and the reconstitution of rubber from isoprene, as demonstrated by Bouchardat[60,61] and Wallach[62]), the relation between these two materials remained controversial for a long time. Since the distillation of rubber yielded much more dipentene than isoprene, it seemed natural to assume that dipentene was the "building block" of rubber, particularly because no isoprene could be isolated from plants. The two questions then to be answered concerned the chemical structure of the dipentene and the nature of the process that led to the dipentene-rubber transformation.

These problems were the subject of extensive studies carried out by Harries from 1902 to 1905.[100,101] At first Harries used the action of nitrogen oxides, nitric acid, and permanganate to obtain fragments he called "nitrosite c" and to which he assigned the formula $(C_{10}H_{15}N_3O_7)_2$.[100] Later, after he had established ozonolysis as a powerful method for the study of unsaturated compounds, he applied this technique to *Hevea* rubber.[101] His enthusiasm is reflected in the exultant sentences:

> I believe that this method has solved in principle the problem of rubber degradation. I shall shortly fill in the remaining gaps.

Harries found that ozonolysis of rubber yielded a single product, levulinic aldehyde

$$\underset{\underset{O}{\|}}{CH_3C}CH_2CH_2\underset{\underset{O}{\|}}{CH}$$

and he concluded that rubber consists of dimethylcyclooctadiene residues

$$\ldots \underset{\underset{\ldots CH-CH_2-CH_2-C}{\|}}{\overset{\overset{CH_3}{|}}{C}} -CH_2-CH_2-CH \ldots \underset{\underset{\ldots CH-CH_2-CH_2-C \ldots}{\|}}{\overset{\overset{CH_3}{|}}{C}-} CH_2-CH_2-\underset{\underset{CH_3}{|}}{\overset{\|}{C}} \ldots$$

and that "the polymerization must be due to a loose addition of the dimethylcyclooctadiene molecules,* since otherwise the easy breakup could not be explained." He wrote that this breakup is analogous to the hydrolysis of cellulose to glucose and speculated that rubber might be formed in the *Hevea* tree by a reduction of pentose sugars to C_5H_8 residues which "condense in the nascent state to a complex $(C_5H_8)_x$." Harries also believed that the difference between *Hevea* rubber and guttapercha[†] may be due to a difference in the positioning of the methyl groups on the cyclooctadiene.

In his publications Harries referred several times to his desire to discover a method by which rubber could be obtained from agricultural products available in Europe:[101a]

> I should like to refer to an objective which must attract the rubber chemist: I showed that levulinic acid is formed from rubber or guttapercha. It is also known that the same acid may be obtained from starch. This physiological relation is of great significance for technology; if one should succeed to recover rubber by a suitable reduction of levulinic acid, agriculture could utilize starch in a more lucrative manner.

How strange that we should encounter here after 95 years a revival of Kirchhoff's dream (see p. 37) to convert starch into rubber!

Harries' interpretation was challenged by Pickles[102] on a number of grounds. He objected to the "vague and unnecessary conceptions of polymer-

---

*J. Thiele [*Ann.*, **306**, 87 (1899)] had suggested that in compounds that contain carbon-carbon double bonds "the strength of the affinity is not fully used and on each atom there is still an affinity residue or a partial valence."

†In the Malay language "gutta" is "sap" and "percha" is the name of the tree from which guttapercha is obtained.

ization" and pointed out that if the dimethylcyclooctadiene
together by forces between the double bonds bromination
degradation for which there was no evidence. He there
rubber is a polymer of the type

$$\ldots CH_2-\underset{\underset{CH_3}{|}}{C}=CH-CH_2-CH_2-\underset{\underset{CH_3}{|}}{C}=CH-CH_2 \ldots$$

He was in error in assuming that a ring has to form to eliminate unsatisfied end-group valences and he grossly underestimated the size of the molecules. These, however, seem to be minor defects in a historically important contribution.

An influential monograph published in 1913 by Dubosc and Luttringer[102a] summarizes the conventional wisdom of that time concerning rubber structure. It dismisses Pickles' chain molecules and accepts without reservation Harries' dimethylcyclooctadiene rings joined to one another by partial valences. It claims in support of Harries' structure (p. 334)* that heating the unsubstituted cyclooctadiene leads to a rubberlike product. Yet this argument was disproved in the same year by Lebedev,[102b] who showed that the polymer of butadiene is different from products obtained from cyclooctadiene. Harries, himself, who had originally expressed a poor opinion of Pickles' suggestion,[64a] conceded in 1914[102c] that his structure might, after all, be correct.

As noted before, nineteenth-century scientists were fascinated with rubber, largely because of its unusual physical properties. To quote Charles Goodyear, the inventor of the vulcanization process[103a]:

> There is probably no other inert substance the properties of which excite in the human mind an equal amount of curiosity, surprise and admiration. Who can examine and reflect upon this property of gum-elastic without adoring the wisdom of the Creator?

Two observations of rubber elasticity with far reaching significance were reported by Gough in 1805.[104] His first experiment on a rubber band was described as follows:

> Hold one end between the thumb and forefinger of each hand and bring the middle of the piece into slight contact with the edge of the lips . . . extend the slip suddenly and you will perceive a sensation of warmth. . . . The increase in temperature may be destroyed in an instant by permitting the slip to contract again. . . .

To forestall all criticism of the qualitative nature of this report, Gough added

> . . . we are not inquiring about proportions but endeavoring to establish the certainty of the fact   the following simple experiment . . . seems to afford no

---

*The page numbers refer to the original French edition.

inconsiderable insight into the plan which nature pursues in producing this phenomenon. . . .

In the second experiment he found that

> if one end of a slip of Caoutchouc* be fastened to a rod and a weight be fixed to the other extremity . . . the thong will become shorter with heat and longer with cold.

More than a hundred and thirty years would pass before the significance of Gough's observations began to be clarified. His explanation, at a time when the nature of heat was not understood, was as follows: pores in the rubber attract the "caloric fluid." Rubber extension reduces the volume of the pores, leading to evolution of "caloric"; the retractive force arises "from the mutual interaction of Caoutchouc and Caloric which attraction causes an endeavour to enlarge the interaction of the former." The contraction on heating "is occasioned by the absorption of the caloric fluid . . . in the same manner that ropes . . . are obliged to contract . . . by absorption of water."

In 1857 W. Thomson (better known by his later title, Lord Kelvin) published a theoretical paper that dealt with thermal effects accompanying distortions of solids.[105] Unaware of Gough's experiments, he predicted that "an india-rubber band suddenly drawn out (within the limits of perfect elasticity) produces cold . . . (for it is certain that an india-rubber band with a weight suspended by it will expand in length if the temperature be raised)." Thomson, however, encouraged Joule to carry out experimental studies and his calorimetric measurements were made with extraordinary skill.[106] They confirmed Gough's qualitative observation that rubber heats up on stretching, that a loaded rubber band contracts on heating and provided quantitative data for these effects. (Because of the many deferential references to Professor Thomson in Joule's paper, some writers who reviewed this work seem to have overlooked the fact that Joule's experiments contradicted Thomson's predictions [505a]). On the other hand, Joule showed that Gough was wrong in claiming that the density of rubber increases when it is stretched at constant temperature; on the contrary, he found a slight expansion of the specimen. It should also be noted that because vulcanization of rubber had now been discovered Joule was able to compare the behavior of raw and vulcanized rubber samples.

Starch

The improvements in the elemental analyses of organic compounds achieved by Gay-Lussac and Thénard[107] made them conclude that sugars and starch contain hydrogen and oxygen in the same proportion as water and led them to designate such substances as "carbohydrates." The observation that starch

---

*Derived from the word used for rubber by South American natives.

may be converted by acid to a sugar is generally credited to experiments carried out by Kirchhoff in 1811. An "Excerpt from a letter of Academician Nasse to Professor John concerning Kirchhoff's new preparation of sugar,"[108] published in 1812, contains the following quaint passage:

> The discovery of this preparation of sugar has much in common with the accidental discovery of porcelain. Kirchhoff, who has always considered chemistry as a technical discipline, aimed at producing rubber, which is at present very expensive, but obtained sugar instead.

Three years later Kirchhoff[109] described his process including the purification of the "sugary syrup" by boiling it with skimmed milk or beaten egg white, followed by filtration through a woolen cloth. He also reported the liberation of sugar during the germination of seeds.

Another experiment that led to an entirely unforeseen result was destined to play an important role in the study of starch over the next hundred years. In 1814, three years after the discovery of the element iodine, Colin and Gaultier de Claubry[110] reported on their study of the interaction of iodine with a random assortment of biological materials. Nothing remarkable occurred when iodine was added to fats or proteins. But on reading this paper we cannot fail to sense the surprise and excitement of the authors when they observed that on mixing iodine with starch "the color is a magnificent blue if the two substances are in the right proportions."

By 1879 Brown and Heron[111] wrote that they had a list of 400 publications that dealt with starch, and their paper contains an excellent bibliography of early work in this field. Several factors favored studies of starch degradation. Enzymatic catalysis of the process made it possible to work under mild conditions and to avoid destruction of the sugar formed. In addition, three independent analytical methods were available, based on the color of the starch-iodine complex, measurements of the optical activity, and the reducing power of the reaction products.

Biot and Persoz[112a] in 1833 followed the effect of acid on starch by changes in optical acitivity and reported that "first the acid aided by heat tears the membrane of the starch granule releasing a substance which we call dextrin" since "no organic substance known up to now produces such a strong rotation of the plane of polarized light." Later in the same year[112b] they described the conversion of starch to dextrin by an enzyme which they called diastase. For a long time dextrin was believed to be an intermediate in the conversion of starch to sugar, but Musculus[113] found thirty years later that the enzymatic reaction produces sugar and dextrin concurrently and that diastase acts only on substances that form the blue iodine complex. During all this time it was assumed that the sugar obtained in the enzymatic breakdown of starch is glucose and only O'Sullivan's study in 1872[114] showed that this sugar is a disaccharide, $C_{12}H_{22}O_{11}$, which was named maltose.*

---

*O'Sullivan acknowledged that Dubrunfaut[114a] had reported twenty-five years earlier the same disaccharide as the product of the enzymatic degradation of starch. His finding was ignored.

The detailed studies of the diastase-catalyzed reaction was clearly hampered by the heterogeneity of both the starch and the enzyme. O'Sullivan[115] found that the maltose yield decreased with an increase in the reaction temperature. He believed that a single maltose was split off from the starch molecule and this led him to wonder whether the size of the starch molecule could increase with temperature or whether "the character of the transforming agent changed." In an ambitious investigation Brown and Heron[111] used both optical activity and reducing power to characterize the changes that accompany starch degradation and interpreted their results as consistent with starch molecules that contain at least ten maltose residues; these split off one maltose unit at a time with equilibrium being attained in a system that contains 80% maltose and 20% dextrin. In a later paper Brown and Morris[116] suggested $[(C_{12}H_{20}O_{10})_3]_5$ for the starch molecule.

In 1889 Brown and Morris[117] attempted a cryoscopic determination of the molecular weight of starch. Although the data corresponded to values between 20,000 and 30,000, the experimental uncertainty was too large to accept these numbers with confidence. However, the value of 6400 obtained for dextrin was considered reliable, and because the authors assumed that the dextrin represented a fifth of the original starch molecule they felt that a molecular weight of 32,000 could be assumed for the starch. It is interesting to observe how the attitude changed in the next ten years. By 1899 the influence of colloid chemists caused Brown and Millar[118] to conclude that "the method of freezing is not applicable to non-crystalline products." Instead, the dextrin was lightly oxidized to produce a carboxyl group at the end of the chain. The carboxyl content corresponded to 40 glucose residues in the "dextrinic acid" and this was taken as indicating a chain of 100 maltose residues in the original starch.

Although there had been various indications that starch consists of fractions that differ from one another in structural properties other than chain length, there was a tendency to dismiss "amylocelluloses" as impurities. Only in 1906 did Maquenne and Roux[119] show that starch, in fact, consists of two fractions with very different properties. Amylose (identical with the "amylocellulose") produced a more deeply colored iodine complex than the original starch and could be completely converted to maltose. The other compound, named amylopectin by Maquenne, yielded, after enzymatic breakdown, a dextrin which gave no color reaction with iodine.

Cellulose

Gay-Lussac and Thénard[107] believed that starch and wood have the same elemental composition but Payen[120] showed in 1839 that woods of different origin contain a higher proportion of carbon and a higher H/O ratio. He also made the important observation that extraction of beech wood with dilute nitric acid yields a residue analyzing 44% carbon and 56% water (similar to the composition of starch), whereas the original sample contained 54% C, 6.2% H, and 39.8% O. He wrote: "In fact, wood contains a substance isomeric with starch which we call cellulose and a material filling the cells, the real ligneous substance." Payen found cellulose in a variety of samples, such as apple seeds and nut shells, and noted that it was present in an almost pure state in cotton fibers.

Compared to studies of starch, cellulose investigations were handicapped by the insolubility of the material in water and the drastic conditions required for its degradation. Braconnot[121] demonstrated in 1819 that when linen rags are treated with sulfuric acid "starch sugar" is formed. Pelouze,[122] in describing the conversion of cellulose to glucose, wrote that he was "sure that the reaction will become the basis of a new industry." Yet the nature of the sugar produced by the hydrolysis of various "celluloses" remained for a long time poorly defined, with the various authors specifying sweetness, fermentability, and optical activity. The relation of different hydrocarbons to one another became the subject of a heated controversy between Payen and Frémy.[123] Payen[120] wrote that "cellulose, starch and dextrin, which appear to be so different, constitute the same substance in different states of aggregation," whereas Frémy insisted that "differences in the properties of celluloses are not due to varying states of aggregation . . . but to isomeric states of these substances."[124]

An excellent account of early cellulose studies was compiled by Hess,[125] and views from earlier vantage points were given by Cross, Bevan, and Beadle[126] and Tollens.[127a] As better procedures became available to characterize sugars, it became apparent that the different "celluloses" contained a variety of sugar residues.[128] Flechsig[129] found that acid hydrolysis of cotton yielded only glucose (characterized by optical activity and reducing power) and Schulze[128] proposed in 1891 "to reserve the designation of cellulose, without qualifying adjectives, to the constituent of cell walls which is resistant to dilute acid and alkali, is soluble in ammoniacal copper solution and yields grape sugar (dextrose) on hydrolysis."

A question arose then naturally: What was the relation between these dextrose residues and cellulose? In particular, it seemed puzzling that starch and cellulose, both consisting of the same sugar residues, should have such different properties, so that "if cellulose is an aggregate of simple molecules, the nature of their binding must be entirely different than in starch."[130] This problem should have been largely disposed of by Skraup and König,[131] who in 1901 isolated cellobiose* octaacetate from the products of the acetolysis of cellulose and concluded that "cellose is the simplest polysaccharide of cellulose, just as maltose is the simplest polysaccharide of starch. It follows that cellulose and starch are basically different substances and that cellulose cannot be considered a more highly polymerized starch." It is ironic that a quarter of a century later Hess[125] should have dismissed the clear implication of this work when he wrote in his monograph (p. 494) that "cellobiose had been considered

---

*Originally named cellose.

for decades a key to the cellulose structure since it was assumed that this compound sugar is preformed in cellulose. New experiments showed this assumption to be in error." We shall return to this problem when discussing developments of a later era.

Opinions among investigators differed as to whether cellulose should be considered an aggregate of small molecules or whether the sugar residues should be thought of as covalently bonded to each other. Tollens[127b] proposed formula (I) for the cellulose chain. In a later extensive discussion of the "hydrolysis and acetolysis of cellulose" Ost[132] wrote: "If the oxygen atoms form ether-like links between the glucose residues one may either conclude that there is an open chain . . . or that the chain is closed into a ring." On the other hand, Cross, Bevan, and Beadle[133] proposed structure (II), which implied that in cellulose these biose units are somehow aggregated. Green and

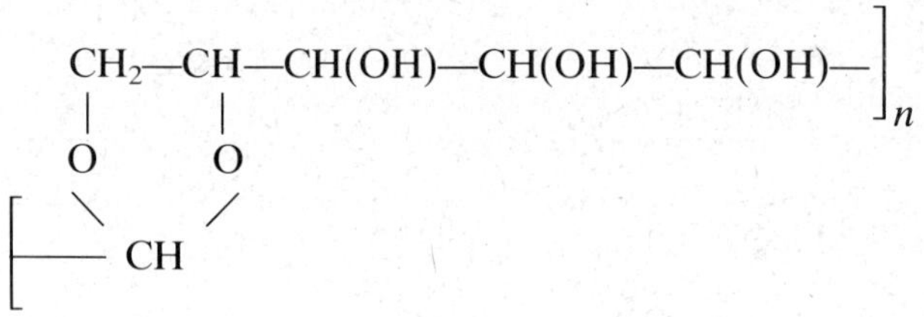

$$\left[\begin{array}{c} CH_2-CH-CH(OH)-CH(OH)-CH(OH)- \\ |\quad\quad | \\ O\quad\ O \\ \diagdown\ \diagup \\ CH \end{array} \right]_n$$

I

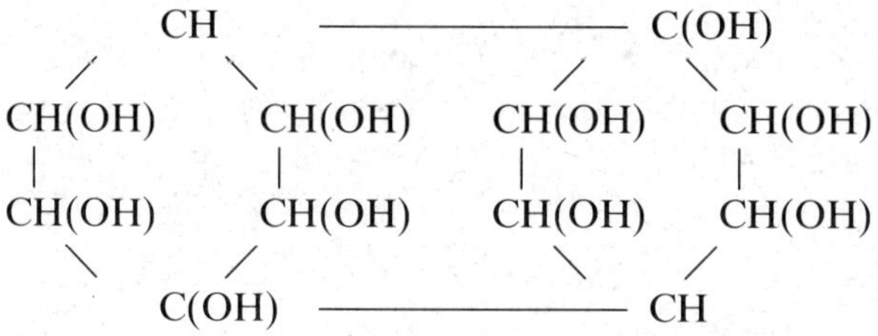

$$\begin{array}{cccc}
CH & \text{———} & C(OH) \\
\diagup\ \diagdown & & \diagup\ \diagdown \\
CH(OH)\ \ CH(OH) & CH(OH)\ \ CH(OH) \\
|\quad\quad | & |\quad\quad | \\
CH(OH)\ \ CH(OH) & CH(OH)\ \ CH(OH) \\
\diagdown\ \diagup & & \diagdown\ \diagup \\
C(OH) & \text{———} & CH
\end{array}$$

II

$$\begin{array}{c}
CH(OH)-CH-CH(OH) \\
|\quad\ \diagdown\ \ \diagdown \\
\quad\quad O\ \ O \\
\quad\quad \diagup\ \diagup \\
CH(OH)-CH-CH(OH)
\end{array}$$

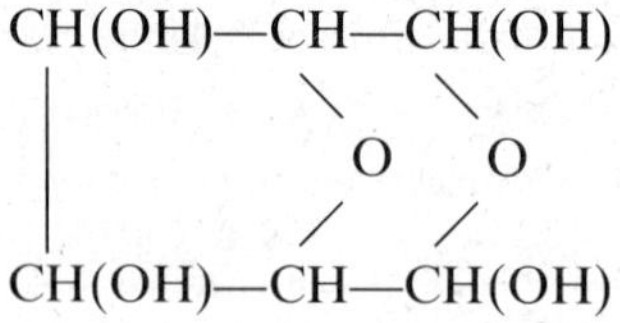

III

Perkin[134] proposed structure III with the following revealing comment:

> With reference to the objection that so simple a structure is at variance with the physical properties which seem to indicate a body of high molecular weight . . .

the suggested formula is only intended to represent cellulose in its unpolymerized form in which it may exist in ammoniacal cupric solution. The cellulose of fibers may be regarded either as a physical aggregate of simple molecules . . . or as a chemical polymer . . . the former hypothesis appears to us the more probable.

This passage is important since it raises for the first time bulk physical properties as an argument for large molecules.

In the last twenty years of the nineteenth century it was believed that by brief exposure to concentrated sulfuric acid cellulose is converted to "hydrocellulose" in which one molecule of water is bound for every two sugar residues.[135] Ost showed[132] that this concept is wrong and that the treatment which presumably led to the formation of hydrocellulose results instead in the gradual breakup of a chain molecule. His argument went as follows:

> Because of the much lower viscosities of hydrocellulose in ammoniacal copper solution (as compared to corresponding solutions of cellulose) one must conclude that hydrocellulose has a lower molecular weight. It may be objected that cellulose solutions are colloidal and that the high viscosity is due to swelling of the disperse phase . . . but dilute solutions of hydrocelluloses and their esters are true solutions. . . . The best argument is their increased reducing power. . . .

A relation between solution viscosity and the chain length of solute molecules was also suggested by other investigators of this period (see pp. 104, 105). However, Ost seems to have been first to use the concentration of end groups (responsible for the "reducing power") as a measure of polymer chain degradation.

For those who accepted the concept of cellulose as a chain molecule the problem of the length of the chain had to be addressed. This was done first in 1900 by Nastukoff,[136] who carried out ebullioscopic measurements on acetylated cellulose. He concluded that the chain contains about forty glucose residues.

Proteins

If the study of rubber, starch, and cellulose presented great difficulties to the chemists of the nineteenth century, the problems faced in any attempt to unravel the nature of proteins were much more formidable. Although it was recognized that the "protoplasm" of living cells contains "albumen-like"* material that plays a central role in the life process, thus providing a powerful incentive for its study, it was far from clear whether the techniques available to the chemist of that period would permit any significant progress in this confus-

---

*The word "albumen" is derived from the Latin "album ovi" (egg white) and the German literature used for a long time "Eiweisskörper" or "Eiweiss-Stoffe" for proteins.

ing field. Excellent histories of protein research are contained in a monograph by Fruton[137] and in a symposium published by the New York Academy of Sciences.[138]

In 1841 Liebig[139] published a long paper on "the nitrogen-containing foodstuffs of the plant kingdom," in which he claimed that "herbivores feed on vegetable albumin, fibrin and casein which have exactly the same composition and generally the same properties as their blood, their albumin, their muscle fibers." He was so sure of the identity of "vegetable casein" and "animal casein" that he commented that "it is incomprehensible that the identity of the two materials had not been noted previously." (Liebig's belief that animals cannot synthesize protein but obtain it ready-made from their vegetable food was accepted for sixty years until Loewi showed[140] that dogs can survive on food in which all protein was hydrolyzed.) In addition, Liebig claimed that these proteins contain "the same ratio of organic elements but undoubtedly in different sequence so as to account for their different properties."* Yet more careful analyses established differences in the composition of species such as egg albumin, casein, and serum albumin. Also, at an early date, salting-out procedures were developed for the separation of proteins in complex mixtures. Nevertheless, Kolbe[141] had some justification when he complained in one of his vitriolic attacks on those who wrote as if Kekulé's work had solved all major problems of organic chemistry "that albumin, casein, fibrin . . . also belong among organic compounds and that we are completely ignorant about the constitution of these most important constituents of animals and plants . . . is of little concern to them."

One of the conceptual obstacles to the development of this field was the widespread conviction that proteins undergo a chemical change when isolated from a living organism and that a study of a "dead protein" may have little relevance to the properties of the physiologically significant "living protein."[137] In fact, Kekulé wrote in his influential textbook that "organic chemistry has nothing to do with the study of chemical processes in the organs of plants and animals." Some workers believed that the isolated proteins were fragments of giant structures in the living organism; others suggested chemical transformations such as

$$\begin{array}{ccc} & O & OH \\ & \parallel & \vert \\ NH_2 & CH & NH-CH \\ \vert & \vert & \longrightarrow \quad \vert \quad \vert \\ -CH-C- & & -CH-C- \\ & \vert & \vert \end{array}$$

*"das nämliche Verhältniss an organischen Elementen, und zwar, wie man nicht Zweifeln kann, in einer andern Ordnung miteinander vereinigt, was die Verschiedenheit in ihren Eigenschaften erklärt."

characterizing the change from a "living" to a "dead" protein.[142]* These ideas may seem fanciful, but to the nineteenth-century scientists phenomena such as the rapid enzymatic breakdown after death of proteins that are stable in the living organism were difficult to account for without the assumption that life somehow changes the chemical structure of these substances.

By 1850 three amino acids—glycine, leucine, and tyrosine—had been isolated from protein digests but no methods were available for the quantitative determination of these constituents. Thus Hlasiwetz and Habermann,[143] who in 1873 introduced the acid hydrolysis of proteins for their study, believed that leucine, tyrosine, aspartic acid, glutamic acid, asparagine, and glutamine accounted completely for the composition of casein. They wrote that other proteins consisted of the same amino acid residues, although in different proportions. Impressive progress in this field was achieved only when Emil Fischer, one of the most brilliant scientists in the history of organic chemistry, concentrated on this problem. He clearly considered this move the most daring of his scientific career when in 1906 he said in a lecture that summarized this work:[144]†

> While careful colleagues fear that a rational study of this class of substances faces insurmountable difficulties, others, to whose numbers I count myself, are inclined to feel that one should at least attempt to lay siege to the virginal fortress. . . .‡

---

*Such notions have a long history. In the translation of A. F. Fourcroy's *Elements of Chemistry and Natural History to Which is Prefixed the Philosophy of Chemistry* (5th ed., Edinborough, 1800) we find in Vol. 1, p. 72–73: "When vegetables and animals are deprived of life, *or when their products are removed from the individuals of which they made a part* (my italics) movements are excited in them which destroy their texture and alter their composition."

†In a footnote to the published text of this lecture Fischer wrote: "I greatly regret that the press has exaggerated fantastically the content of this lecture. One may note from this paper . . . that I am not guilty of overvaluing the results obtained." It is hardly surprising that Fischer was annoyed. The Vienna *Neue freie Presse* carried on January 8, 1906, a report of Fischer's lecture under the heading "Artificial preparation of proteins." It included the following assertions: "Knowing the constitution of protein, it must be possible to build it up from its components just as the discovery of the constitution of indigo led to its artificial preparation. . . . Once embarked on this path, chemistry will produce many more proteins than nature succeeded in making, just as several tens of thousands of the 130,000 known carbon compounds exist only as a result of chemical synthesis."

‡Some readers may suspect that the wording of this passage betrays some "male chauvinism." In this context the following passage in Fischer's autobiography (Emil Fischer, *Aus meinem Leben*, Springer, Berlin, 1922, p. 139) is of some interest: ". . . an American lady, Helen Abott, introduced herself as a fellow chemist. . . . She declared that she intended to work scientifically in Würzburg and was surprised that women could not attend lecture courses. . . . She made quite knowledgeable comments and showed that her theoretical background was not bad. When she left, we held a 'council of war' whether we should obtain for her from the Senate of the University access to the laboratory. Some were enthusiastically in favor but the thoughtful elements could not repress the fear that she could introduce confusion into the hitherto harmonious circle. I wrote to her according to the majority view that we could not accept her and received a polite but fairly forceful reply in which she castigated Germany's backwardness concerning the status of women.

Fischer introduced a new approach to the characterization of protein hydrolyzates in which the amino acids were converted to esters that could be separated by fractional distillation. This led rapidly to the discovery of a number of new amino acids. Assuming that proteins have essentially the structure of polypeptides (a term he coined in his 1906 paper), he set himself the task of obtaining chain molecules by the laborious step-by-step synthesis. In this he realized that the asymmetry of amino acid residues is an essential feature of protein structure and that with synthetic amino acids the racemate had to be resolved before the appropriate enantiomer could be used. He insisted that only by a carefully controlled build-up of polypeptide chains could meaningful results be obtained. In a famous passage he stated:

> If one should succeed by chance through a brutal reaction to prepare a protein and if—which is even less likely—the artificial product could be shown to be identical with a natural substance, little would be gained for the chemistry of proteins and nothing for biology.

Eventually Fischer reached a polypeptide chain of eighteen amino acid residues (three leucines and fifteen glycines with the sequence LeuGly$_3$Leu Gly$_3$LeuGly$_9$)[145] and wrote: "The properties are close to those of proteins and had we first found them in nature, we should not have hesitated to consider them proteins." He doubted that proteins could attain molecular weights of 12,000–16,000, as suggested by some workers, and in this he followed the widespread reluctance of the time to accept the existence of very large molecular structures. Of equal importance was the comment in his 1906 paper that "amide formation may not be the only way the protein molecule is enchained. On the contrary, I consider it most probable that piperazine rings also occur." In addition, he considered it possible that hydroxyls in the polypeptide side chains may participate in intramolecular ester or ether linkages. Although Fischer's polypeptides were degraded by some preparations of proteolytic enzymes, suggesting that their structure was similar to that of proteins, the resistance of these polypeptides to pepsin was puzzling. Because nothing was known at the time about pepsin specificity, it was easy to conclude that this result pointed to the existence in proteins of some linkages distinct from the peptide bond.

In one case, that of hemoglobin, elemental analysis offered a means for the estimation of a "minimum molecular weight" on the assumption that the molecule contains a single iron atom. On that basis Zinoffsky[146] proposed in 1886 the empirical formula $C_{712}H_{1130}N_{214}S_2FeO_{245}$ which corresponded to a molecular weight of 16,700. This result was important in that it was accepted

---

She married later Arthur Michael, but I understand that they separated." One wonders whether this last comment is meant to justify Fischer's refusal to offer Miss Abott a job or if he intended to suggest that a woman chemist cannot be expected to be successful in matrimony. It is also curious that Fischer should have recalled in such detail an 1890 incident when he wrote his memoirs twenty-eight years later.

even by those who in general favored the colloid concept that large particles are aggregates of much smaller molecules.[147]*

The nature of enzymes excited a great deal of controversy and, as we shall see, the question was finally settled only well after the period we are considering here. Once again Liebig[148] advanced an extreme view in an article that dealt with "molding, rotting and fermentation," in which he considered an enzyme a decomposing substance transferring "to another substance with which it is in contact . . . the same changes which it undergoes." Considering that this paper was published in 1839 at a time when chemistry was still in a rather rudimentary state it is surprising that this view remained influential for a long time. Fischer was impressed by the stereospecificity exhibited by the enzymatic attack on sugars and concluded that it pointed to an involvement of the asymmetric amino acid residues of a protein. Because the complexities of metabolic sequences were still unknown, he found it plausible to write:[149]

> Among the agents utilized by the living cell the various proteins play a leading part. They are optically active and since they are formed from the carbohydrate of the plant one may assume that the geometry of their molecules is similar to that of the hexoses as far as asymmetry is concerned. With this assumption it is not hard to understand that yeast cells with their asymmetric agent can attack only those sugars whose geometry is not too far from that of glucose.

From this statement it was only a short step for Fischer to use in his next paper[150] the famous metaphor "that enzyme and glucoside have to fit each other like a lock and a key." Yet, although Fischer's argument for the protein nature of enzymes is suggestive, a much more convincing argument was advanced in a paper which remained without impact and does not appear ever to have been cited in histories of this subject. In 1898 Wróblewski[151] reviewed the controversy concerning the nature of enzymes and reported that the enzymatic activity of diastase is destroyed by pepsin. Apparently Wróblewski's standing in the scientific community of that time was too low for any attention to be paid to this striking observation.†

In 1900 Hardy[152] reported the electrophoresis of coagulated egg-white particles showing that these particles bear a positive charge in acid solution

---

*Nevertheless, in one of his last public lectures, Emil Fischer[309] expressed his belief that few proteins exceed a molecular weight of 4000 and continued: "It is true that one has derived from the iron content of oxyhemoglobin, which forms good crystals, a molecular weight of 16000 but against such calculations one may always raise the objection that the existence of crystals does not by itself guarantee chemical purity, since one may deal with isomorphous mixtures such as are so frequently encountered among silicates. . . . In studying substances of high molecular weight, molecular physics should thus restrict itself to synthetic products of known structure. I shall therefore continue experiments on the build-up of giant molecules." At the end of his lecture Fischer suggested that, just as physics is concerned with the details of atomic structure, it is the task of synthetic organic chemistry to clarify the limits of molecular size.

†This is the conclusion expressed in a letter addressed to me by J. S. Fruton, who pointed out that Wróblewski never gained a professorship and was widely considered "to be mad."

but a negative charge in basic media. He defined an "isoelectric point" at which the particles would not move in an electric field, but a quantitative study of this condition by Michaelis and his collaborators[153,154] could be carried out only after Sørensen's definition of pH and his proposal of pH buffers. About the same time the clarification of the Donnan effect on osmotic pressure[155] laid the basis for osmometric determinations of the molecular weights of proteins.

One of the most characteristic properties of proteins—indeed a property that served for a long time as one of the distinguishing marks of these substances—was the coagulation of heated protein solutions. The cause of this phenomenon was unknown, but it was noted by Michaelis and Rona[156] that heat-induced characteristic changes, which were referred to as "protein denaturation," led to coagulation when the pH was adjusted to the isoelectric point. Chick and Martin[157] found in 1912 that the rate of this denaturation has an unusually high temperature coefficient and is accompanied by an increase in the viscosity of the solution. This observation was said at the time to indicate that "the volume of the dispersed phase has increased at the expense of the continuous phase" and was to suggest at a later date the remarkable structure of globular protein molecules.

6. Colloids

Toward the end of the nineteenth century a great deal of interest was directed to "colloid phenomena." The doctrines of the scientists working in this area are summarized in monographs written by the leaders of this field, Ostwald,[158]* Freundlich,[159] Zsigmondy,[147] and Pauli.[160] They uncovered a great variety of interesting phenomena, but their influence undoubtedly produced a substantial delay in the acceptance of the existence of macromolecules. Ostwald, the most extreme proponent of the tenets of "colloid science" went to great lengths in insisting on its autonomy. In the introduction to his book [158] he wrote:

> The author could not suppress his criticism regarding "purely" chemical interpretations of colloid phenomena. . . . One of our leading physical chemists believes that colloid chemistry is only "a portion of physical chemistry" and that physicochemical laws derived from "real solutions" may also be applied to colloid systems. . . . But even an average physical chemist will be surprised when looking objectively at the properties of colloids, the enormous variety of phenomena, so that he will be more likely to believe that the properties of "real solutions" are unrelated to those of colloid systems.

In 1861 Thomas Graham published his celebrated paper "Liquid diffusion applied to analysis."[161] He argued that differences in "the diffusive power possessed by all liquid substances" are analogous to differences in volatility so that fractional diffusion might be used for the separation of chemical species in a way similar to fractional distillation. Following up this idea, he found that substances characterized by very slow diffusion ("hydrated silicic acid, hydrated alumina, starch, dextrin, albumen") are also unable to crystallize and called such materials colloidal (glue-like). He observed that a polysaccharide gum could not diffuse through starched paper, which was easily penetrated by sucrose, and he interpreted this difference by assuming that

> sugar . . . can separate water from any hydrated colloids. The sugar thus obtains the liquid medium required for diffusion. . . . Gum, possessing as a colloid an

---

*Wolfgang Ostwald, whose father, Wilhelm Ostwald, was one of the early pioneers of physical chemistry and the founder of *Zeitschrift für physikalische Chemie*.

affinity for water of the most feeble description is unable to separate water from the gelatinous starch and so fails to open the door for its own passage by diffusion.

We may note that nowhere did Graham suggest that a low diffusion rate may be the result of a large particle size. On the other hand, his contention that substances he designated as colloids interact very weakly with the solvent medium was to exert a strong lasting influence on the development of "colloid chemistry."

Another concept that pervaded this field was molecular aggregation. This had first been stressed by the botanist Nägeli largely on the basis of microscopic observations of fibers, cell membranes, and starch granules. His books, published between 1858 and 1877, are not easily accessible, but a review of Nägeli's theories by Katz[162] contains extensive quotations that convey the flavor of his arguments.* He was impressed with the fact that hair or plant fibers could be bent, stretched, or swollen without change in their birefringence and concluded that

> these optical properties reside in molecular groupings which we designate as micelles and are independent of the inter-micellar separation. . . . Each micelle behaves like a small crystal.

Nägeli's concept of micelles was taken later to justify the assumption of such aggregates in solution. This view was supported by the authority of Graham who, unaware of Nägeli's work, wrote[161] on the basis of a presumed low reactivity of colloids:

> The inquiry then suggests itself whether the colloid molecule may not be constituted by the grouping together of a number of smaller crystalloid molecules and whether the basis of colloidality may not really be this composite character of the molecule.†

The concept of molecular aggregates in solution seemed legitimate, particularly after Krafft demonstrated the very small boiling point elevation in solutions of soaps[163] and dyes.[164] Krafft drew far-reaching conclusions from his observations[163]:

> Colloidal molecules rotate in small closed paths, they have therefore a motion which differs in principle from that of gas molecules; this allows an easy explanation of the behavior of colloid bodies. Molecules whose state consists mainly in rotation can exert only a negligible pressure compared to gas molecules and therefore also a negligible freezing point depression or boiling point elevation. In

---

*Extensive extracts from Nägeli's work pertaining to micelles are contained in C. Nägeli, *Die Micellartheorie*, Ostwald's Klassiker No. 227, Akad. Verlagsges, Leipzig, 1928.
†Note the ambiguity with which the term "molecule" is used here. This haunts the entire literature in this field.

applying my theorem to. . . starch, it will be unnecessary to make the abnormal claim that such bodies have a molecular weight of 30,000 to 50,000.

Many of the scientists of this era believed that the systems they considered colloidal consisted of two phases. On this basis, supported by the authority of the influential textbook by Wilhelm Ostwald, Paterno[165] concluded that the small freezing-point depressions he observed in solutions of albumin and gelatin must be due to impurities. This conviction was at times so strong that it led workers to strangely cramped rationalizations when experimental data refused to agree with preconceived ideas. Thus when Reid[166] found an osmotic pressure in protein solutions he wrote:

> These observations. . . do not prove that the proteins of serum and egg white exert an osmotic pressure though . . . these complex mixtures contain something in true solution which gives a readable pressure in an osmometer.

In fact, Reid thought that he found no osmotic pressure in solutions of proteins which had been carefully purified. The view that freezing-point depressions, boiling-point elevations, and osmotic pressures of colloid solutions must be due to "crystalloid" impurities was strongly championed in the monograph on colloid chemistry published in 1909 by Wolfgang Ostwald.[158] This opinion, however, was not unchallenged. Rodewald and Kattein[167] found that osmotic pressures of starch-iodine solutions remained stable for many weeks, a result inconsistent with the assumption that the effect is due to impurities consisting of small diffusible molecules. Their data corresponded to a particle weight of 38,000, and it is interesting that the authors of this study considered the possibility that the sample might contain particles of different weights so that osmotic pressure would yield an average particle weight. The argument of Rodewald and Kattein was cited approvingly in Freundlich's "Kapillarchemie"[159] and we can see that the leaders of the colloid school were by no means in agreement on some fundamental points.

Some years after the publication of the Rodewald and Kattein paper Moore[168] wrote

> there exists at the present time not a particle of experimental evidence that colloidal solutions are *qualitatively* different form crystalloid solutions. . . . There is no evidence that a visible suspended particle does not exert the same osmotic pressure as a molecule of a dissolved crystalloid.

This point was settled by Perrin's elegant experiments[169] which showed that the equilibrium distribution in a gravitational field of particles large enough to be observed in a microscope follows the same law as the distribution of gas molecules in the atmosphere.

Of course, an acceptance of the validity of the laws governing colligative properties for polymer solutions had no bearing on the question whether the osmotically active particle is a molecule or a molecular aggregate. There was almost universal conviction that large particles must be considered aggregates. This was the view championed by Noyes in his presidential address to the American Chemical Society in 1904.[170] Similarly, when Gladstone and Hibbert[171] found a molecular weight of 6504 (!) from the freezing-point depression of *Hevea* rubber solutions they could accept this result only by assuming the presence of molecular aggregates. Moore[168] reported that the osmotic pressure of gelatin solutions increases more rapidly than proportionately to the absolute temperature, an indication that molecular aggregates dissociate when the temperature is raised. This result conforms with our modern understanding of gelatin, but it is difficult to see how Moore could have found that the osmotic activity of serum proteins increases when albumin and the globulins are separated from one another.

Some attempts were made by proponents of the colloid school to establish other methods for the estimation of particle or molecular weight. Pauli[160] believed that he could estimate the molecular weight of proteins from their titration and conductance behavior. However, to use this approach he had to estimate the number of ionizing groups carried by the protein molecule. For instance, on page 121 of his monograph he concluded that globulin is divalent so that its molecular weight is 10,000; he criticizes Hardy for assuming that the protein is pentavalent, leading "to the altogether too high molecular weight of 25,000."

Another, much more fruitful approach was based on the measurements of diffusion coefficients. In 1888 Nernst[172] showed that the diffusion coefficient D is equal to the ratio of Π_o, the osmotic pressure exerted by one molecule of solute in unit volume, and the frictional coefficient, f, representing the force required to sustain a unit volocity of the solute molecule in the viscous medium. This formulation can easily be shown to be equivalent to the familiar Einstein relation[175] $D = kT/f$, where k is Boltzmann's constant and T, the absolute temperature. Nernst commented on the very large values of f obtained from the diffusion coefficients of small molecules but rather surprisingly made no attempt to interpret them. In 1902 Thovert[173] reported diffusion coefficients in water for a large number of organic compounds with molecular weights ranging from $M = 32$ for methanol to $M = 500$ for raffinose. He found that MD^2 was constant for all these materials, a result used by Herzog,[174] who estimated the molecular weight of egg albumin as 17,000 from its diffusion coefficient and Thovert's constant. It is surprising that neither Thovert nor Herzog realized that no function of MD^2 and the viscosity is dimensionless, so that this treatment could not be correct. The proper approach was pointed out by Einstein,[175] who suggested that as long as it can be assumed that molecules behave as rigid spheres the frictional coefficient should be given by Stokes' law $f = 6\pi\eta a$, where η is the viscosity of the medium and a, the radius of the sphere. Thus, provided Einstein's assumption is correct that the macroscopic laws of hydrodynamics can be applied legitimately to the motion of a molecule moving through a medium composed of other molecules, the volume of a spherical solute particle can be obtained from its diffusion coefficient. Einstein realized that the dimensions of a solute particle may change on dissolution. To take care of this he showed that the viscosity η of a dilute suspension of spheres is related to the viscosity η_o of the suspending medium by $\eta = \eta_o (1 + 5/2)\phi)$, where ϕ is the fraction of the volume occupied by the spheres.* Assuming that a sucrose molecule would behave as a sphere, Einstein's relations led to a diffusion coefficient of sucrose in water within 8% of the correct value. Surprisingly, no one seems to have tried to use this approach in the immediately following years for the estimation of the molecular weight of globular proteins. It is also astonishing that as late as 1921 Nernst[177] quoted both the Thovert and Einstein approaches to the problem of deducing molecular weights from diffusion coefficients without indicating a preference.

Strangely enough, a more complex process, the diffusion of a variety of gases through rubber membranes, was studied by Mitchell, a Philadelphia physician, as early as 1831.[177a] He found very large differences in the diffusion rate by estimating, for instance, that CO_2 diffuses through rubber thirty times as fast as CO and noted that the gases which passed most easily through the rubber barrier were "generally highly soluble in water or other liquids." Graham[177b] embarked in 1866 on a detailed study of this phenomenon demonstrating a correlation between the solubility of a gas in the rubber phase and its rate of permeation. He found that a rubber balloon filled with nitrogen expands in contact with air because oxygen flows into the balloon faster than nitrogen flows out and he suggested dialysis through rubber as a means of enriching the oxygen content of air.

Colloid chemists were captivated by the idea that the phenomena they observed would lead them to an understanding of the life process. Graham[161] wrote:

> The colloid possesses ENERGIA. It may be looked upon as the probable primary source of the force appearing in the phenomenon of vitality.

Pauli[178] expressed a similar sentiment:

> "The colloid chemistry of proteins provides then an approach to the obscure region which scientists contemplate today only with a silent longing for the promised land—the physical chemistry of living matter.

Because surface energies and adsorption equilibria were central concerns of colloid chemists, many characteristic biochemical phenomena were interpreted in these terms. Such concepts were invoked to explain the oxygen

---

*It is little known that Hatschek[176] considered the problem of the viscosity of a fluid with suspended spheres apparently unaware of Einstein's work. Although he obtained the wrong coefficient of ϕ (9/2 instead of 5/2), he concluded correctly that the viscosity of such a suspension depends only on ϕ and not on the size of the spheres.

binding by hemoglobin[179] immunochemical reactions,[180]* and it was even claimed that the fertilization of the egg is a "colloid chemical process."[181]

The phenomenon of reversible gelation exhibited by solutions of gelatin or agar exercised a particular fascination. Graham[182] had demonstrated as early as 1862 that the rate of diffusion of salts in gelatin solutions is not significantly reduced when the solution gels. The nature of gels was then studied extensively by Hardy.[183] He found that the ternary system, water-ethanol-gelatin, separated into two phases during the gelation process and reasoned that a similar phase separation into gelatin-rich and gelatin-poor phases must also occur during the gelation of the two-phase, gelatin-water system, although "the optical characteristics of the two phases do not differ sufficiently to permit microscopical investigation of the surface of separation." He supported this concept by describing how he could squeeze out a dilute solution from an agar gel but did not allude to a similar experiment with gelatin gels. Hardy's conclusion that reversible gels are two-phase systems was widely quoted and seems to have been generally accepted. Ostwald[184] supported it since it seemed consistent with his experiments which indicated that liquid-liquid emulsions have viscosities higher than either component; thus the failure of gels to flow was only a consequence of the high degree of subdivision of the discontinuous phase.

The proponents of colloid chemistry insisted that rubber vulcanization cannot be understood in chemical terms alone but must be viewed as a colloid phenomenon. The most extensive work in this field was carried out by Weber.[185] His experiments led him to the following conclusions:

1. Sulfur adds to rubber; there is no hydrogen sulfide formed.
2. A continuous series of addition products is formed; there is no indication of a definite stoichiometry.
3. Prolonged mastication of the rubber is without effect on the kinetics of the sulfur addition but affects profoundly the physical properties of the vulcanizate.

He interpreted these findings in terms of Harries' dimethylcyclooctadiene structure for rubber, believing that after sulfur addition the rubber, as a typical colloid, "passes into the pectinized state." Somewhat related to Weber's work were the observations of Schidrowitz and Goldsbrough,[186] who found a correlation between the viscosity of rubber solutions and the ultimate stress and elongation of the vulcanization products obtained from these rubbers. They made no attempt to interpret these results.

---

*It is startling to read that Landsteiner, who later became the great pioneer of immunochemistry, in 1908 expressed his skepticism concerning Michaelis' suggestion that immunochemical reactions involve "specific chemical affinity."

7. Beginnings of a Polymer Industry

As we have seen in the preceding account, few chemists were inclined to accept the concept of the very large molecules, which we now call polymers, before the first world war. Also, during this time of the dramatic rise of a chemical industry it would have been difficult to foresee the dominant role that would eventually be assumed by the production of polymers. Nevertheless, three industrial developments falling into this period foreshadowed two important principles by which man may obtain polymeric materials with useful properties:

1. The chemical modification of macromolecules synthesized by living organisms led to the conversion of *Hevea* rubber, which has limited usefulness in its natural state, to vulcanized products which, among numerous other applications, made possible the development of the automobile. Similarly, the chemical modification of cellulose produced a useful plastic and the first synthetic fibers.
2. The controlled condensation of phenol with formaldehyde resulted in the first industrial manufacture of a fully synthetic polymeric material.

Rubber

In spite of the glowing account of the multiple uses of rubber suggested by de la Condamine's report published in 1751[94] it took many years before these hopes were realized.[93,187] The use of solutions of rubber in turpentine as a cement was suggested by Herissant and Macquer in 1763. In 1770 the famous chemist Joseph Priestley wrote, "I have seen a substance excellently adapted to the purpose of wiping from paper the marks of a black pencil" and this led to the use of the name "india rubber" for the material to which the French referred by its native name of caoutchouc. In 1823 Charles Macintosh patented a process in which rubber was used to bond two layers of fabric for the production of a waterproof sheet; the success of his enterprise is best proved by the use of "mackintosh" as a synonym for raincoat.* Yet Macintosh's fabrics (and rubber shoes imported from Brazil) suffered from the fact that they became sticky at slightly elevated temperatures and were brittle in the cold. Consequently rubber products were frequently returned by irate customers, and this led to a series of bankruptcies. Numerous attempts were made to overcome this deficiency before the discovery of the vulcanization process by Charles Goodyear.[188,189]

Goodyear made it perfectly clear in his account that the discovery was due to an accident. He had tried a number of procedures to eliminate the stickiness of rubber and had patented some processes that proved useless. In 1838 he received a suggestion from Nathaniel Hayward that incorporation of sulfur into rubber and exposure to sunlight ("solarization") would remove the stickiness; unfortunately the change was limited to the surface of the rubber sheets. Then, in January 1839, he noticed that one of these sulfur-coated rubber specimens "being carelessly brought in contact with a hot stove charred like leather. He endeavored to call attention . . . to this effect as remarkable since gum-elastic always melted when exposed to a high degree of heat." This chance observation inspired Goodyear to define conditions that led to products with most desirable properties. He noted that "while the inventor admits that these discoveries were not the result of scientific chemical investigations, he is not willing to admit that they were the result of what is commonly termed accident; he claims them to be the result of the closest application and observation."†

In spite of the excitement generated by this discovery, practical difficulties in his attempts to apply the process made Goodyear delay a patent application for four years. A patent granted to him on June 15, 1844[190] claimed "combining the said gum with sulfur and white lead to form a triple compound—the formation of a fabric of india rubber by interposing layers of cotton batting between those of the gum (i.e., the triple compound of gum, sulfur and white lead)—and exposing this india rubber fabric to a high degree of heat." An example proposed by Goodyear used 20 parts of sulfur and 28 of white lead (basic lead carbonate) for 100 parts of rubber; with the composition heated to 270°F.

The delay in the patent application was to cost Goodyear dearly. One of his samples came to the attention of Thomas Hancock in England, who had long attempted to improve the properties of natural rubber. He eventually discovered how the vulcanization had been accomplished and on May 21, 1844, was

---

*For an obituary of Charles Macintosh see *Proc. Roy. Soc. London*, **5**, 486 (1843). It should also be noted that silk coated with rubber was used in France before 1800 for hydrogen-filled balloons. Also, a company in Paris used rubber for the manufacture of suspenders and garters, starting in 1803. ("Synthetic Rubber, the Story of an Industry," International Institute of Synthetic Rubber Producers, New York, 1973.)

†Ironically, Faraday had heated rubber with sulfur 14 years earlier but noticed only an evolution of hydrogen sulfide.[6b] He suggested that the reaction could be used to prepare H_2S, to test for hydrogen, or to increase the carbon content of organic compounds.

awarded an English patent. It is not clear whether he knew at that time of a French patent issued to Goodyear on April 16. At any rate, Hancock was not required under English law to swear that he was the first inventor but merely that he was the first to introduce the invention to England. Because Hancock's patent application preceded Goodyear's by eight weeks, Goodyear's claim was rejected. In his memoirs[188] Goodyear comments bitterly: "This may be correct according to English patent laws but is not according to the idea of justice entertained by the inventor."

Hancock's account[191] is much more sophisticated than Goodyear's.* He conceded that his success was conditioned by his acquaintance with a sample received from America, which "was not much affected by solvents heat or oils." He first suspected that the new secret process involved "some new solvent which was very cheap in America" and claimed that he eventually discovered conditions that led to vulcanization without subjecting the American sample to chemical analysis, commenting that "it is a singular fact that although sulfur had long since been compounded with rubber . . . its true value for producing such a result had never been dreamt of." Although the fact that Hancock never mentions Goodyear in his book may detract from our appraisal of his character, it should be noted that he invented the rubber masticator and was first in effecting vulcanization in steel molds, thus being responsible for two developments which proved essential to the growth of the rubber industry.

The impact of the discovery of the vulcanization process is clearly evident from statistical data on rubber consumption:[93] 25 tons in 1830, 150 tons in 1840, 750 tons in 1850, and 6000 tons in 1860. Further dramatic increases occurred after the invention of the pneumatic tire† and the frantic efforts to increase the supply of natural rubber in the Belgian Congo and South America led tragically to some of the worst crimes of man against man.[192]

Goodyear's original patent includes the use of white lead, the first material used as an accelerator in the vulcanization process. Later, litharge and magnesium oxide were used for this purpose. The observation that addition of carbon black leads to great improvement in the physical properties of vulcanized rubber was made in 1904‡ but consumer resistance to the black color delayed for many years the widespead use of this reinforcing agent.[193,194]

---

*His book also contains the full text of his patents.

†Solid rubber tires were being manufactured in Britain in 1846. The pneumatic tire was patented in 1888 by John Boyd Dunlop, a veterinary surgeon in Belfast, who built it for his son's bicycle to reduce vibration. This tire was fastened to the rim of the wheel. The detachable tire was patented by C. K. Welch in 1890. (M. E. Lerner, *Encyclopedia Americana*, 1974, Vol. 26, p. 777.)

‡It appears that S. C. Mote of the India Rubber Gutta Percha and Telegraph Works, Silverton, England, never published his discovery of rubber reinforcement. In the monograph *Rubber* (P. Schidrowitz, Methuen, London), published in 1911, there is no mention of carbon black. According to H. J. Stern's *Rubber, Natural and Synthetic* (MacLaren, London, 1954, pp. 119–120, the importance of the discovery was realized only in 1912, after the American Goodrich Company purchased rights to the Silvertone process for tire manufacture. Stern[194a] wrote a detailed account of Mote's work.

Cellulose Nitrate and Regenerated Cellulose

In 1833 Braconnot[195] described the nitration of various cellulose samples and Liebig, in an addendum to his paper, noted the tendency of cellulose nitrate to decompose violently when heated. Schönbein[196] improved the nitration process in 1847 by using sulfuric acid as a catalyst. After the discovery that cellulose nitrate was soluble in a mixture of ethanol and ether such solutions (often referred to as collodion) were widely used for the production of lacquers.[197]

The idea for producing artificial fibers seems to have been advanced first by Robert Hooke in his *Micrographia*, published by the Royal Society in 1665. This book contains the following passage:*

> And I have often thought that probably there might be a way . . . to make an artificial glutinous composition resembling, if not full as good, nay better, than that Excrement or whatever other substance it be out of which the Silk-worm with-draws its clew. . . . This hint may, I hope, give some Ingenious inquisitive Person an occasion of making some trials. . . .

Two centuries had to pass before Ozanam† found a way to put "this hint" into practice. He discovered that silk is soluble in ammoniacal copper solutions and wrote:

> The solubility of silk lends itself to important applications. . . . One may imitate nature by dissolving silk as in the caterpillar and resolidify it by evaporation; thus one could extrude it in the form of a fabric instead of weaving it or else . . . using spinnerets of varying size one could produce fibers of any thickness and length; one could also utilize old and used silk and the perforated cocoon from which the silk moth has emerged.

One year after publication of Ozanam's paper Swan took out a patent[198] in which he claimed the production of fibers by the extrusion of cellulose nitrate solutions into a coagulating medium. His patent bore the title "Manufacture of carbon fibers for incandescent electric lamps." The preparation of the cellulose nitrate fiber, which was only the first step, was followed by "treatment with a deoxidizing agent until it is no longer in a condition to burn explosively . . . and carbonizing it in a suitable manner."

In the year after Swan's patent was issued Count Hilaire de Chardonnet deposited with the French Academy of Sciences a sealed envelope which was opened during the session of November 7, 1887, to reveal his report "On an artificial textile substance similar to silk."[198a] Chardonnet realized that fibers prepared by the extrusion of a cellulose nitrate solution into water would not be

---

**Micrographia* was republished by Dover, New York, in 1961. The passage quoted is on page 7.
†Ozanam, *Compt. Rend.*, **55**, 833 (1862).

usable because of their high flammability unless they were subjected to denitration. Originally, this was to be accomplished by hydrolysis in a nitric acid bath. This procedure, however, led to a weakening of the fibers and Chardonnet eventually arrived at a process in which the denitration was effected by treatment with acid sulfide solutions.[199,200] In his original report Chardonnet wrote that his fibers had a tensile strength comparable to that of silk (25–35 kg/mm^2) and his product became generally known as "artificial silk." In 1890 he began the manufacture which made him the father of the artificial fiber industry.

The utility of cellulose nitrate (pyroxyline) was substantially broadened when the Hyatt brothers discovered the plasticizing effect of camphor.[201] The inventors specified that the mixture of nitrocellulose and camphor had to "contain sufficient moisture to prevent the pyroxyline from burning or exploding" during the compounding operation. The chemical instability of cellulose nitrate was, of course, a serious impediment to its use. A highly readable account of these problems[202]* cites a letter from the 1860s which recounts explosions of collodion billiard balls when accidentally touched with a lighted cigar. Nevertheless, numerous applications of the plasticized cellulose nitrate (celluloid), for example, in dentures, toothbrushes, combs, and dolls, was developed. For some time the most important application was in the manufacture of stiff collars and cuffs which became fashionable around 1885. In the early years of the twentieth century film for the fledgling motion picture industry was also made of celluloid for want of a better material. According to Gloor,[203] the life of a reel was on the average only fifty projections, because it yellowed and cracked at the reel slots. Worse still, a number of serious fires broke out in motion picture theaters.

A development of much more lasting significance was the discovery by Cross and his associates[204–206] that cellulose may be converted by alkali and carbon disulfide to the water-soluble xanthate from which cellulose may be regenerated by various reagents. Fibers obtained in this way were referred to as "viscose" and it was immediately realized that they had two crucial advantages over cellulose nitrate in not requiring a denitration procedure and use of the expensive ether solvent in its manufacture. The quality of the product was greatly improved by the introduction in 1905[207] of a solution of sulfuric acid and sodium bisulfate for the coagulation bath which resulted in fibers of high luster, elasticity, and strength. According to Hottenroth,[199] twenty-two plants were producing viscose fibers as early as 1906 (seven in Germany, six in France, four in Switzerland, three in Italy, and two in England). Around 1911 large-scale production started at the Elberfelder Glanzstoffabriken in Germany, by Samuel Courtauld in England, and by the American Viscose Company in the United States.

---

*Reference 202 also describes the fascinating story of "Parkesine," the first commercially successful cellulose nitrate.

Bakelite

The acid catalyzed condensation of phenol with methylal* (used as a source of formaldehyde, which was not easily available at the time) was reported in 1872 by Baeyer,[208] who noted that when resorcinol was used in place of phenol the product "burns in a manner suggesting a composition similar to that of a component of wood." Some insight into the nature of the reaction was provided by ter Meer,[209] who obtained dimethoxydiphenylmethane from anisole and methylal, and by Fabrinyi[210] and Claus and Trainer,[211] who reported the formation of ethylidene diphenol from acetaldehyde and phenol. Later Emil Fischer suggested to Kleeberg that he repeat Baeyer's experiments with formaldehyde,[212] which in the meantime had become a cheap industrial chemical. Using 10 grams of phenol and 20 mL of a 40% formaldehyde solution, Kleeberg obtained an insoluble infusible mass and concluded that it consisted of a mixture of condensation products that lacked phenolic groups. In 1894 it was shown that the alkaline condensation of phenol with formaldehyde led to methylolphenol.[213]

In a remarkable lecture on February 5, 1909[214] Baekeland traced the history of research on phenol–formaldehyde condensation and early attempts to arrive by this reaction at a useful product. He referred, somewhat contemptuously, to Kleeberg who "could not crystallize his mass. . . . he described his product in a few lines, dismissed the subject and made himself happy with the study of nicely crystalline substances." He then reviewed the patent literature that dealt with soluble acid-catalyzed condensation products, which had been suggested as shellac substitutes but were too brittle to be useful. As for the base-catalyzed condensation, which could lead to insoluble infusible products, earlier workers had used rather low temperatures to avoid the formation of a spongy mass due to the liberation of formaldehyde in the final stages of the condensation process. This was avoided by Baekeland by the expedient of applying appropriate pressure during the curing of the resin after mixing with a suitable filler. Among the applications for the product Baekeland cited linings for metallic pipes or containers, valve seats, billiard balls, phonograph records, and numerous parts for electrical appliances. He also tried to formulate the chemistry of the process by which his resin was produced. It may seem strange today that he does not seem to have considered the possibility that the infusible resin may be constituted of an infinite molecular network—an idea which, as we have noted, had already been suggested by Kekulé (see p. 28). Instead, he proposed, "until we have something better," the ring structure

$$\begin{array}{c} \overline{[-C_6H_4OCH_2-]_6} \\ \underline{-CH_2O-} \end{array}$$

*Dimethoxymethane, $CH_2(OCH_3)_2$.

Baekeland had patented his "heat and pressure" process in 1907[215] and in a fine tribute to the U.S. Patent Office[216] he wrote that "United States patent examiners with very few exceptions are able, earnest and fair. They compare very favorably with the best of all other countries." In subsequent papers[217] he described the conversion of the soluble, acid-catalyzed product (designated as Novolak) into infusible Bakelite by heating it under pressure in the presence of formaldehyde, its polymers, or hexamethylene tetramine.

Baekeland wasted no time in founding companies for the industrial exploitation of his invention. The Bakelit-Gesellschaft in Germany and the General Bakelite Company in the United States started production in 1910.

Part 2. 1914–1942

8. The Beginnings of a Synthetic Rubber Industry

In August 1906, at a meeting of the British Association for the Advancement of Science held at York, Professor W. R. Dunstan presented a lecture[218] on "Some Imperial Aspects of Applied Chemistry" in which he expressed the opinion that

> the preparation of artificial rubber from isoprene is virtually complete although it is still necessary to work out an industrial process. It will then depend on the cost whether this will endanger the culture of natural rubber as in the case of quinine.

This unfounded optimistic estimate provided the incentive for the management of Farbenfabriken in Elberfeld to initiate a program of research aimed at synthetic rubber production.[218a] A detailed proposal, which referred to Dunstan's lecture and to an endorsement by the highly respected A. von Baeyer, was submitted by F. Hofmann in February 1907, and on its basis Hofmann was charged with the direction of this research program. In his recollections [218b] Hofmann stated quite candidly that he was entirely ignorant of the structure of rubber when he embarked on his work. Since Harries was generally considered the leading expert in this area, Hofmann tried first to synthesize dimethylcyclooctadiene, which, according to Harries, was the building block of rubber. He failed in this attempt and decided to try instead the polymerization of isoprene, as suggested by Bouchardat in 1879.[61] He was determined to obtain the monomer in a purity as high as possible and therefore rejected the common procedure of cracking terpenes to isoprene. A number of synthetic routes were explored [218a] before success was achieved with a six-step process, starting with *p*-cresol:

$$\underset{\underset{CH_3}{|}}{\overset{OH}{\bigcirc}} \rightarrow \underset{\underset{CH_3}{|}}{\overset{OH}{\bigcirc}} \rightarrow \underset{\underset{COOH}{|}}{\overset{CH_3}{\underset{|}{CH_2CHCH_2CH_2}}} \rightarrow \underset{\underset{CONH_2}{|}}{\overset{CH_3}{\underset{|}{CH_2CHCH_2CH_2}}} \rightarrow$$

$$\rightarrow \underset{\underset{NH_2}{|}}{\overset{CH_3}{\underset{|}{CH_2CHCH_2CH_2}}} \rightarrow \underset{\underset{{}^+N(CH_3)_3}{|}}{\overset{CH_3}{\underset{|}{CH_2CHCH_2CH_2}}} \rightarrow \overset{CH_3}{\underset{|}{CH_2{=}CCH{=}CH_2}} + 2\,NH_4^+$$

Using this procedure, Hofmann obtained 30 grams of isoprene in March 1909.[218a,b;219] (Butadiene was obtained in the same manner from phenol.) However, attempts to polymerize isoprene by procedures described in the literature proved disappointing. Eventually Hofmann found a simple solution to his problem. Heating isoprene in an autoclave led to complete conversion to a rubbery product. The September 1912 patent, which described this process,[220a] states flatly that this product "is indistinguishable from natural rubber." The scope of the claim is rather amusing: "In carrying out this reaction, inert diluents or catalytic agents of neutral, alkaline or acidic nature may be added to the isoprene." Three months later Farbenfabriken was granted a second patent[220b] which first describes natural rubber latex and then states:

> We made the surprising discovery that one can obtain artificially a product similar to natural latex by heating isoprene in the presence of various viscous liquids. In these viscous liquids the artificial rubber is emulsified and can then be separated by customary procedures. The technical advantages are due to the physical properties of the artificial rubber. The product exhibits high toughness and elasticity and low stickiness so that its fabrication is easier.

As an example, the patent specified the addition of 300 g of isoprene to 500 mL of an aqueous solution that contained 7 g of egg albumin (!), starch, or gelatin. The system was then to be shaken for *several weeks* at 60°C.

In spite of the extreme slowness of the polymerization, the main problem was generally seen to be the high cost of monomer preparation. A very large amount of work was devoted to the improvement of methods of monomer synthesis.[221,222] In Russia two processes were developed for producing butadiene from alcohol: the Ostromislensky process

$$CH_3CH_2OH + CH_3CHO \xrightarrow{\text{catalyst}} CH_2=CHCH=CH_2 + 2H_2O$$

and the Lebedev process in which ethanol was dehydrogenated and dehydrated in a single step. The cheapest diene at that time proved to be 2,3-dimethylbutadiene, which could be obtained in two steps from acetone:

$$2CH_3COCH_3 \rightarrow CH_3-\underset{\underset{OH}{|}}{\overset{\overset{CH_3}{|}}{C}}-\underset{\underset{OH}{|}}{\overset{\overset{CH_3}{|}}{C}}-CH_3 \rightarrow CH_2{=}\overset{\overset{CH_3}{|}}{C}-\overset{\overset{CH_3}{|}}{C}{=}CH_2 + 2H_2O$$

This monomer was selected for the production of synthetic rubber by Germany during the first world war.

When war broke out a German government commission, under the chairmanship of Fritz Haber, concluded that there would be no need to construct synthetic rubber plants[93,218b] on the grounds that the war would be too short for such an enterprise to be worthwhile. Later, when the rubber supplies were depleted, desperate efforts were made to obtain rubber from America. Rubber disguised as coffee was shipped to European neutral countries in the hope of evading the British blockade and the large submarine Deutschland was sent twice to the United States (landing on July 18 and November 1, 1916) to pay for rubber and strategic metals with its cargo of dyestuffs.* The British blockade, however, eventually forced Germany to produce two grades of poly (2,3-dimethylbutadiene) ("methyl rubber"): a soft grade for tires, hoses, and gas masks and a hard grade for casings of storage batteries used in submarines. The soft rubber was produced from monomers pretreated with oxygen† and polymerized at 65°C; the hard rubber was made by polymerization at 30°C, using some preformed rubber as initiator. Both processes involved reaction times of several months. In spite of the low efficiency of the process 2500 tons of rubber were produced between 1916 and 1918.‡

The incentive to develop synthetic rubber was influenced profoundly by the wild fluctuations in the price of natural rubber. This price reached an all-time high of $2.88 per pound in 1910, having more than doubled in two years, but after the end of the first world war it had fallen below $0.50 because of the large expansion of rubber plantations in Southeast Asia. It is then not surprising that

---

The New York Times, July 10, p. 1; August 9, p. 1; August 15, p. 1; November 11, p. 12.

† The acceleration of isoprene polymerization in the presence of oxygen was reported in a thesis written by Lautenschläger in 1913. Later, Staudinger[224] commented: "This acceleration of polymerization by oxygen should be important in nature. It may play a role in the formation of resins, asphalt, etc. . . .It may also be of importance to keep oils which polymerize easily in the absence of air under carbon dioxide. . . . "

‡This "methyl rubber" was of inferior quality. According to Whitby and Katz, however, who wrote an interesting early history of synthetic rubber,[222a] tt could be improved by addition of carbon black much more than natural rubber, with the "soft" grade having its tensile strength increased sixfold! Whitby and Katz commented that "Had carbon black attained at the time of the war the position of a recognized rubber compounding ingredient. . . the story of synthetic rubber during the war might have been different." (See also Ger. Pat. 578 965, June 1, 1933.)

the outlook for a synthetic rubber industry, competitive in peacetime, was seen very differently at different times. Thus Dubosc and Luttringer wrote optimistically in 1913:[223]

> The struggle (for the rubber market) which was limited to America and Asia will have in synthetic rubber a rival with which the natural product will have to count. If the synthesis can be rapidly and economically industrialized, the price will inevitably fall in spite of the growing consumption.

Six years later Staudinger summarized the outlook for synthetic rubber much more skeptically in a lecture before the Swiss Chemical Society.[224] He concluded:

> When I view the large volume of work devoted to the synthesis of rubber, I feel that chemistry has gone astray. . . . Each country should contribute the goods for which it is best suited. . . . Let us leave it then to the tropics to provide us with raw materials formed by its plentiful solar energy.

Staudinger's attitude was determined by the high expense associated with the then known syntheses of dienes, although he conceded that petroleum might in the future become a cheaper source of rubber monomers.

With the passing of the war emergency in Germany and with the price of natural rubber far below the cost of any synthetic product which could be envisaged at the time, all efforts toward the production of synthetic rubber ceased.* In the middle of the postwar decade, however, a cartel agreement of natural rubber producers resulted once again in a dramatic rise in the cost of rubber from a low of $0.17 per pound in 1924 to a high of $1.21 in 1925. An additional incentive toward a renewed attempt to develop a synthetic rubber industry was provided by Staudinger's work on the nature of polymers, which implied that research in this field could now be planned more rationally. Although the rubber price soon collapsed to reach its all-time low of $0.035 in 1932, the intensive effort of Germany to be self-sufficient in strategic materials provided a new driving force for synthetic rubber development.

The 1912 patent[220b] on isoprene polymerization aimed at producing a rubber latex by using "egg albumin, starch or gelatin, for latex stabilization (in analogy to the natural rubber latex which contains protein in the aqueous phase, as already found by Faraday in 1826). This process would be classed today as a suspension polymerization. An important advance was achieved when the advantages of a true emulsion polymerization were appreciated, and a patent claiming the emulsion polymerization of dienes was granted in 1930.[225] (Emulsion polymerization was also used by Carothers as early as 1931).[229] Late in 1933 a patent described the emulsion copolymerization of dienes with

---

*An editorial in the October 1925 issue of Industrial and Engineering Chemistry (p. 992) says "work on synthetic rubber, begun when promise of large returns was alluring, has been more or less laid aside."

styrene or α-methylstyrene[226]. The wording of this patent is of interest since it shows that the difference between the properties of a blend of two homopolymers and a copolymer of a similar overall composition came as a surprise at the time this patent was applied for (1929):

> It may be mentioned that the formation of mixed polymerization products with pronounced valuable rubber-like properties . . . is most surprising since polymers of phenyl substituted olefins are generally resins and not rubber-like . . . rubber-like products are not obtained when mixing polymerized products from butadiene or isoprene and from polymers of phenyl substituted olefins, nor are they obtained when styrene is polymerized in the presence of completely polymerized butadiene or vice versa.

The styrene-butadiene copolymers were marketed under the trade name Buna S.*

In the remaining years before the outbreak of the second world war a number of synthetic rubbers were developed.[93] Butadiene-acrylonitrile copolymers (Buna N), developed in Germany, were found to be superior to natural rubber in their resistance to oil, gasoline, and aging at elevated temperatures, so that their production remained economically viable after the collapse of the price of natural rubber. In the United States three synthetic rubbers were introduced. The condensation of 1,2-dichloroethane with sodium polysulfide to a rubbery polymer was discovered by accident when J. C. Patrick tried to convert dichloroethane to glycol.[227] It was patented in 1927 and sold under the name of Thiokol.† A synthetic rubber produced by polymerizing 2-chlorobutadiene was patented in 1931 and sold by Du Pont in 1932 under the name Duprene."‡ Thus it may be considered the first commercially successful synthetic rubber. According to Nieuwland and collaborators,[228]

> The Du Pont Company has for some time been interested in acetylene reactions and the possibility of manufacture of synthetic rubber, because of the well known limitations of natural rubber and especially because of the lack of adequate supply in this country.

The last remark is surprising since at that time self-sufficiency was generally not an important concern in the United States. A cooperative program between the Chemistry Department at Notre Dame University and the DuPont Labora-

---

*Buna stood for butadiene polymerized by sodium (Na); S stood for the styrene comonomer. The development of Buna S in Germany was determined by the expectation of war. Since Buna S was more than five times as expensive as natural rubber, Schacht, the Minister of Finance, favored stockpiling of natural rubber. However, Hitler did not believe that this would be adequate in a long war and ordered that the synthetic rubber program should be pushed regardless of cost.[226a]

†Thiokol was patented by J. C. Patrick and N. M. Mnookin (Br. Pat. 302,270 December 13, 1927). The outstanding property of polysulfide rubbers is their solvent resistance. Thiokol was first produced in 1929 and was the first synthetic rubber manufactured in the United States.

‡Changed to "Neoprene" in 1937.

tories led to the cuprous chloride-catalyzed conversion of acetylene to vinyl acetylene[228] and addition of hydrogen chloride[229] resulted in the formation of 2-chlorobutadiene which was named chloroprene:*

$$2CH\equiv CH \rightarrow CH\equiv C-CH=CH_2 \xrightarrow{HCl} CH_2=C-CH=CH_2$$
$$\phantom{2CH\equiv CH \rightarrow CH\equiv C-CH=CH_2 \xrightarrow{HCl} CH_2=C}|$$
$$\phantom{2CH\equiv CH \rightarrow CH\equiv C-CH=CH_2 \xrightarrow{HCl} CH_2=C}Cl$$

This diene which polymerized much faster than isoprene yielded a rubber much more resistant to oxidation than natural rubber. It also turned out to be the first synthetic rubber that crystallized on stretching.

The discovery of "butyl rubber" has been described in some detail by Howard.[227] When he visited the IG Farbenindustrie in Ludwigshafen in April 1932 he was given a laboratory demonstration which he describes as follows:

> Isobutylene was kept in open glass beakers packed with dry ice. . . . Then the catalyst (BF_3) was introduced. . . . The reaction was more like an explosion . . . there was a slight puff and the liquid in the beaker changed into a sponge of Vistanex.

This was the trade name adopted at the time for polyisobutene. According to the patent,[227a]

> the strong cooling prevents the undesired too rapid reaction which leads to products not having the desired high molecular weight and the loss of isobutene by sudden overheating.

Molecular weights only up to 5000 were claimed. Because the material could not be vulcanized, it was not used as a rubber. Two other uses were found. It proved to be an excellent additive to automobile oil, counteracting its tendency to gel at low temperatures. Also, because it could be easily pyrolyzed to isobutene, it could be used as a reserve aviation fuel.† A cooperative venture of IG Farbenindustrie and the Standard Oil Company of New Jersey, the "Joint American Study Company" (Jasco) was given the task of developing the material for commercial use. Jasco worked out a method of introducing 1–2% of diene residues into the polyisobutene chain to enable it to be vulcanized, yet

---

*According to J. W. Hill,[229a] the discovery of chloroprene was entirely accidental. Carothers planned to reduce divinylacetylene to hexatriene, hoping to obtain a rubber from that substance. In the fractional distillation which aimed at the separation of divinylacetylene from monovinylacetylene a few milliliters of a liquid with an intermediate boiling point were collected. This turned, unexpectedly, into a rubbery material that proved, on analysis, to have a high chlorine content. Carothers realized immediately that HCl derived from the cuprous chloride catalyst used for the oligomerization of acetylene must have added to vinylacetylene.

†German Patent 641,284 (M. Otto and M. Müller-Conradi to IG Farbenindustrie, January 27, 1937) suggested the use of polyisobutene as an adhesive.

to minimize the unsaturation in the vulcanized product for optimum oxidation resistance.[230,231] Since the possibility to copolymerize isoprene with 2-butene was demonstrated in 1932[232] it is not clear why it should have taken Jasco several more years to arrive at isobutene-diene copolymers* and why the investigators at the IG Farbenindustrie, who had originally discovered the isobutene conversion to a high polymer, should have failed to obtain this result before the start of the second world war.

---

*In 1937 v. Rosenberg submitted a report to the Ludwigshafen laboratory in which he described his observations during a trip to the United States from April to September of that year. (BASF archive #K0212.) He found that "Standard Oil of New Jersey study next to the polymerization of pure isobutene also copolymers, e.g., with styrene or vinyl chloride, on a laboratory scale. So far, nothing useful has come of it." Not a word about diene copolymers!

9. The Impact of X-Ray Crystallography

In 1917 Ambronn[233] published a study of the optical properties of cellulose samples which had been obtained by denitrating cellulose nitrate film and stretching the specimen to double its initial length. He observed that when such films were swollen by liquids of increasing refractive index the birefringence decreased, passed through a minimum, and increased once again. He concluded that this behavior could be explained only by the presence in the sample of rodlike optically anisotropic particles which tended to be oriented with their long axes parallel to the direction of stretch. He wrote:

> A much better approximation to a complete orientation of these rods or micelles is realized in many cell walls of plants, specially in the extended fibers. . . . Such cells exceed often a length of 20 cm and their walls exhibit a birefringence six times as large as quartz. From this one could estimate the birefringence of cellulose rods since these membrances behave over small dimensions like crystals. Their overall structure represents, of course, because of the radial concentric arrangement of the optical symmetry axes, a cylindrical crystal aggregate.

We may recall here Nägeli's statement, when observing plant fibers under the microscope forty years earlier (cf p. 48): "each micelle behaves like a small crystal." Ambronn, however, was writing at a time when powerful new methods had been discovered for the study of crystals and this prompted him to make an important proposal:

> It would be most desirable if the spatial anisotropy of the particles and their transition from a random arrangement to a greater degree of orientation could be proved by a different method. One may consider in this context the transmission of X-rays through these samples. . . . I do not believe that the stretching produces sufficient orientation to lead to a Laue diagram as in single crystals. . . .

Success would be more probable in adopting the method of Debye and Scherrer to isotropic samples. Unfortunately, I lack the means to carry out such experiments.

Stimulated by Ambronn's suggestion, Scherrer[234] carried out a study of X-ray scattering by a variety of "colloid systems". He found that gold or silver sols yielded, even with particle sizes of the order of 20 Å, diffraction patterns similar to those obtained by macroscopic samples of gold or silver. He also noted that

> organic colloids (albumin, casein, cellulose and starch) appear amorphous; the colloid particles therefore consist of individual molecules or of groups of irregularly oriented molecules.

However, R. O. Herzog, head of the Institute for Fiber Research at the Kaiser-Wilhelm-Institut in Berlin-Dahlem, found it difficult to reconcile Scherrer's negative results with Ambronn's evidence of the optical properties of cellulose[235] and had the experiment repeated in his laboratory. The first was carried out on crumpled fibers of cotton or ramie as well as on cellulose pulp, as suggested by Ambronn, and sharp diffraction rings were observed. Then,

> so as to determine whether the fiber has a regular arrangement along its long axis, the method of Debye and Scherrer was combined with that of Laue* by irradiating a bundle of parallel ramie or flax fibers instead of the crumpled up fibers. The diffractions do not form lines but a series of four-point groups at the corners of rectangles whose center is the intersection of the X-ray beam. . . .[236]

Herzog understood that this implied some orientation of crystallites and was anxious to obtain a complete analysis of the phenomenon. He was fortunate in having just added to his staff M. Polanyi, one of the most brilliant scientific minds of his time. In a memoir published forty-two years later Polanyi recollected:[237]

> There was excitement about the diagram and I was asked to solve the mystery. In the next few days I made my first acquaintance with the theory of X-ray diffraction of which, owing to wars and revolution and my exclusive interest in thermodynamics and kinetics, I had heard little before.

He then concluded that the pattern of the diffraction spots corresponded to an orientation of the crystallites with one of their axes parallel to the fiber axis.[235] This suggestion was confirmed by showing that a similar pattern of spots lying on hyperbolas was obtained from a cold-drawn copper wire.[238] A full theory

---

*Laue used polychromatic X-ray diffraction by a single crystal, whereas Debye and Scherrer used powdered crystals and monochromatic X-rays.

of "fiber diagrams" was published by Polanyi in 1921[239a,c] and by Polanyi and Weissenberg in 1922.[240]*

Meanwhile Scherrer had been urged by Ambronn to repeat his experiment on cellulose and had received from him ramie samples that served for his optical measurements. This time he also obtained a fiber diffraction photograph and interpreted it in terms of crystallite orientation.[241] The same result had been published, unknown to the European investigators, in Japan as early as 1913 by Nishikawa and Ono,[242] who also studied samples of asbestos and silk fibers.† Neither Scherrer nor Nishikawa and Ono, however, went beyond a qualitative interpretation of these fiber diagrams.

The orthorhombic unit cells of cellulose proposed by Polanyi[239b] measured $7.9 \times 8.45 \times 10.2$ Å and could accommodate only four glucose residues. No more detailed interpretation in terms of molecular structure was feasible at the time; not only could no precise measurements of the relative intensity of diffraction spots be obtained but, as we shall see later, the structure which was generally accepted for glucose by organic chemists was wrong. Nevertheless, Polanyi[239b] was able to conclude that the size of the unit cell and the symmetry relations satisfying the diffraction pattern were compatible with two alternatives: The cell contained either two disaccharide segments of long plysaccharide chains or two cyclic disaccharides. No preference was expressed for either structure. In his memoir[237] Polanyi wrote

> unfortunately I lacked the chemical sense for eliminating the second alternative . . . that the elementary cell of cellulose contained only four hexoses appeared scandalous. . . . I was gleefully witnessing the chemists at cross-purposes with a conceptual reform when I should have been better occupied in definitely establishing the chain structure.

The silk fiber, which had been shown first by Nishikawa and Ono[242] to yield a typical X-ray diffraction fiber diagram, consists of two protein components. Silk fibroin is the crystalline constituent to which the fiber owes its strength; sericin, the second component, is responsible for its adhesive properties and can easily be washed out by a hot soap solution. The diffraction pattern of silk fibroin was studied by Brill.[243] The results were surprising. Although fibroin was known to contain residues of amino acids with bulky side chains such as tyrosine, the unit cell was clearly too small to contain them. Brill concluded that the fibroin had to be composed of a mixture of proteins, a crystalline component made up of glycine and alanine residues, and one or more amorphous proteins containing the bulky amino acids. From the size of the unit cell he concluded that the molecular weight M of the crystalline protein had to be given by $500 \leq nM \leq 660$, where n was the number of molecules in the unit

---

*This paper introduced the cylindrical camera so that the diffraction spots, which lay on hyperbolas when a flat film was used were now arranged on parallel layer lines.
†Nishikawa and Ono had no means of producing monochromatic X-rays.

cell. Nowhere in the body of the paper does he refer to the alternative interpretation in which the unit cell might contain only segments of long chains; it is therefore strange that he should have listed in the "conclusions" the possibility of "polypeptides built up from alternating glycine and alanine, so that the unit cell would contain glycylalanine segments of four chains." Brill's work was supervised by Polanyi and one has the impression that Polanyi suggested this passage, which was added to the manuscript as an afterthought.* The possibility that fibroin might consist of a single species with the polypeptide chain running through crystalline and amorphous regions was a concept lying far in the future.[243a] (See also p. 245). As Polanyi recalled,[237]

> A failure . . . was my treatment of silk fibroin . . . in the elementary cell there was only room for glycine and alanine. But I could not make up my mind what to make of the other decomposition products. I did not recognize the importance of the question.

It would appear that little thought was given before 1925 to Polanyi's suggestion that the unit cell of cellulose (and, by analogy, unit cells of other crystalline fibers) might contain sections of long chain molecules rather than small molecular species. Herzog was quite emphatic[244] in his assertion that a molecule could not be larger than the crystallographic unit cell. Yet some doubt must have persisted because he wrote, somewhat enigmatically, that "the chemical resistance and the related insolubility of these fiber substances must be due to other causes, possibly some common structural principle" *(Aufbauprinzip).* The mechanical strength of fibers was visualized as due to crystallites embedded in a puttylike matrix *(Kittsubstanz).*

A somewhat different model adopted later emerges from the following formulation of Herzog and Weissenberg:[245]

> With certain substances, which are referred to as high molecular, e.g., polysaccharides, proteins, rubber, etc., only colloidal solutions are obtained while X-ray diagrams yield unit cells so small that no large molecule could be accommodated in them. The question then arises whether cohesive forces do not attain in such cases the order of magnitude of valence forces so that any distinction between the two ceases to be meaningful and the building block of the crystal structure, which must be smaller than the unit cell, loses its independent significance.

This point of view was also championed by Mark[246] at the 89th meeting of the Society of German Natural Scientists and Physicians, which was held in Düsseldorf in 1926 and which led to a confrontation of Staudinger, who defended

---

*This suspicion is strongly supported by the fact that Brill's dissertation (University of Berlin, 1923) does not mention the possibility that the crystallites of silk fibroin might contain polypeptide chains. In fact, Brill compared the diffraction diagram of the silk fibroin with a rotation photograph obtained with alanylglycine anhydride crystals and concluded that their similarity indicated a close chemical relationship.

the concept of long chain molecules against his generally skeptical colleagues. Speaking about "The X-ray Structure Determination of Organic High Molecular Substances," Mark stated

> "The micro-building block" of a substance must be smaller than the unit cell; if one is to link a group of atoms which is larger, than it will extend throughout the crystal. . . . In the case of hexamethylene tetramine, dissolution in water disintegrates the crystal easily into its "micro-building-blocks." The failure to do so up to now with high molecular substances indicates that their lattice forces are quantitatively and qualitatively comparable to their intramolecular forces: the entire crystallite behaves as a large molecule.

These quotations are noteworthy in several respects. The expression "high molecular" (*hochmolekular* in German) did not distinguish between a molecule in the modern sense, that is, an assembly of atoms joined by covalent bonds, and an aggregate of smaller molecules held together by secondary valence forces. This was justified by the assumption that the forces responsible for molecular association are as strong as the valence bonds and "that any distinction between the two ceases to be meaningful." It is understandable that such a notion might have been held in the case of cellulose and silk fibroin in view of the known tendency of compounds that carry hydroxyl or amide groups to associate. It is less easy to understand how such a strong association could have been assumed in the case of rubber. Finally, it is strange that Weissenberg's nomenclature of "micro-building-blocks"[247] should have been adopted while ignoring his explicit statement that the grouping of strongly bonded atoms constituting such a building-block need not be contained within the unit cell nor extend throughout the crystal lattice but could also be chainlike or netlike.

A year before the Düsseldorf meeting a paper was published by Katz[248] under the title "What is the cause of the peculiar extensibility of rubber?" It should have shaken accepted notions of the interpretation of the diffraction patterns we have discussed. *Hevea* rubber had been known to yield a diffuse X-ray scattering halo typical of liquids but attracted no further attention from crystallographers. Apparently Katz hoped to find some clue to the understanding of rubber elasticity when he decided to look for any changes in the X-ray scattering pattern with the rubber samples under tension.* What he found was

---

*R.H. Marchessault wrote to me about a meeting with A. Weidinger who had been an assistant of J. R. Katz and who described the circumstances that led to the discovery of rubber crystallization on stretching: "Katz had built an X-ray tube with 9 portholes and he had instructed Weidinger never to turn on the instrument unless samples were available for each port. One Friday afternoon, Weidinger was preparing to put the instrument on when there were only 7 occupied ports. Katz entered the laboratory and was peeved at this wastefulness of X-rays and immediately looked for some material to fill the two empty ports. His eyes fell on a rubber band with which he immediately filled one of the ports and after some cogitation, decided to place a second extended rubber band in the final port. The result, of course, was a landmark discovery in polymer science, viz. in diffraction, unstretched rubber yields an amorphous halo, but stretched rubber shows sharp diffraction spots indicative of crystallinity and orientation."

so unprecedented that according to his testimony he did not dare for many months to report his results:

> ... fibers of amorphous rubber extended sixfold exhibit in addition to the amorphous halo a line spectrum characteristic of a body containing crystallites whose axes are oriented in the direction of the stretch.

The phenomenon was so surprising that Katz felt compelled to make sure that his rubber sample did not contain a crystalline impurity, yet the data clearly excluded this possibility. The intensity of the fiber diagram increased with an increase in the extension of the specimen, whereas the intensity of the diffuse halo decayed, but the orientation of the crystallites was high even at small extensions. Katz noted that these results could not be interpreted in terms of crystallites present in the relaxed rubber and merely oriented by the tensile stress (in analogy with Ambronn's model for the behavior of cellulose). He was compelled to conclude that "molecules or parts thereof" are being oriented and ordered into a three-dimensional repeat pattern. This revolutionary notion was supported by two extracrystallographic arguments:

1. Joule's observation sixty-six years earlier[106] (see p. 36) that a rubber band heats up during extension could be interpreted as due to the release of the latent heat of crystallization.*
2. The increase in the elastic modulus of rubber at high extensions can be interpreted as a consequence of crystallization.

Katz estimated the volume of the crystallographic unit cell of rubber as 338 $(\text{Å})^3$, which could accommodate no more than three isoprene molecules. His paper contains a passionate attack on all those who would insist that molecules cannot be larger than the unit cell:

> What can it mean that we find such a small unit cell in a substance such as rubber which is so certain to be high molecular? It is well known that Herzog and his collaborators Polanyi and Brill repeatedly defend the theory that a unit cell cannot be smaller than the molecule, so that one may derive from its size the molecular weight. This principle leads to the paradoxical conclusion that substances generally regarded as high molecular (cellulose, silk fibroin) . . . should have at most a molecular weight of 600! . . . Katz protested against this concept in a discussion with Herzog. . . . He suggested that swellable high molecular substances seem to have a peculiar molecular structure: Their molecules could be built up from building blocks of low molecular weight which are connected by

---

*As we shall see later, there is also a second contribution to this "Joule effect." It is probable, however, that the contribution pointed out by Katz is dominant.

secondary valences or in some other manner. The crystallographic unit would then possibly measure only the building block but not the whole much larger molecule. Herzog has resolutely rejected this concept . . . if we had to assume that rubber is merely an isoprene trimer then there is a high probability that the crystallographic unit cell measures merely the building block of the associated molecule. . . .

It is a pity that we have no independent record of the confrontation between Katz and Herzog at the Innsbruck Meeting of Natural Scientists to which the foregoing passage refers. The "associated molecule" in Katz's paper does not appear to be substantially different from the formulation used by Herzog and Weissenberg and it is then far from clear what the argument was about. From our vantage point it is surprising that Katz, in concluding that "molecules *or parts thereof*" are being oriented when rubber is stretched, did not state clearly that this implied the existence of chain molecules.

A few years after Katz's original observation of the crystallization of *Hevea* rubber under tensile stress thermodynamic data were obtained for the phenomenon.[248a] The heat of crystallization was determined calorimetrically and the melting point was determined as a function of stress. Finally, it was shown that an increase in the melting point with increasing tension was a consequence of the reduced entropy of the disordered chains as they were being extended. Thus the entropy of melting (ΔS_m) decreased and the melting temperature (T_m), given by the ratio of the enthalpy and entropy of melting, ($T_m = \Delta H_m/\Delta S_m$) had to increase. Moreover, it is possible to calculate the heat of fusion from the dependence of the melting point on the applied tension.

While these studies were being carried out in Europe a crystallographic investigation of cellulose fibers was in progress at the University of California. After two preliminary reports, published by Sponsler in the *Journal of General Physiology* in 1925, Sponsler and Dore[249] in the following year summarized their findings, which for the first time showed that the small unit cell of cellulose is compatible with a long chain composed of glucose residues. They seem to have based their considerations from the beginning on a chain structure, since "a satisfactory formula for cellulose must account for its physical properties." It should be realized that at that time only rather crude data on the relative intensities of diffraction spots could be obtained so that no means existed to interpret diffraction photographs in terms of atomic coordinates within the unit cell. Sponsler and Dore built models of possible polysaccharide chains consisting of glucose residues, using available data on atomic radii and maintaining tetrahedral bond angles. They found that the five-membered ring structure of glucose residues (I), which had been generally accepted since Emil Fischer's classical work, was incompatible with the observed fiber repeat distance but that the crystallographic data fitted the six-membered ring structure (II), which had just been proposed on two different grounds:[250,251]

The molecular model also showed that the glucose units had to have the β structure (with the ring inverted in neighboring glucose residues) to generate a straight chain and meet the symmetry requirements of the lattice. On the other hand, Sponsler and Dore considered that the relative intensities of higher order reflections from phases perpendicular to the fiber axis were incompatible with the accepted structure for the skeleton of cellobiose (III), the disaccharide generally assumed to be the building block of cellulose, but reached the

1,4 linkage
III

erroneous conclusion that the chain skeleton was constructed by joining alternately two carbons in the first and two carbons in the fourth position of the glucose ring (IV).

```
       C                              C                    C
       |                              |                    |
       C—O       O       C—C          C—O       O
      /   \    /   \    /   \        /   \    /   \
     C     C          C     C     C          C
    / \   / \        / \   / \   / \        / \
   O   C—C          O—C   O     C—C        O
                      |
                      C
```

       1-1 linkage   IV   4-4 linkage

These authors, however, felt that evidence of the preexistence of cellobiose residues in cellulose was not sufficiently conclusive to invalidate their proposed structure.*

The first demonstration on crystallographic grounds that a synthetic polymer consists of long chain molecules was provided for polyoxymethylene. For some years Staudinger and his students had studied oligomers and high polymers of formaldehyde since

> their study may possibly provide new points of view regarding the constitution of other high molecular substances, such as cellulose. In both cases we have insoluble polymerization products so that their molecular weight cannot be determined but they must be considered high polymeric due to their physical and chemical properties. . . .[252]

Staudinger was interested in having these substances characterized not only by chemical techniques but also by crystallography and induced J. Hengstenberg to adopt this problem for his doctoral dissertation. Hengstenberg showed[253,254] that, in addition to reflections that depended on the length of the molecule, X-ray diffraction from the oligomers also contained reflections due to a subcell that enclosed sections of the chain. As the length of the oligomers increased, these subcell reflections became more prominent and one could easily arrive, by an extrapolation of the data for the oligomers, at the diffraction pattern exhibited by the high polymer. Whereas mixtures of relatively short oligomers formed a mixture of crystals of the various species, fraction with a mean degree of polymerization of 45 formed a single crystalline phase and it was concluded that chain ends must be distributed at random in the crystals.†

---

*According to the H. Mark's recollection, Polanyi's conclusion[239b] that the X-ray diffraction pattern of cellulose is compatible either with a cyclic disaccharide or a polysaccharide chain met with contemptuous amusement among organic chemists: What was the use of a method which could not distinguish between such alternatives? Mark remembers that this skepticism was reinforced by the fact that the structure proposed by Sponsler and Dore disagreed with the chemical evidence for cellobiose fragments in the cellulose molecule.

†As Staudinger frequently emphasized, this means that the chain length of polymers cannot be obtained from crystallographic data.

A most remarkable phenomenon was described by Staudinger et al.,[255] who found that formaldehyde could be polymerized under conditions that resulted in the formation of macroscopic single crystals. (No comparable process was observed until the polymerization of crystals of diacetylene derivatives forty years later.)[255a] These single crystals of polyoxymethylene (with a length of 0.06 mm) served to confirm Hengstenberg's structure.[256]*

H. Lohmann, one of Staudinger's students, found[257] that not only polyoxyethylene but also polyoxypropylene yielded sharp X-ray diffraction patterns.† Staudinger and Lohmann realized that this presented them with a dilemma since they could not imagine how stereoregularity, required for crystallization, could have been achieved in their polyoxypropylene. They thought that this difficulty could be avoided by postulating for the polyoxypropylene the structure

$$\begin{array}{ccccccc}
O & & CH_3 & & O & & CH_3 & & O \\
/ \;\backslash & / & & \backslash & / \;\backslash & / & & \backslash & / \;\backslash \\
C & & & & C & & & & C & & & & C \\
| & & & & | & & & & | & & & & | \\
C & & & & C & & & & C & & & & C \\
\backslash \;/ & \backslash & & / & \backslash \;/ & \backslash & & / & \backslash \;/ \\
O & & CH_3 & & O & & CH_3
\end{array}$$

"where the methyl groups can be placed in the gaps so that the molecules have a regular structure enabling them to crystallize."

One member of the group working on X-ray crystallography at the Kaiser Wilhelm Institut was H. Mark, who had been trained as an organic chemist but who rapidly became a leading expert in this new discipline. According to Polanyi,[237] "his experimental skill bordered on genius." While in Berlin-Dahlem Mark worked on crystals of metals and small organic molecules (hexamethylene tetramine, pentaerythritol, triphenylmethane), but in 1926 he accepted a position with the IG Farbenindustrie laboratories, where he was to work in close collaboration with K. H. Meyer, another organic chemist fascinated by the power of crystallographic techniques. This collaboration proved to be most productive. Already in 1928 four important papers were published in which the crystal structures of cellulose,[258] silk fibroin,[259] chitin,[260] and

---

*In an important publication, which has been ignored in the literature, H. W. Kohlschütter and L. Sprenger reported [Z. Phys. Chem., B, **16**, 284 (1932)] that trioxane single crystals kept in vacuum are partially converted to highly oriented fibers of polyoxymethylene. Complete conversion of trioxane single crystals to oriented polyoxymethylene was observed when the trioxane was exposed to formaldehyde vapor. This is the first example of a solid-state polymerization in which a parent crystal imposes the direction of chain propagation. The oriented polymerization of trioxane was rediscovered by S. Okamura in 1961 when he initiated the reaction by γ-irradiation.[611]

†Dr. N. Bikales has informed me that he has confirmed the formation of a crystalline product when propylene oxide was polymerized under Lohmann's conditions. Lohmann seems, therefore, to have reported the first stereospecific polymerization.

Hevea rubber[261] were shown to result from the packing of long chain molecules.* In the paper on cellulose[258] it was proved that Sponsler and Dore had misinterpreted the diffraction diagram, which was, after all, compatible with a chain in which the glucose residues are joined to one another by an oxygen link that connects the first carbon of one glucose ring to the fourth carbon of the next.† Thus there was no contradiction between the cellulose structure and the chemical evidence that it consists of cellobiose residues.

The paper on the cellulose structure contains two arguments which led to a long and often acrimonious controversy with Staudinger. Meyer and Mark objected to Staudinger's writing about cellulose "molecules," since "one will not speak of a petroleum molecule but apply the term 'molecule' only after separation of the various paraffins." They suggested, instead, the somewhat awkward designation "principal valence chain" *(Hauptvalenzkette)*. Because Staudinger had stated frequently that samples of "macromolecules" consist of chains of varying length, it is not easy to understand today why this semantic question should have produced so much excitement. The second argument was much more important. Meyer and Mark claimed that "in our case the osmotically active particle in 'solution' is not the isolated chain of glucose residues but their aggregate." In this they were perpetuating Herzog's conviction[262] that "in colloidal dispersions the size of particles is identical with that of the original crystallites," and Mark's contention[246] that "if one wants to link a group of atoms longer than the unit cell, then it extends throughout the crystal." Thus they believed that in estimating the size of crystallites from the breadth of X-ray diffraction spots they had also obtained the size of the micellar aggregate that would remain stable in solution. They estimated that cellulose crystallites consisted of an aggregate of 40–60 chains, each containing 30–50 residues.

With the experimental techniques available at the time it was not possible to decide from the diffraction diagram whether the cellulose chain segments were parallel or antiparallel in the unit cell. Meyer and Mark concluded that it would be difficult to conceive of a biosynthetic mechanism in which neighboring cellulose molecules would have opposite orientations and they opted, on that ground, for the parallel arrangement. Some years later Meyer and Misch[263]

---

*It is remarkable that the management of IG Farbenindustrie was so enlightened that it enabled Meyer and Mark to carry out fundamental studies, many of which could have no relation to the business of the company—in particular, Mark's pioneering studies of the electron diffraction by gases. Equally remarkable is the fact that Mark, a research director who supervised 50 scientists, managed to publish almost 100 papers in the five years he spent with IG Farbenindustrie.

†Freudenberg, using methods of organic chemistry, published in the same year as Meyer and Mark two papers in which he arrived at similar conclusions concerning the chemical nature of cellulose. In the first [Freudenberg and Braun, *Ann.*, **460**, 288 (1928)], written without knowledge of crystallographic work, he disproved the notion that cellulose was an association colloid. In the second [*Ann.*, **461**, 130 (1928)] he stated unambiguously that the cellulose "backbone is made up of cellobiose junctions, the polysaccharide is built up of long chains."

found that regenerated cellulose had the same structure as mercerized cellulose.* They argued that

> on coagulation the filament should contain equal numbers of chains oriented in any direction. Thus, on drawing the fiber, the number of chains oriented in the two opposite directions should be equal. One must then assume that crystallites formed on subsequent deswelling should contain chains of opposite orientation. Since the lattice of these crystallites is the same as that of mercerized ramie, the latter must have the same arrangement.

Modern work has shown that none of these arguments was quite correct.[†] Meyer and Mark were right in assigning to the chains in native cellulose a parallel orientation, but they were wrong in assuming that a living cell cannot produce a crystalline polymer in which the chains are antiparallel (in α-chitin the chains are antiparallel). Meyer and Misch were correct in assigning the antiparallel structure to the crystallites of regenerated cellulose but they were wrong in assuming that the orientation of the cellulose chains cannot change during mercerization. A related observation was made by Hess et al.,[263a] who found that acetylation of cellulose fibers under mild conditions produces a cellulose triacetate with a crystal structure that changes when the polymer is dissolved and reprecipitated. It is now known[‡] that the chains retain the parallel orientation of native cellulose during the acetylation of the fibers, whereas they are antiparallel in the reprecipitated material.

In 1932 Mark published a paper[264] in which he estimated the tensile strength of an "ideal" cellulose fiber on the basis of crystallographic data and the energy of covalent bonds which would have to be broken during the mechanical failure of a specimen. With a cellulose chain occupying a cross section of 25 (Å)^2 and covalent bond energies of 70–90 kcal/mol he arrived at a tensile strength of 800 kg/mm^2 for the "ideal" fiber, as against experimental values of 100 kg/mm^2.* A few years later Meyer and Lotmar[265] estimated the elastic modulus of cellulose by using force constants for bond elongation and bond-angle distortion obtained from spectroscopy and found reasonable agreement with experiment. These two publications must be considered pioneering contributions to an understanding of the physical properties of polymers on the basis of their molecular structure.

---

*In "mercerization" stretched cellulosic fibers are exposed to concentrated sodium hydroxide solutions. It was known that this changes the crystal structure but a reversal of chain orientation was considered most improbable.

[†]See Gardner and Blackwell, *Biopolymers*, **13**, 1975 (1974); Minke and Backwell, *J. Mol. Biol.*, **120**, 167 (1978); Kolpak and Blackwell, *Macromolecules*, **9**, 273 (1976); Sarko and Muggli, *Macromolecules*, **7**, 486 (1974); Stipanovic and Sarko, *Macromolecules*, **9**, 851 (1976).

[‡]E. Roche, H. Chanzy, M. Boudelle, R. H. Marchessault, and P. Sundararajan, Macromolecules, **11**, 86 (1978).

*He assumed that the work of breaking a chemical bond could be estimated by substituting for the dependence of the force between two covalently bonded atoms on the interatomic distance a step function which drops to zero at an interatomic separation of 4–5 Å.

82   1914–1942

In the case of *Hevea* rubber, crystallography was able to settle the question of whether it is *cis*- or *trans*-polyisoprene:

$$\text{cis-polyisoprene} \qquad\qquad \text{trans-polyisoprene}$$

Meyer and Mark[261] showed that the unit-cell dimensions are compatible only with the cis structure;* moreover, to arrive at the observed identity period of 8.2Å and to avoid improbably small distances between one of the methylenes and the methyl side chain, internal rotations of about 50° in opposite directions around the two bonds next to the double bonds had to be assumed.[266] This appears to be the first case in which crystallography was used to arrive at such a well documented conclusion concerning a rather complex conformation of a polymeric chain.

We have noted already[229] that polychloroprene was also found to crystallize when stretched. In this case the X-ray diffraction pattern showed that the chain had the trans structure. Stretched poly(vinyl chloride) indicated a repeat period of 5.2 Å[266a] showing that the chain could not be planar but had to be puckered

because a planar chain should have been characterized by a repeat distance half as large. Of particular interest was the observation by Brill and Halle[266b] that stretched polyisobutene exhibits the very long repeat period of 18.5Å. This could be explained only by assuming a helical structure for the chain backbone and it is undoubtedly the first case in which such an assumption was made.†

---

*Guttapercha was later shown to have the trans structure[266] (pp. 133–138 of the English edition, Vol. II, pp. 123–127 of the German edition).
*Fuller, Frosch and Pape [*J. Am. Chem. Soc.*, **62**, 1905 (1940)] analyzed the polyisobutene structure in detail and proposed a helix with eight monomer residues in a repeat of 18.63 Å.

In 1939 Bunn[266c] carried out a crystallographic study of the structure of a high molecular weight paraffin which was estimated to have about 3000 methylene residues. He had the benefit of previous crystallographic studies of homologs of the paraffin series and his work constitutes a new level of sophistication made possible by the application of Patterson's method[276] of Fourier synthesis. This enabled Bunn to locate the carbon atoms in the chain and led to a C—C bond length of 1.53Å and a bond angle of 112°, significantly larger than the tetrahedral angle assumed in older work.

The paper by Meyer and Mark on silk fibroin[259] proved of special importance because it described for the first time polypeptide chains in the fully extended conformation to be designated later as the β structure. According to D. Crowfoot Hodgkin[267] reports of this work induced W. H. Bragg to ask W. T. Astbury to take X-ray photographs of wool and hair which he intended to use in a Christmas lecture at the Royal Institution on "The imperfect crystallization of common things." Astbury and his collaborators[268,269] found that hair, wool, and even tips of porcupine quills yielded similar diffraction patterns. More surprisingly, when wet hair was heated it could be extended to double its original length and the diffraction photograph changed strikingly. Astbury designated the two crystalline forms as α and β and proposed the following chain geometries for the two structures:*

*In the Spiers Lecture delivered by Astbury to the Faraday Society in 1938[270] he quoted the following ditty by A. L. Patterson:

Amino-acids in chains
are the cause, so the X-ray explains
of the stretching of wool
and its strength when we pull
and show why it shrinks when it rains.

Collagen, another fibrous protein which occurs in tendon, cartilage, and fish scales, was also shown by Herzog and Gonnell[271] to yield a typical fiber diffraction photography. Astbury was interested in this material[270] but did not attempt to interpret the X-ray data. Gelatin, a degradation product of collagen, was found to heat up when stretched and since this behavior was similar to that of rubber, Katz and Gerngross[272] decided to study its diffraction in the relaxed and stretched forms. They found that the diffraction ring of the relaxed gelatin broke up into arcs on stretching the sample and that the pattern obtained from the stretched specimen was identical to that of collagen. They wrote:[273]

> This shows for the first time how an originally amorphous substance—gelatin—can acquire on stretching the characteristic X-ray diffraction pattern of a previously known other substance, collagen.

A careful study of gelatin gels by X-ray diffraction carried out by Herrmann, Gerngross and Abitz[274] led to an important new concept: to account for the mechanical strength of the gel these authors postulated that polypeptide chains are ordered only along part of their length in the crystallite, which is "fringed" at both ends. The disordered chains in these "fringes" are mutually entangled providing a mechanical link between the crystallites. This model disposed of the need to assume several protein components in such semicrystalline materials as silk fibroin or of Herzog's "putty" in which crystallites are imbedded. The "fringed micelle" model of semicrystalline polymers was to exercise a lasting influence. Herrmann et al. also applied this model to an interpretation of the behavior of rubbers.*

It had been known for a long time that many proteins present in solution in the body fluids can be crystallized, and after Sumner's demonstration in 1926[484] that the enzyme urease is a protein a number of enzymes were obtained in the crystalline form. Initial attempts to obtain X-ray diffraction photographs from these crystals were disappointing with only a few poorly resolved spots seen on the photographs.† The great breakthrough occurred in 1934 when Bernal and Crowfoot[275] found that more than a hundred sharp

---

*The concept of a "fringed micelle" was partly anticipated by Meyer and Mark[258] in their paper on the crystal structure of cellulose when they wrote "A portion of cellulose is amorphous. Since it can be degraded to glucose just as the crystalline part, we have no doubt that it also consists of glucosidic chains. We consider it even probable that chains which are well ordered in the interior of the crystallites become gradually disordered at their surface. The chains need not be restricted to the size indicated by the X-ray diagram. . . ." This passage differs from the model of Herrmann et al. only in that it does not invoke entanglements between disordered chains to explain the mechanical behavior of the semicrystalline polymers. The suggestion that a chain might pass through several crystallites is due to A. Frey-Wyssling [*Protoplasma*, **25**, 261 (1936)].

†The first X-ray diffraction experiments on protein crystals were carried out by Herzog and Jancke [*Naturwiss.*, **9**, 320 (1921)], who obtained only a diffuse pattern. As a result protein crystals were for some time referred to as "pseudocrystals." It seems that E. Ott [*Kolloid. Beih.*, **23**, 108 (1927)] was first to find that a protein (ovalbumin) yielded a diffraction pattern "quite different from that of amorphous substances"

diffraction spots can be obtained from a pepsin crystal if it is irradiated while immersed in its mother liquid. It now became apparent that past failures to obtain highly resolved X-ray diffraction photographs were due to a gradual loss of the water contained in protein crystals which resulted in a collapse of the lattice. The new information was striking in that it revealed a very large unit cell that could accommodate a number of pepsin molecules whose molecular weight had been found by ultracentrifuge analysis to be close to 40,000. Also, these molecules appeared as compact bodies separated by spaces filled with water. Finally, the important conclusion was reached that

> the arrangement of atoms inside the protein molecule is of a perfectly defined kind, though without the periodicities characterizing the fibrous proteins.

Soon after this epochal discovery H. Mark and F. R. Eirich visited Bernal's laboratory. Eirich told Mark that M. Perutz, a young friend of his, was interested in working on a peptide problem. Mark agreed to encourage Perutz to join Bernal's laboratory and he recommended him to Bernal. Perutz's uncle, F. Haurowitz, who had long been interested in hemoglobin crystals, suggested that they might be a good subject for crystallographic study."*

After the publication of Patterson's work on the Fourier analysis of diffraction patterns[276] it became possible to interpret the relative intensities of diffraction spots in terms of electron density distributions. An outstanding early example of this kind of analysis was Crowfoot's work on insulin,[277] the results of which showed that the unit cell contained protein with a molecular weight estimated at 37,600. However, the cell had trigonal symmetry and it seemed unlikely that this would be true of a single molecule; it seemed more probable that the unit cell contained $3n$ protein molecules.†

In 1936 Wyckoff and Corey[278] published a paper in which they reported a preliminary crystallographic study of a "crystalline protein that apparently carries all the biological characteristics of the infectious virus" isolated by W. K. Stanley in 1935 and responsible for the tobacco mosaic disease. It is easy to understand that the crystallization of a substance which had been thought of as having the attributes of life produced a profound impression on the scientific community. While early work ignored the nucleic acid contained in the virus, the shape of the particle (a rod with a 15-nm diameter and a length of about 300 nm[279]) was established by electron microscopy. A most surprising aspect of its behavior was the formation of a liquid crystalline phase in aqueous solution.[280] Although a more detailed discussion of virus particles is clearly beyond the scope of this book, it should be noted that this phenomenon stimulated in another era a theoretical analysis of conditions leading to anisotropic solutions of synthetic polymers which eventually assumed also technological importance.

---

*As Mark recalled,[267] Perutz was reluctant because he knew nothing about X-ray crystallography. Mark told him: "You will learn"; and indeed he did, eventually winning a Nobel Prize for his solution of the hemoglobin structure.
†It contains six molecules with an aggregate molecular weight of 34,656.

10. Staudinger's Struggle for Macromolecules

Around 1920 H. Staudinger, professor of organic chemistry at the Federal Institute of Technology in Zürich (ETH), decided to devote in the future most of his efforts to the study of what were then called high molecular compounds (*"hochmolekulare Verbindungen"*). He was at that time 39 years old and had acquired a fine reputation, particularly for his studies of ketenes and aliphatic diazo compounds, which had earned him a chair at one of the world's most prestigious universities. He stuck to his decision: in the list of his publications in fields other than that of macromolecules only some 20% refer to work published after 1922, and although he entered the macromolecular field at an age when many scientists have passed the peak of their productivity, three-quarters of his publications are in that area. In fact, after he left Zürich and accepted a professorship in Freiburg, Germany, he concentrated all his energies on "high molecular compounds." This was undoubtedly an act of great courage since the "establishment" of organic chemistry in the academic world viewed this field with unconcealed contempt. As Staudinger recalls in his memoir,[281] it was common to refer to attempts to deal with these materials, which could not be purified by distillation or crystallization, as "grease chemistry" and he quotes an admonition by H. Wieland: "Drop the idea of large molecules, organic molecules with a molecular weight higher than 5000 do not exist. Purify your rubber. . . then it will crystallize." Frey-Wyssling[282] recalls a lecture by Staudinger to the Zürich Chemical Society in 1925 at which he was attacked by a number of prestigious scientists until he exclaimed in frustration, quoting Martin Luther's famous pronouncement, "Here I stand and can do no other." Mark[283] recalls the hostile reception which Staudinger's "farewell lecture" received at the ETH in the following year with the crystallographer Niggli insisting that molecules could not be larger than the crystallographic unit cell.

To place Staudinger's struggle to gain acceptance for the concept of very long chain molecules in proper perspective it is appropriate to survey the opinions of his most important opponents, who championed the notion that cellulose, starch, protein, and rubber are aggregates of small molecular species exhibiting "colloid behavior." These views were passionately held by leading scientists such as Herzog, Karrer, Pringsheim, Hess, Bergmann, and Pummerer.

Herzog's position is puzzling. He knew about Polanyi's 1921 paper[239b] in which it was clearly stated that the small unit cell of cellulose crystals could be compatible with a long chain structure but he generally ignored this observation and wrote on numerous occasions[244,262,284,285] that substances like cellulose and silk fibroin must consist of molecules smaller than the crystallographic unit cell. Only the paper published with Weissenberg in 1925,[245] which I quoted in Chapter 9, contained a footnote referring to Polanyi and Brill as having alluded to the possibility that molecules "extend over the whole crystal," but nowhere do the authors mention chains. Herzog was also convinced[262] that in high molecular compounds "the size of particles in colloidal dispersions is identical with that of the original crystallites." Herzog and Krüger[285] reported that cellulose nitrate in acetone solution had a diffusion coefficient that corresponded to a particle diameter of 20 nm, close to the size of crystallites observed by X-ray diffraction. Cellulose in cupric ammonia solution[286] seemed to have particles with a diameter of only 5 nm, which suggested to the authors some disaggregation of the colloid. Significantly, the gradual decrease in the viscosity of this solution was interpreted as due to a continuous breakup of the colloidal aggregate, with no consideration given to the possibility of a chemical change.

Karrer, Staudinger's colleague at the ETH, insisted for many years that starch and cellulose were colloidal aggregates of small molecules. He used the following surprising argument when he entered this field:[287] Starch is intrinsically insoluble in water but can be dispersed in it, just as silver can form colloidal solutions in which each particle of the dispersed phase is an aggregate of hundreds of silver atoms. When starch is methylated the aggregate is broken up, just as true solutions are obtained after converting silver to silver nitrate. This is proved by the disappearance of the blue starch–iodine color, "which is well known to be a colloid phenomenon." In 1921 Karrer and Nägeli[288] stated that starch is an anhydride of maltose and continued, confidently,

> One has to be surprised that the view of hundreds or thousands of glucose molecules joined by glucosidic bonds to long chains could have remained unchallenged for several decades. . . . It is also improbable that a plant in converting sugar to a reserve substance from which it might soon have to be recovered would perform such complex work as would be required in the build-up of a polyglucoside. . . .

Karrer was influenced to some extent by experiments reported by Pringsheim and Langhans in 1912[289] concerning the behavior of the "Schardinger dextrins," now known to be cyclic oligoglucosides. They claimed that the species containing six glucose residues is split on acetylation into two halves, which on deacetylation reconstitute the original particle. A similar behavior was reported for the species with four glucose residues. In 1926 Pringsheim[290] stated,

> Tetraamylose is not held together by glucosidic links between two diamyloses but by supermolecular forces which stabilize the "basic unit" diamylose. This point of view ,extended to polyamyloses and cellulose, has become the centerpiece of the modern chemistry of polysaccharides. . . . The action of molecular valences must depend on the structure of the building blocks. . . . This is all we can state at this time; the laws governing such forces are unknown.

For a diagnostic method to differentiate between valence bonds and molecular aggregates Pringsheim suggested the measurement of optical activity which would be altered by the formation of covalent bonds but not by molecular aggregation.

Hess and his collaborators[291] summarized their beliefs on the structure of cellulose in a paper that extended over 144 pages. They observed concentration-independent melting-point depressions in solutions of cellulose acetate and reasoned that "these constant depressions show that the solutions do not obey the osmotic law." Their conviction that cellulose is a colloidal aggregate of anhydroglucose was based largely on the study of the optical activity of cellulose solutions in ammoniacal cupric solutions. Since the data could be fitted by assuming a simple association equilibrium between one glucose and one cupric ion, they concluded that the cellulose molecule could not be a chain containing many glucose units. "It is a lucky accident," they wrote "that the behavior of cellulose in cupric solution furnishes a means for the recognition of its structural unit which remained hidden in Raoult's method." The fallacy of this reasoning was pointed out by MacGillavry[292]—glucose residues in a polyglucoside chain may bind copper independently of one another—but the argument was lost on Hess.[293] Hess et al.[291] expressed their beliefs as follows:

> Polymerization and depolymerization involves . . . an isomerization which produces a change in the molecular weight. . . . If this change concerns on the whole physical rather than chemical properties, then we are dealing with association and dissociation. It is conceivable that the difference between physical and chemical properties of two substances related by polymerization is very small so that a structural change cannot be proved chemically . . . but because of such a limiting case . . . the different definitions of polymerization and association cannot be abandoned.

One of the arguments used by Hess against a structure of cellulose in which long chains of glucose residues are connected by β-glucosidic linkages[125] was the low yield of cellobiose obtained on hydrolytic degradation. He believed that this might even have resulted from a condensation of two glucose residues formed in the primary process. In this he ignored a thoughtful paper by

Freudenberg,[293a] who had discussed the yields of cellobiose to be expected under two assumptions, (1) that a glucose chain with β-glucosidic linkages is split at random with all fragments remaining in solution; (2) that cellulose is split at random but that the cellobiose which is formed crystallizes and is then protected from breakdown. The latter assumption led to a 67% yield of the biose,* which was considered in reasonable agreement, with experimental data obtained under optimal conditions.

The work of Meyer and Mark[258] on the cellulose structure did not immediately change the views that Hess had held so firmly for so many years. In a paper published with Trogus[294] he concedes the utility of the crystallographic approach but complains that "the underestimation of the chemist's work causes the numerous contradictions based on X-ray data in the cellulose field."

Similar, possibly even more extreme views were held by Bergmann, who believed[295] that high molecular natural products involve "three-dimensional ordering of simple units which are unstable when isolated." In his view, the formation of an anhydride from cellobiose led to such instability that caused the anhydride to associate to the water-insoluble cellulose. He believed that

> it would be wrong to think of high molecular substances which have not been observed in molecular dispersion as consisting of individual molecules. The units which recur periodically in complex carbohydrates are in no sense molecules. Solid bodies do not contain molecules.

This last statement is truly astonishing, coming from a highly regarded scientist at a time (1926) when crystallography had already yielded detailed structures of a number of organic molecules.† Bergmann held similar views concerning proteins[296] and believed that special enzymes had to produce a dissociation of their molecular aggregates before other enzyme systems could effect proteolysis.

All the workers mentioned above claimed to have found confirmation of their views by obtaining low molecular weights of starch derivatives, cellulose derivatives, or proteins by cryoscopic measurements in solvents that would produce a disaggregation of the colloid particle. In the case of carbohydrates splitting of the glycosidic bonds during the derivatization procedures used at that time was underestimated; in other cases the high-melting solvents used undoubtedly led to solute degradation. In a typical example Herzog[284b] reported a low cryoscopic molecular weight for solutions of silk fibroin in resorcinol (melting at 110°C)—an approach Meyer and Mark[259] justly charac-

---

*This result is equivalent to a prediction on the basis of Keller's theory [*J. Chem. Phys.*, **37**, 2584 (1962)] that one third of the substituents of a long chain will remain unreacted if any group lying between two reacted nearest neighbors is unable to react.

†As Meyer[296a] commented: "X-ray crystallography in inorganic chemistry led to the abandonment of molecules and to the notion that crystals are built up directly from atoms This concept applicable to ionic lattices was wrongly extended to organic crystals."

terized as "brutal." Thermal degradation must also have been the cause of the low molecular weight (1400) which Pummerer et al.[297] obtained for rubber from the melting-point depression of camphor (mp 176°C). They could detect no melting-point depression of rubber in benzene solution and concluded that benzene is unable to overcome the tendency of the rubber molecule to associate.

Staudinger's interest in polymeric materials seems to have been stimulated first by his curiosity concerning the structure of rubber. During his years as associate professor at the Technical University in Karlsruhe (1907–1912) he took an interest in the research project of L. Lautenschläger, who wrote a thesis on "Autoxidation and Polymerization of Unsaturated Hydrocarbons."[298]* This thesis does not foreshadow in any way Staudinger's later concept of "high molecular compounds" as composed of long-chain molecules. Polymerizations were subdivided into (1) those leading to open chains (styrene dimer and diisobutene were given as examples), (2) those leading to ring compounds (cyclodimers of diphenyl ketene and cinnamic acid, cyclotrimers of acetaldehyde, and photodimer of anthracene), and (3) those leading to high molecular compounds. This carried the obvious implication that "high molecular compounds" [*Hevea* rubber, "metastyrol," poly(vinyl bromide), and "fossil resins"] are not constituted of "open chains," a conclusion supported by Lautenschläger's acceptance of Harries' contention that *Hevea* rubber is "probably a multiply polymerized dimethylcyclooctadiene." High molecular compounds were defined operationally as products that cannot be vaporized in a high vacuum and the opinion was expressed that "their molecular weight is so high that its determination appears hopeless."

Staudinger's speculation that "high molecular compounds" consist of covalently bonded long-chain molecules was first expressed in a lecture on rubber, presented to the Swiss Chemical Society in 1917; a paper based on that lecture was published in Switzerland in 1919.[224] A detailed exposition of his point of view appeared then during the following year.[229] Citing as typical examples the polymerization products of formaldehyde, vinyl bromide, and styrene, he commented:

> . . . . one may assume that in a primary step unsaturated molecules combine in a manner similar to that in the formation of four - or six-membered rings but that for some reason, possibly steric, ring formation does not take place so that a large

---

*According to the acknowledgment contained in this thesis, the work was stimulated by C. Engler, who was the head of the department and acted as thesis advisor. Thanks are expressed to Staudinger "for his great kindness and helpfulness." A paper based on this work was published only eighteen years later[298b] when much more was known about polymerization; therefore it has a flavor different than Lautenschläger's 1913 dissertation. Curiously, it does not list Engler as a coauthor. It may be noted that Lautenschläger was a pharmacist and completed the thesis research in a single year, during which he studied the kinetics of polymerization and oxidation of seventeen compounds, many of which he had to prepare. The introduction to the thesis represents a fascinating statement of views current at the time.

number, possibly hundreds of molecules, assemble* until equilibrium between the large molecules has been attained, which depends on temperature, concentration and solvent.

As examples of the proposed chain molecules Staudinger included formulas for polyoxymethylene, polystyrene, and polyisoprene. The first two were still referred to by the then common designations—paraformaldehyde and metastyrol.

Staudinger foresaw that objections might be raised to his proposal on the ground that "unsaturated atoms with free valences would be present at the ends of the long chains." Staudinger dismissed this concern, writing that when several hundreds of reactive molecules combine "leaving only two unsaturated positions" the reactivity would have been reduced several hundredfold. This notion, that very long chain molecules become rather unreactive was to become a recurrent theme not only in Staudinger's work but also in that of other early workers in the high polymer field.

It should be emphasized that the 1920 paper which marks Staudinger's formal entry into polymer chemistry contained no new experimental data. It stated a doctrine for which evidence had yet to be obtained. The first task seemed to be to dispose of the notion that "secondary valence forces" have to be invoked to account for "colloidal behavior" (i.e., high solution viscosity). A convenient example was *Hevea* rubber, for which Harries had postulated that small cyclic molecules are held together by "residual valences" between carbon-carbon double bonds. He had suggested[300] that

> it would be important to reduce rubber since the product could probably be distilled in a high vacuum without decomposition allowing an unambiguous determination of its constitution.

The experiment was carried out by Staudinger and Fritschi[301] in 1922. No distillable hydrocarbons were obtained and solutions of this product were as viscous as those of the original rubber. It was also noted that the reduced rubber was not altered by the action of sulfur chloride, thus disposing of Ostwald's contention that rubber vulcanization is an adsorption phenomenon.

One of Staudinger's most fruitful ideas was to substitute for the study of such natural products as polysaccharides and rubber the study of "model polymers" which would lend themselves to an easier experimental approach. We have already referred to the studies of polyoxymethylene, conceived of as a model of cellulose because of its insolubility.[252,254] Unlike cellulosic chains, formaldehyde oligomers with a wide range of chain lengths could be prepared as pure compounds enabling Staudinger to study their properties as a function of the degree of polymerization. Similarly, polyoxyethylene was studied as a model of

---

*The expression used by Staudinger, "*lagern sich zusammen*" was the same as that employed by his mentor, Engler, twenty-three years earlier[72a] when speculating about polymerizations (see p. 24).

starch since both consisted of polar, water-soluble chain molecules.[257] Because studies of *Hevea* rubber were hampered by the sensitivity of its molecules to oxidative breakdown, polystyrene, a much more inert hydrocarbon polymer, was selected as a model polymer.[302] The one case in which Staudinger's selection of a "model" was clearly misleading was that of poly(acrylic acid), which was thought of as an analog of a protein.[303] However, although this study was carried out at a time when the structure of proteins was a mystery it marked the beginning of the exploration of polyelectrolytes, pinpointing, at least qualitatively, a number of the characteristic properties of their solutions.

Staudinger's studies of polyoxymethylenes were initiated in Zürich and continued after his move to Freiburg. An extended summary of this work (by six of Staudinger's students) was published in 1929.[304] It is the only instance in which a systematic study was carried out of the changes in physical and chemical properties of a homologous series that extended from oligomers to high polymers; it therefore represents an important milestone in the understanding of long chain molecules. In the early studies commercial "paraformaldehyde" was treated in different proportions with acetic anhydride to obtain polyoxymethylene diacetates. The first five members of the series were separated by fractional distillation, whereas pure oligomers containing 8, 10, 12, 14, 15, 16, 17, and 19 formaldehyde residues were obtained by repeated recrystallization. A comparison of the densities, melting points, and crystal structures of these oligomers and the insoluble high polymer led to the conclusion that they were built on the same structural principle. This had four important consequences:

1. The polymer was not an association colloid but consisted of chain molecules with normal valence bonds.
2. The chains did not have free valences at their ends, as suggested by Staudinger in 1920,[299] but were terminated by characteristic functional groups. (Later similar polyoxymethylenes terminated with hydroxyl or methoxyl groups were prepared.)
3. By determining the end group concentration it was possible to estimate the average chain length of the insoluble high polymers.
4. As already mentioned in the preceding chapter (p. 78), crystallographic studies[253,254] led to the assumption that chain ends are randomly distributed within the crystal lattice (see, in particular, Staudinger and Signer[305]) so that crystallographic data cannot furnish information about the polymer chain lengths.* This assumption was strongly reinforced when single crystals of polyoxymethylene were prepared.[255] A

---

*E. Ott [*Z. Phys. Chem., B*, **9**, 378 (1930)] claimed later that he could determine the chain length of a polyoxymethylene with 60 formaldehyde residues by crystallography. He gives no data how such a species could have been isolated and his claim is altogether not credible.

chain running through the entire crystal would have corresponded to a degree of polymerization of more than ten thousand, whereas end group analysis indicated very much shorter chains.

Although Staudinger embarked on a study of polyoxymethylene, considering it a cellulose model, he was not prepared in 1929 to apply the conclusions reached with polyoxymethylene to a chain structure of cellulose. In the paper cited above[304] we find the following passage: "with cellulose the question is still undecided whether individual molecules such as glucose or cellobiose anhydrides are joined by normal or coordinate valences."* This statement is curious, coming a year after Meyer and Mark's convincing crystallographic evidence[258] of the chain structure of cellulose.†

Some other points raised in Staudinger's summary paper are of special interest. For the first time he presents the concept of a polymerization conditioned by the crystal lattice forces of the polymer arguing that long polyoxymethylene chains are insoluble, so that they could not have formed in solution before being added to a growing polymer crystal. He goes further in claiming that long polyoxymethylene chains are intrinsically unstable:

> Thus, when Bergmann assumes that the existence of these insoluble solid compounds depends on strong crystal lattice forces he is right in a sense, only they are not lattice forces between individual groups but between macromolecules.

He also noted that chains with hydroxyl end groups are much less stable than those in which the chain ends are methylated but failed to understand the cause for this difference.

Staudinger's demonstration that long chain molecules could crystallize also disposed of a long-standing conviction of organic chemists that very large molecules are likely to be amorphous, a concept endorsed by the great Emil Fischer.[309] It was also believed that the solubility of organic compounds decreases with increasing molecular size, so that substances with apparent molecular weights of the order of 100,000 had to be colloidal aggregates if they were soluble. Staudinger explained that this reasoning ignored the fact that branched molecules are generally more soluble than those with a linear structure.[310] Moreover, the solubility of polymers such as polystyrene is favored by their amorphous nature. Originally Staudinger believed[302] that unsaturated valences at the ends of polystyrene chains react with one another to form large

---

*Staudinger's use of "coordinate valences" included hydrogren bond formation.
†Staudinger[306] similarly claimed in 1930 a trans structure for *Hevea* rubber, completely ignoring the incontrovertible crystallographic evidence of Meyer and Mark[259] for the cis structure. In the book which Staudinger published in 1932[307] he repeated his assignment of the trans structure for rubber. He referred in a footnote to the work of Meyer and Mark but gave the impression that their cis assignment was based merely on an analogy with the relative melting points of low molecular weight cis and trans isomers. One is left with the impression that Staudinger was influenced by his increasingly adversary relation to Meyer and Mark.[308]

rings and these were held responsible for the failure of the polymers to crystallize, but later, he was the first to realize[311,312] that vinyl polymers in which the asymmetric centers have neither the same nor a regularly alternating configuration cannot be expected to fit into a crystal lattice.

In a review published in 1931[313] Carothers commented that the solubility of high polymers is "surprisingly great." He suggested that this is due "no doubt" to the presence of chains of varying length, with each different species "behaving independently." Originally, Staudinger[314] assumed that this polydispersity is the reason why polystyrene would not crystallize "according to the experience of organic chemists that mixtures crystallize more poorly than pure substances." His students therefore tried to fractionate amorphous polymers, hoping to obtain a crystalline product. It was only after the elucidation of the crystal structure of polyoxymethylene[254] that it was realized that polymer crystallites could contain chains differing in length so that polydispersity did not impede crystallization.

The studies of polystyrene by Staudinger et al.[302] also showed that polystyrenes prepared under varying conditions had different properties (in particular, solution viscosity) and that these substances could be separated into fractions with different molecular weights. This was in striking contrast to the old assumption of the existence of a discrete "metastyrol" species and it was entirely incompatible with the notion of an association colloid.[315] The observations were interpreted convincingly by the existence, in any polystyrene sample, of a "polymer homologous" mixture of chains of different lengths but containing similar residues. It should be pointed out, however, that this conclusion was reached only gradually, as testified by this revealing passage[311]:

> At the outset of studies about the structure of synthetic high molecular hydrocarbons we believed that polyindene and polystyrene are colloids and that differences in viscosity are caused by differences in the degree of subdivision. We assigned initially no special significance to the small melting point depressions assuming that they are due to low molecular weight contaminants which are strongly adsorbed to high molecular substances. Only repeated experiments showed that there is indeed a connection between viscosity and the average molecular weight.

We see that Staudinger was in some ways influenced by the "colloid school," as is also apparent by his adoption of the terms "hemicolloid" and "eucolloid"* for molecules with molecular weights of 1000–10,000 and 10,000–100,000, respectively.

Staudinger's work provided an essential incentive to the German chemical industry to venture into the polymer field. A collection of documents published by Farbwerke Hoechst† provides a revealing insight into this relationship and

---

*In Greek hemi = half, eu = true.
†"*Zur Strukturaufklärung der Makromoleküle, Ein Briefwechsel zwischen Prof. Staudinger und Dr. Kränzlein,*" Farbwerke Hoechst AG, 1966. See in particular Staudinger's letter of June 2, 1927, Voss' report of June 4, 1927, and the Hoechst letter to Staudinger of October 4, 1927.

the early difficulties in the industrial exploitation of Staudinger's scientific work. For instance, in June 1927 Dr. A. Voss of IG Farbenindustrie visited Staudinger and was shown the photopolymerization of vinyl acetate, but in October the Hoechst laboratory of that company wrote to Staudinger that they could not satisfy his request for larger quantities of poly (vinyl acetate) for spinning experiments "since there is no sunshine and we can operate only on a small scale with other light sources." In reporting on his visit to Freiburg, Voss suggested three alternatives:

1. If he was to continue work on vinyl polymerization he would need "at least one collaborator."
2. If Hoechst is to discontinue such work a closer agreement with Staudinger should be established.
3. All interest in this area might be abandoned, "since so far no technologically useful results have been obtained."

In 1930 a meeting of the "Kolloid Gesellschaft" held in Frankfurt[316] clearly signified the victory of the concept of long-chain molecules. Meyer was elected president of the society and Ostwald, the high priest of the colloid concept, said in his introduction:

> A colloid particle held together by principal valences will exhibit, apart from general colloid properties, some peculiarities. . . . We colloid chemists have not missed these special features of "chemical colloids" or "molecular colloids". . . . only in their case can one speak of "true colloid substances" rather than a colloidal state.*

Meyer, Staudinger, Mark, and Herzog spoke about various aspects of macromolecular behavior. Hess softened his stand by stating that it was unclear whether cellulose contains many or only a few glucose residues. Pummerer stuck to cyclic rubber molecules but conceded that the rings were possibly large.

But another problem now emerged. Even if crystallography indicated that cellulose, *Hevea* rubber, and silk fibroin consisted of chain molecules could it be assumed that such molecules were separated in dilute solution? To obtain background information on the magnitude of intermolecular forces Meyer asked Dunkel, a member of his research staff at the IG Farbenindustrie research laboratories, to collect data on the heat of vaporization of a series of compounds from which contributions of various groups present in organic molecules to the cohesive energy could be obtained. This early application of a concept which was widely used at a much later time led to increments to the molar cohesive energy such as 990 cal for methylene, 6500 cal for a hydroxyl,

---

*Here Ostwald refers to the belief, introduced by P.v. Weimarn, that *any* substance can exist in a "crystalloid" or "colloid" state.

and 8970 cal for a carboxyl substitution in a hydrocarbon chain.[317,318] From data of this kind Meyer concluded[318] that

> the principal valence chains exceeding a certain length are firmly held together so that they do not separate in solution but form micelles. . . . The molar cohesion of a chain of 60 glucose residues. . . is calculated as 1,500,000 cal. A chain of 100 isoprene residues corresponds to a micellar energy of about 500,000 cal.

He insisted that this high cohesion between the chain molecules would ensure the persistence of crystallites as micelles in solution, so that

> it is not useful to designate as molecules the principal valence chains, which differ in length; their mean size cannot be determined by osmometry which yields rather the weight of a micelle formed from many chains.

In this formulation Meyer made an obvious error in neglecting the fact that the forces acting between polymer chains may be partly or fully compensated by the interaction between the chain molecules and the molecules of the solvent. Staudinger drew attention to this feature,[310b] but he arrived at the incorrect conclusion that "the solvation energy is not proportional to the length of the molecule, otherwise high and low molecular substances of a polymer homologous series would have the same solubility.* Staudinger denied vigorously the existence of micelles in dilute solutions of hydrocarbon polymers. He pointed out that the molecular weight of polystyrenes did not change significantly when the aromatic rings were hydrogenated, although such a chemical change might have been expected to alter the tendency of the chain molecules to associate.[319] On the other hand, Staudinger believed that the viscosity observed at higher solution concentrations reflected an association between the macromolecules.[320] He also wrote at this stage that[321]

> we should not like to claim that micelles in Nägeli's sense can never exist in solution. . . . For instance, the chain molecules in cellulose crystallites are held together by lattice forces more strongly than the amorphous fraction. . . . A solvent could here (and in similar cases) dissolve the amorphous fraction, while its solvent power might be insufficient to overcome lattice forces. Thus, the crystallites would appear in solution as discrete units.†

The controversy between Staudinger and Meyer over the existence of macromolecular aggregates in dilute solution‡ continued for many years, with some

---

*The fallacy of this argument became clear only after Flory's analysis of the thermodynamics of polymer solutions.
†Three years later (Ref. 307, footnote 2 on p. 483) Staudinger denied most emphatically that cellulose or its derivatives could form molecular aggregates in dilute solution. His change of mind was caused by his interpretation of the viscosity of polymer solutions, discussed in the next chapter.
‡Carothers[313] sided with Staudinger, writing that "the osmotic unit is the molecule." He based this conclusion on the observation that the molecular weight obtained in different solvents was the same. Later it was found that there are exceptions to this behavior.

unreasonable arguments being used by both antagonists. In their paper on the crystal structure of *Hevea* rubber Meyer and Mark had written[259]

> Under the influence of certain solvents, the rubber micelles disintegrate, either into smaller aggregates of principal valence chains or into the isolated chains. From this point of view rubber assumes an intermediate position between solutions of soap whose micelles are in constant equilibrium with free fatty acids and cellulose or starch whose micelles cannot be split up by any solvent.

This passage was referred to repeatedly by Staudinger as implying an analogy between rubber and soap—a meaning that was certainly not intended. On the other hand, Meyer[322] attacked Staudinger's contention that "the transformation of polystyrene into hydropolystyrene of the same particle size is proof against the existence of micelles" with the far-fetched argument that the tendency of oleic acid to associate remains unaltered when it is converted to stearic acid.

Even in their *Hochpolymere Chemie*, published in 1940, Meyer and Mark[323] insisted that for polymers

> a separation into independently moving molecules takes place only at extreme dilution such as cannot be obtained either in solutions used in practice nor in those employed for scientific study.*

Later studies were to show that polymer aggregation in dilute solutions is neither as general as assumed by Meyer and Mark nor generally absent as claimed by Staudinger. The question was, of course, of fundamental importance because on its answer depended the possibility of the experimental determination of the average molecular weight of high polymers.

Staudinger expressed his antagonism to Meyer in a particularly extreme manner on the occasion of the publication of Meyer's monograph on polymeric compounds.† He had a sheet glued to the cover of the volume in the Freiburg University library which included the following passages:

> This book is not a scientific work but propaganda. . . . Meyer takes essentially the results of the Freiburg laboratory without citing them. This scientific expropriation is disguised to the uninitiated by the distortion and the arrogant criticism of the Freiburg researches. . . . According to Meyer, only two of the thirteen recent publications worth a mention are German. This could be taken by the outside world as an indication of the decline of German science.

---

*The interested reader may consult a review of this subject.[324] Molecular association is revealed by a dependence of the apparent molecular weight on solvent and temperature; recently nuclear magnetic resonance spectroscopy has been demonstrated as a powerful tool for the study of polymer association.[325]

†The second volume of *Hochpolymere Chemie* does not cite Mark as coauthor.

This devastating judgment of a valuable book—which today seems incomprehensible—was shared by Kern,* who added the suspicion that Meyer (now a professor in Geneva) continued to receive confidential information from the IG laboratory in Ludwigshafen. A most serious accusation in wartime!

---

*W. Kern, letter to Staudinger on the occasion of his 60th birthday (Staudinger Archiv., Deutsches Museum, Munich).

11. New Methods for the Determination of the Molecular Weights of Macromolecules

The Ultracentrifuge

A powerful new research tool, the ultracentrifuge, extended the study of macromolecules in two important respects. It made it possible, for the first time, to characterize particle weights unambiguously over an enormous range, from a few thousands to millions of molecular weight units. At the same time it provided information not only concerning the average particle weight but also about the particle weight distribution. This development had its roots in two different traditions, involving the hydrodynamics of the motion of rigid particles through viscous media and the study of the kinetic equilibrium of particles in thermal motion.

In 1845 Stokes had shown that for spheres of radius a moving through a fluid at a velocity v sufficiently slow to avoid turbulence, the resistance of the medium is $f_0 v$, where the "frictional coefficient" f_0 is given by $6\pi\eta a$, η being the viscosity of the fluid. Thus if F is an external force exerted on a spherical particle the particle will accelerate until $F = f_0 v$, so that with η known the radius of the sphere may be obtained from the steady-state velocity. If F is the gravitational force acting on a sphere of density ρ suspended in a medium with a density ρ_0 in the terrestrial gravitational field g then $(4/3)\pi a^3 g(\rho - \rho_0) = 6\pi\eta a v$ and the radius of the particle will be related to the velocity of its steady-state fall by $a = \sqrt{9\eta v / 2g(\rho - \rho_0)}$. This principle was used in 1911 by Svedberg and Estrup[326] to estimate the size of colloid particles and elaborated in 1916 by Odén,[327] who followed the weight of the particles deposited on a pane below a colloid suspension and derived from the data a distribution function of the radii of the particles. A few years later Svedberg and Rinde[328] refined Odén's method for the study of gold sols but noted that it was limited to relatively coarse-grained particles; for finer particles the driving force would have to be increased by using a centrifugal field in place of the terrestrial gravitation. But then "determination of the weight of sediment becomes almost impossible. A more promising procedure is the determination of the variation of the concentration of the disperse phase with height."

The second approach, depending on the distribution of particles in a gravitational field, has its origin in the observations recorded by Brown in 1828 and 1829.[329] In an article entitled "On the particles contained in the pollen of plants and on the general existence of active molecules in organic and inorganic bodies"* he described an investigation originally motivated by a desire to find the mechanism by which pollen reaches the pistil of plants to effect fertilization. When he found that the pollen particles were engaged in a rapid random motion he examined the pollen of a large number of plant species to satisfy himself that the phenomenon can be observed with all of them. Next he found that other small particles of vegetable origin behave in the same way and discovered "the very unexpected fact of the seeming vitality retained by these minute particles long after the death of the plant." He reported that even "the dust and soot deposited . . . especially in London, is entirely composed of such molecules" and, in fact, the same kind of random motion was observed in finely ground particles in "rocks of all ages, including those in which organic remains have never been found." Brown stressed that he had no explanation for this phenomenon. He ruled out convection currents caused by evaporation by observing particles suspended in a water droplet immersed in oil.

In spite of the care with which Brown had carried out his experiments the "Brownian motion" was generally assumed to be an artifact caused by "unequal temperature of the strongly illuminated water, its evaporation, air currents, etc."[330]† Gouy,[331] who subjected Brown's observation to a most critical analysis, wrote that Maxwell insisted that no motion, as reported by Brown, could be verified. Gouy carefully excluded vibrations, convection currents, and other external influences as causes of the particle motion. He found that the velocity depended only on the size of the particles, which increased as they became smaller. He concluded that "Brownian motion, alone among physical phenomena, reveals a constant internal state of agitation in the absence of external forces. One cannot avoid seeing in it the result of weakened thermal motions of molecules."

In 1905 Einstein[332a] published a paper "About the molecular motion of particles in resting fluids as required by the kinetic theory of heat" in which he prefaced his physical analysis by stating that "it is possible that the motion to be treated below is identical with 'Brownian motion'; data accessible to me are too inaccurate to pass judgment on this question". He introduced the crucial

---

*Brown used the term "molecule" for all the microscopically observed particles.
†A summary of early views on Brownian motion is given in A. Einstein *Investigation on the Theory of the Brownian Movement*, English translation, Dover, New York, 1956, pp. 86–88.

assumption that suspended particles should be subject to the same physical laws as dissolved molecules and concluded his paper with this appeal: "May an investigator soon succeed in resolving the suggested question which is of such importance to the theory of heat." The challenge was accepted by Perrin in what may be regarded as one of the most spectacular experimental studies of this era. In a long paper published in 1909, "Brownian motion and the reality of molecules"[333] he started with the assumption that "a large particle composed of many molecules will behave like a very large molecule so that its mean energy will have the same value as isolated molecules." This implied, as already suggested by Einstein,[332b] that the distribution of particles suspended in a fluid and subjected to the force of gravity should be of the same form as the distribution of gas molecules in the terrestrial atmosphere; that is, the particle concentration c should decrease with height h according to $-d\ln c/dh = Vg(\rho - \rho_0)N/RT$, where V is the volume of the particle and N is Avogadro's number. To prove this assumption Perrin had to fractionate microscopic resin globules with radii ranging from a fraction of a micrometer to several micrometers and estimate their size by various procedures. Microscopic observation of their distribution in a suspension led to an estimate of N of about 7×10^{23} in reasonably close agreement with values obtained by a variety of techniques. This work earned Perrin a Nobel Prize in physics in 1926.

Svedberg was invited in 1923 by the University of Wisconsin to lecture on colloid research and during his visit he and Nichols built a centrifuge with optical equipment such that a colloid solution could be photographed while it was being spun.[334] This centrifuge generated only 150 times the terrestrial gravitational field ("150 g")—minute compared to those achieved later—but this was sufficient to establish the feasibility of the basic idea and the radii of gold sol particles as small as 20 nm could be obtained from their sedimentation velocity. After this initial success Svedberg devoted his energy to the construction of centrifuges with gravitational fields that would be sufficient for the measurement of the molecular weights of proteins. By 1925 fields of 5000 g were achieved and these proved sufficient for the determination of the molecular weights of hemoglobin[335] from its equilibrium distribution in the centrifugal field. On the other hand, it was realized that much higher fields would be needed for the study of the sedimentation velocity of protein molecules. This aim was attained by 1927 with a centrifuge field of 100,000 g, which allowed Svedberg and Nichols[336] to obtain the molecular weight of hemoglobin from the sedimentation constant s (defined as the sedimentation velocity in a unit field) and the diffusion coefficient D by $s/D = M(1 - \bar{V}_{sp})/RT$, where $\bar{V}_{sp}$ is the partial specific volume of the protein. The value obtained in this way[68,350] was in excellent agreement with that derived by the equilibrium distribution. In the same year Svedberg received the Nobel Prize for this work.[337]*

---

*Eventually a centrifuge was built which could be operated at gravitational fields up to 750,000 g [K. O. Pedersen, *Nature*, **135**, 304 (1935)]. This represented an enormous engineering feat; it should be realized that every gram of material in the centrifuge exerted a force equal to that of three

Svedberg's efforts were entirely devoted to the study of proteins. Here the solute distribution could be determined by photography, using visible light in the case of hemoglobin and ultraviolet light for other proteins† For many other macromolecules, however, the photographic method was not applicable. Here a principle exploited by Thovert[339] for the measurement of diffusion coefficients proved useful. A beam of light passing through a column perpendicular to a gradient of refractive index is bent by an angle proportional to this gradient. Various optical arrangements can then be used to apply this principle for the determination of solute distribution in the ultracentrifuge cell.[340]

DuPont was the first industrial company to acquire an ultracentrifuge and in 1933 Kraemer and Lansing[341] published the first study in which this instrument was used for the characterization of a synthetic polymer, poly(ω-hydroxyundecanoic acid). This material had an average molecular weight of 25,200, as determined by end-group analysis, and the molecular weight, obtained by equilibrium centrifugation, was 27,000. On the other hand, the values obtained from the sedimentation velocity and the diffusion coefficient were twice as high. At this stage Kraemer and Lansing did not appreciate the distinction between the number- and weight-average molecular weight and concluded that the result obtained from sedimentation and diffusion was unreliable.‡ In a later thoughtful study of cellulose acetate and cellulose nitrate[342] they explained for the first time the distinction between a number-average and a weight-average molecular weight and how the average obtained depends on the experimental method employed. They also introduced the term "degree of polymerization" (DP) as a measure of the length of high polymer molecules.

A very extensive study of polystyrene by ultracentrifuge equilibrium and ultracentrifuge sedimentation was carried out in 1934 by Signer and Gross[343] in Svedberg's Uppsala laboratory. They noted that the data yielded no evidence of the formation of micellelike aggregates. However, frictional coefficients calculated from the data increased much more slowly with the molecular weight than expected on the basis of hydrodynamic theory for rigid rods—the Staudinger model for macromolecules. Signer and Gross concluded that

---

quarters of a ton in the gravitational field of the earth. Svedberg[338] records that a centrifuge operated at 900,000 g exploded after a few runs. In fact, the danger of such an accident led to the shielding of investigators from the ultracentrifuge by a protecting wall. As late as 1936 only two ultracentrifuges existed outside Svedberg's laboratory; one at the Lister Institute of Preventive Medicine in London, the other in the DuPont laboratory in Wilmington, Delaware. In that year the Rockefeller Foundation made a grant to the University of Wisconsin for an ultracentrifuge to be operated by J. W. Williams (Rockefeller Foundation Annual Report, 1936, p. 187).

†I refer to the results of these studies in a later chapter devoted to proteins.

‡Some years later (Ref. 337, p. 418) they wrote that a molecular weight of 26,000 was obtained when extrapolating s and D to zero concentration. It is hard to believe, however, that their sample would have had such a narrow molecular weight distribution to yield almost the same value for a number- and weight-average molecular weight.

the poor agreement of experiment and theory could of course be due to the fact that the dissolved polystyrene does not contain extended particles. We believe that another explanation is more probable, that hydrodynamic theories do not treat adequately chain molecules.

This reasoning was justified by pointing out that the thickness of the polystyrene chain is comparable to the dimensions of the solvent molecules so that macroscopic hydrodynamic theories that deal with rigid particles suspended in a structureless continuum may not be applicable.* This study by Signer and Gross is important in that it represents the first attempt to derive molecular weight distributions from sedimentation data. It also identified the two complicating factors in any interpretation of the distribution of sedimentation coefficients; that is, particle interactions and diffusion superimposed on sedimentation. To overcome these problems the authors recommended that the experiments be carried out at the highest practicable dilution and that a solvent be used, the density of which differs as much as possible from that of the polymer so as to minimize the sedimentation time.

Solution Viscosity

In 1906 Einstein[344a] published a paper in which he set out to estimate Avogadro's number N from his theory of the diffusion coefficient D which predicted for spherical particles $D = RT/6\pi\eta aN$. To estimate the effective radius of a dissolved molecule he developed a theory for the ratio of the viscosity η of a dilute suspension of spherical particles to η_0, the viscosity of the surrounding medium (later designated as the "relative viscosity"). He obtained $\eta/\eta_0 = 1 + \phi$, where ϕ was the volume fraction occupied by the spheres, and applied this result to the diffusion coefficient of sucrose in water, assuming that the sucrose molecule can be treated as spherical and that its motion when surrounded by water molecules can be treated as taking place in a structureless continuum. By this procedure he arrived at the estimate of $N = 4.15 \times 10^{23}$. A few years after the publication of this paper Bancelin[345] compared Einstein's theory with experimental results obtained on suspensions of gamboge particles. His data agreed with Einstein's prediction that the viscosity of a suspension of spheres depends only on their total volume but not on their size. The coefficient of the ϕ term, however, was much larger than predicted ($\eta/\eta_0 = 1 + 2.9\phi$) and Bancelin wrote to Einstein about this discrepancy. Einstein, in fact, found an error in his calculations and amended his result[344b] to $\eta/\eta_0 = 1 + (5/2)\phi$. This yielded $N = 6.56 \times 10^{23}$, in surprising agreement, considering the various assumptions used in its derivation, with values obtained in other ways.

---

*Perrin[333] had already pointed out that Stokes's law must break down when applied to particles smaller than the molecules of the surrounding medium.

In his first paper Einstein noted that the volume of the sucrose molecule derived from the viscosity of sucrose solutions was substantially higher than its volume in the solid state, although little change in the total volume accompanied the solution process. He commented that "the results can only be interpreted by assuming that the sucrose molecule impedes the mobility of the adjoining water" which behaves as if it were attached to the sucrose. With the correction in the second paper he found that the hydrodynamically effective volume of a sucrose molecule was 60% greater than that occupied in the sucrose crystal.

Smoluchowski[346] pointed out in 1916 that for any volume fraction of suspended particles the viscosity should increase with an increasing asymmetry of the particles. Treatment of suspensions of elongated particles, however, is complicated by the fact that the extent of interference with the fluid flow depends on the orientation of the particles relative to the flow lines, and as Kuhn was first to point out[346a] this orientation, due to hydrodynamic forces, will be counteracted by Brownian motion. In particular, the viscosity of the suspension should depend on q/D_r, where q is the velocity gradient and D_r, the rotational diffusion coefficient. Kuhn used a dumbbell model, and theories for ellipsoids of revolution with an axial ratio p were later developed by Eisenschitz[346b] and Simha.[347] They show that for very long rods and at $q/D_r \to 0$ $(\eta - \eta_0)/\eta_0\phi$ approaches proportionality to p^2 as predicted by Mark[361] from a qualitative argument (see p. 107).

The first use of the measurement of the solution viscosity for the characterization of a polymer seems to have been published in 1905 by Axelrod,[348] who noted that rubbers with the highest solution viscosities yield after vulcanization products with the highest strength. Considering the fragmentary knowledge of the nature of rubber at that time, his statement that "the quality of various rubbers depends on their degree of polymerization" is astonishing. The experimental procedure for measuring solution viscosities was refined by Schidrowitz and Goldsbrough,[186] who used the Ostwald viscometer and studied the relative viscosity η/η_0 as a function of solution concentration. They thought that after the initial upward curvatures the plots become linear and they felt free to extrapolate them to what they believed was the viscosity of the rubber, which proved to be a good index of its quality. Early work in this field was reviewed in 1913 by Fol,[349] who criticized Axelrod since "as long as we are unable to measure the molecular weight of colloids, we cannot speak of a higher or lower degree of polymerization." Fol found that the relative viscosity of rubber solutions remained essentially constant when the solutions were heated. A year later Kirchhof[350] reported that solutions of rubber were much more viscous in chlorinated hydrocarbons than in benzene and ascribed this to rubber holding "bound solvent several times its own volume in these disperse phases."

The first studies which related the relative viscosity of the solution of a polymer to its particle weight (determined by osmometry) were carried out by Biltz, who established such a relation for dextrin in 1913[351a] and for gelatin in

1916[351b]. Blitz did not, however, attempt to interpret the dependence of the solution viscosity on molecular weight in terms of a physical model; in fact, he stated categorically that the osmotically active particles in gelatin solutions are molecular aggregates rather than isolated molecules.

Yet the very notion that solution viscosity could depend on the size of the solute particle was fought vigorously by the followers of the colloid school. In 1913 the Faraday Society held a meeting on "The Significance of Viscosity in the Study of the Colloidal State." Ostwald[352] gave the keynote address in which he stressed once again the multiplicity of "colloid phenomena" and never referred to Einstein's theory of the viscosity of suspensions. He suggested that phenomena such as gelation, protein denaturation, and the dramatic increase in the viscosity of liquid sulfur when it is heated to a critical temperature might all reflect the same underlying principle. Hatschek, who presided at the meeting, proposed[353] for the relative viscosity the relation $\eta/\eta_0 = \sqrt[3]{1/(1 - \phi)}$ and claimed that it represented well data of glycogen solutions. In 1925 Hess et al.[354] announced that they had found different cellulose samples to exhibit different viscosities in ammoniacal copper solutions, although the optical activity was always the same, but they made no attempt to interpret this observation and did not seem to realize that it was incompatible with their concept of cellulose as an aggregate of small molecules. In 1929 Karrer and Krauss[355] suggested that the dependence of the viscosity of starch solutions on the rate of shear indicates the presence of "structured particles" surrounded by a membrane and as late as 1931 a similar model was proposed for cellulose solutions by Hess et al.[356] Their argument was based on the observation that purification reduced solution viscosity but did not alter the size of cellulose crystallites estimated from the X-ray diffraction pattern: Thus it was reasoned that the purification process destroyed a "foreign skin" that stabilized cellulose micelles in solution and was responsible for its high viscosity.

The first attempts to formulate a theory of polymer solutions based on the understanding that polymers consist of long chain molecules were made in 1929 by Fikentscher and Mark[357] and in the following year by Staudinger and Heuer.[358] Both papers started from a consideration of Einstein's treatment of the viscosity of suspensions. Einstein had shown that if the suspended particles were rigid spheres the viscosity of the system depended only on the volume fraction of the disperse phase, being completely independent of particle size. Yet both Mark and Staudinger had no doubt that an increasing size of polymer molecules led to increasing solution viscosities and they set out to account for this relationship.

The paper of Fikentscher and Mark starts with the observation that the viscosity of a polymer solution depends not only on the chain length of the polymer molecules but also on the interaction of the polymer with the solvent medium. They proposed a model in which the solute molecule was represented by an ellipsoid of revolution, whose long axis was proportional to the length of the polymer chain, with the axial ratio independent of the particle length in any given solvent medium. The equatorial radius was thought to increase with increasing solvation of the polymer. This model led to the prediction that $(\eta - \eta_0)/\eta_0\phi$ should be proportional to the square of the polymer chain length. No attempt was made to confirm this result by experiment; for high polymers this would have been possible at that time only with the use of an ultracentrifuge which was not available to Fikentscher and Mark. But even in the absence of experimental data on polymer chain length this model seems from our vantage point rather implausible. Taking the long axis of the equivalent ellipsoid as proportional to the chain length implied that the molecular chains were fully extended, an assumption which was justly criticized when it was made later by Staudinger. Assuming that the equivalent ellipsoids had an axial ratio independent of the length of the molecular chain implied that the amount of solvent bound to the polymer increased as the square of this chain length, an assumption that could hardly be justified.

According to Mrs. Staudinger,[359] her husband decided to use solution viscosity for the determination of the molecular weights of polymers after "it turned out to be impossible to get an ultracentrifuge for the laboratory since these experiments were looked on with some skepticism by the authorities." In his model[358] the polymer molecule was considered to behave like a rigid rod with a radius r and a length L spinning around an axis perpendicular to its long dimension. Thus the volume swept out by one molecule was $\pi(L/2)^2 \cdot 2r$ and the volume swept out by all molecules in a unit volume of a polymer solution with a concentration c of polymer molecules was $(Nc/M)\pi L^2 r/2$. Staudinger then generalized Einstein's equation by assuming that the "specific viscosity" $\eta_{sp} = (\eta - \eta_0)/\eta_0$ is proportional to the volume swept out by the suspended (or dissolved) particles, no matter what their shape, and since L was proportional to M he concluded that $\eta_{sp}/c = K_m M$, where the concentration was expressed in the molarity of monomer residues and K_m was a characteristic constant for a given type of polymer.

This theory of the solution viscosity of polymers was tested first on polystyrene fractions whose average molecular weight was obtained by cryoscopy.[358] Staudinger and Heuer observed that η_{sp}/c tended to increase with increasing polymer concentrations and attributed this effect to a molecular association. They believed, however, that for polystyrenes with a molecular weight ranging from 1700 to 10,000 this complication could be neglected in 2.5% solutions and viscosity data on such solutions in benzene yielded K_m values between 1.7×10^{-4} and 3.3×10^{-4} with a mean value of 2.5×10^{-4}. (Later[360] the value 1.8×10^{-4} was accepted.) Staudinger and Heuer were sufficiently confident of their theoretical analysis to use this K_m for the estimation of molecular weights of polystyrenes up to 200,000. They predicted that K_m values characteristic of other polymer-solvent systems would similarly yield molecular weight data but assumed that for any given polymer K_m would vary with the solvent because binding the solvent molecules should change the effective thickness of the chain.

A summary of Staudinger's theory as it eventually emerged after measurements on a number of polymer-homologous series is contained in a paper published in 1933.[360] The original formulation was modified in two respects: It was assumed that end groups made characteristic contributions to η_{sp}/c. Moreover, with the exception of cellulose and its derivatives, the specific viscosity of sufficiently dilute solutions containing the same concentration of backbone atoms were thought to have the same specific viscosity for all polymers of the same chain length. Staudinger now also specified the dilution at which solution viscosities were to be measured: He designated as "sol-solutions" systems in which the total volume was larger than the volume swept out by the rotating rodlike particles contained in it. Solutions with higher polymer concentrations were referred to as "gel-solutions."* In "sol-solutions" η_{sp}/c was considered to be essentially independent of concentration and only in these solutions was the relation of η_{sp}/c to the molecular weight of the polymer assumed to apply.

Staudinger's suggestion concerning the relation between the "reduced viscosity" η_{sp}/c and the molecular weight of chain molecules was vulnerable on three counts. First, the extrapolation from the behavior of low polymers to very long chain molecules was overly optimistic. Second, even if the molecule could be treated as a rigid rod, it was difficult to see why its rotating motion should be restricted to a plane. Mark[361] was first to draw attention to this feature by pointing out that if rotation in three dimensions were allowed the reduced viscosity would be proportional to the square of the length of the rodlike particles.† The third objection concerned the very concept of chain molecules viewed as rigid, fully extended particles. At a meeting of the Bunsengesellschaft held in 1934 Mark[362] summarized the extensive experimental data concerning rotation around single bonds which were incompatible with Staudinger's assumption of a single conformation of polymers such as polystyrene. It is instructive, however, to cite the arguments used by Staudinger[363,364] in favor of his model:

1. He believed that the easy crystallization of long chain molecules could not be understood if the chains were randomly coiled in solution but had to assume an extended shape in the crystal.
2. He felt that the small change of volume that accompanies the melting of polymers was incompatible with a transition from a parallel packing of extended chains to randomly coiled chains.
3. He pointed out that the viscosity of unsaturated chain molecules changes very little on hydrogenation.
4. He cited the work of Signer and Gross[365] on the flow birefringence of polystyrene and cellulose nitrate solutions which the authors inter-

---

*Although Staudinger's interpretation of the viscosity of polymer solutions was incorrect, it should be noted that his "critical concentration" that separated "sol-solutions" from "gel-solutions" was conceptually analogous to the modern c^* at which polymer coils are forced to interpenetrate.
†In fact, W. Kuhn [*Kolloid Z.*, **74**, 147 (1936)] and R. Simha (*J. Phys. Chem.*, **44**, 25 (1940)] showed that this is the case for highly elongated particles.

preted as indicating that the polymer molecules are highly extended particles.

5. He cited Stewart,[366] who had reported that liquid fatty alcohols and fatty acids exhibit an X-ray diffraction peak that corresponds to a spacing linear in the length of the chain/suggesting that the aliphatic chains retain in the liquid state the fully extended zigzag shape.

Although some of these arguments must have seemed rather convincing, others could have been acceptable only to prejudiced observers. Thus Signer and Gross found that the flow birefringence of cellulose nitrate followed the behavior predicted for rigid rods but wrote that the behavior of polystyrene solutions is more complicated. They stated explicitly that this could be due possibly to the flexibility of the chain, adding that "this will not be discussed here." Even more surprising is Staudinger's[363] claim that Eucken and Weigert[367] had shown that a carbon-carbon single bond in ethane was rigid, when, in fact, the thermodynamic data of these authors indicated only a small barrier of 315 ± 60 cal/mol for the hindered rotation around this bond.* Finally, Staudinger frequently used a circular argument. The viscosity law must hold because the chains are rodlike; the rodlike nature of the chains is proved by the experimental confirmation of the "viscosity-law." He frequently compared his "law" to "van't Hoff's osmotic law." When this analogy was questioned Staudinger conceded that no rigorous derivation of his viscosity rule had yet been obtained but suggested that this was analogous to "Raoult's rule" concerning the relation between melting-point depression and the molecular weight of the solute which had to await van't Hoff's analysis to be considered a law.[364] Staudinger's low regard for theoretical considerations is revealed in a footnote[368]: "New objections are based on theoretical arguments. These need not be considered further since the viscosity law has been experimentally proved."†

The record, as indicated by the discussion[369] following Staudinger's presentation at a 1931 meeting of the Faraday Society[360] shows that the scientific community was deeply divided concerning the merits of Staudinger's arguments. Although the validity of his "law" proved later to have been an illusion, there can be little doubt that its acceptance at the time advanced the progress of

---

*K. S. Pitzer [$J.$ $Chem.$ $Phys.$, **8**, 711 (1940)] found that this barrier is ten times as high as this early estimate. Even so, conformational transitions are rapid. The controversy between Staudinger, who insisted on rigid chain molecules and Mark, who advanced all the arguments against this concept, induced F. G. Donnan, presiding at a 1931 meeting of the Faraday Society, to comment [$Trans.$ $Faraday$ $Soc.$, **29**, 3 (1933)]: "Indeed, with Professor Mark and Professor Staudinger both present, we may be the witnesses of a conflict of giants."

†Staudinger was not only firmly convinced of the validity of his "viscosity law" but was also skeptical of the validity of procedures for the measurement of molecular weights which had a firm theoretical basis. He wrote (Ref. 307, p. 101): "The osmotic laws are valid for substances with spherical particles. . . . It is still questionable whether the osmotic method is still valid with such an abnormal molecular shape as that of chain molecules of high molecular natural products. . . . Svedberg's ultracentrifuge method . . . should also be used with caution for chain molecules."

polymer science. It provided a simple method of characterizing the chain length of polymers, and although the estimates obtained in this way were incorrect* the method was most useful in the early years of the study of vinyl polymerization.

An attempt to rationalize the molecular-weight dependence of η_{sp}/c for flexible chain molecules seems to have been made first in 1931 by Haller.[369b] He had no doubt that "portions of molecules joined by single bonds can rotate around the connecting axis" and concluded that

> the loose formless structure of chain molecules should influence the surrounding fluid so that it behaves, in some respects, as if it were immobilized. This must be reflected in the solution viscosity.

He then proposed that the chain molecule be represented by a spherical cloud with a uniform density of disconnected chain segments making independent contributions to the dissipation of energy as the cloud was made to rotate by a shear gradient, assuming that the chain segment density was independent of the size of the cloud. This model led to η_{sp}/c proportion to $M^{2/3}$ "a reasonable approximation to Staudinger's equation."†

Haller made no attempt to justify his gratuitous assumption that chain-segment densities may be considered constant within the space occupied by a flexible molecular coil. In 1934 Kuhn[370] published an important paper in which he analyzed the shape of a chain molecule with segments that subtend a fixed bond angle but in which rotation around the bonds is free. Neglecting first the volume of the chain segments, he suggested the replacement of the real chain by a chain with "statistical chain elements" that subtend random angles at their junctions and chosen so that the statistical and the real chain had the same contour length and the same distribution of end-to-end distances. He showed that such chains would on the average occupy an ellipsoidal volume with linear dimensions that would be proportional to the square root of the chain length. Assuming (without saying so explicitly) that the fluid inside this volume would be trapped by the chain, which would, therefore, behave as if it were a rigid ellipsoid,‡ he concluded that the ratio of the hydrodynamically effective volume to the volume of the "dry" chain would increase as the square root of the chain length. This was clearly in disagreement with Staudinger's results (which Kuhn seemed to accept). He traced the source of the discrepancy to the neglect of the volume occupied by the chain segments. He reasoned that

---

*For instance, Signer[369a] obtained in the ultracentrifuge average molecular weights of 30,000, 80,000, and 300,000 for polystyrene samples for which Staudinger estimated 15,000, 31,000, and 135,000.

†On receiving a reprint of Haller's publication, Staudinger expressed surprise about the friendly letter when the paper "was so unfriendly" (Staudinger Archiv, Deutsches Museum, Munich, DII, 13.3).

‡Huggins [*J. Phys. Chem.* **42**, 911 (1938)] showed later that proportionality of η_{sp}/c and the chain lengths as postulated by Staudinger, would be predicted for Kuhn's coils *if they were free draining*. As later shown by Flory, (see pp. 230–231) this assumption is incorrect.

this "excluded volume" would necessarily have the effect of expanding the coiled chain and used an example in which the linear dimensions of the "hydrodynamically equivalent" ellipsoid would be expanded by the 0.112 power of the chain length. How he guessed at this particular figure is obscure, but it led him to predict a reduced viscosity proportional to the 0.84 power of the chain length, strikingly close to the predictions of modern theories of the intrinsic viscosity of chains with a large excluded volume effect.* Thus Kuhn understood the functional dependence of the reduced viscosity–chain-length relation on the magnitude of the "excluded volume effect", although he did not seem to realize the dependency of this effect on the polymer-solvent interaction.†

Einstein's result for the viscosity of a suspension of spheres was restricted to systems so dilute that the hydrodynamic disturbances due to the suspended particles did not interact with one another. An extension of the theory led to $\eta_{sp}/\phi = 5/2 + 14.1\phi$ which was in reasonable agreement with experiment[371] and, as expected, the reduced viscosity of polymer solutions was similarly concentration-dependent. The value extrapolated to zero concentration was designed by Kraemer[372] [who used the equivalent extrapolation of $\ln(\eta/\eta_0)/c$ with c in g/dL] as the "intrinsic viscosity."

Light Scattering

The measurement of light scattering from polymer solutions became an important approach to polymer characterization only at the time discussed in the third part of this book but a modest beginning was made in the period we are

---

*He conceded that other powers were equally probable. Mark[370a] predicted that the exponent should vary from 2/3 in very poor solvents to 3/2 in good solvent media, whereas Houwink[370b] suggested an exponent of 0.6 as giving the best fit for experimental data of Staudinger and Warth.[370c]

†Staudinger's first admission that η_{sp}/c may not be proportional to the first power of the molecular weight of chain molecules was made in a 1940 paper with H. Warth,[370c] in which he found that K_m values for poly(vinyl acetate), poly(methyl acrylate), and poly(methyl methacrylate) decrease as the molecular weight increases. He speculated that this result may be due to chain branching but conceded that chain flexibility could be an alternative explanation of the data. Only in one of his last publications in 1951[370d] did Staudinger write that "results confirm beautifully Kuhn's conclusion that poorly solvated chain molecules contract to a compact coil." This was quite a concession; for many years Staudinger's attitude had been satirized by a rhyme current in his laboratory[370e]: "Denn die Kuhn'schen Knäuel sind dem Herrn ein Greuel" ("Kuhn's coils are a horror to the Lord"). Strangely, Staudinger continued in his lectures to use wooden sticks, about 2 mm thick and 30 cm long, as models of polymer molecules. A bundle of these sticks is exhibited at the Deutsches Museum in Munich. Staudinger's reluctance to agree to the concept of a flexible molecular coil for a macromolecule was based, according to E. Trommsdorff, one of his devoted students, on the conviction that protein specificity could be understood only in terms of a rigid molecular structure. At the time when Staudinger developed his ideas of macromolecules the notion of a specific chain conformation leading to the folding of polypeptide chains to the compact structure of globular proteins lay, of course, far in the future.

now considering. At the 1930 meeting of the "Kolloidgesellschaft" Ostwald,[373] stressing as usual his "colloid" point of view commented:

> Why are many protein solutions turbid? Not because the particles are joined by primary valences, but because the particles are so large that they overlap the region of the wavelengths of light and also have a refractivity sufficiently different from the solvent.

In 1871 Strutt[374a] (later to be known as Lord Rayleigh) published a paper "On the Light from the Sky," in which he showed that light scattering from particles, small compared with the wavelength of light, distributed at random in the atmosphere should lead to scattering inversely proportional to the fourth power of the wavelength. This prediction was confirmed strikingly in a comparison of the spectral intensity distribution of sunlight with that of the sky. In this analysis it was also found that n particles of volume V will lead to a scattering intensity proportional to nV^2. A later paper[374b] raised "the intriguing question whether the light from the sky can be explained by the molecules of the air or whether it is necessary to appeal to suspended particles" and concluded that "even in the absence of foreign particles we should still have a blue sky."* The scattering intensity was shown to be proportional to the square of the difference between the refractive indices of the particles and the suspending medium.

Lord Rayleigh's result predicted that for any weight concentration of suspended particles the intensity of light scattering should be proportional to the volume of the individual particle and Mecklenburg[375] confirmed this prediction by measurements of sulfur sols. It seemed then natural to assume that changes in the scattering from protein solutions reflect changes in the degree of subdivision of "colloid particles" and observations on gelatin solutions were interpreted in this sense.[375a,b] In 1927 Raman[375c] published a paper, "Relation of Tyndall Effect to Osmotic Pressure in Colloidal Solutions," in which he pointed out that light scattering is due both to the fluctuation in the local density and to fluctuation in the local solute concentration. He showed that it may therefore be related to the compressibility and the concentration dependence of osmotic pressure. Raman's result was identical to that obtained by Debye seventeen years later and it is curious that it made no impact and was generally ignored. Two explanations suggest themselves, first, the publication of the paper in the *Indian Journal of Physics*, which would not have been read by those interested in macromolecules and second, Raman's use of the "colloid" nomenclature and his failure to express his result in terms of molecular weights.

In the same year in which Raman's theory appeared, Kraemer and Dexter[375b] reported on the temperature and pH dependence of light scattering

---

*This paper quotes in a footnote a letter received from J. C. Maxwell in 1873 in which it was suggested that the light-scattering intensity could be used to obtain "the density of the ether."

from 1% gelatin solutions. They found that at 17°C the scattered light intensity increased more than fourfold within a small fraction of a pH unit to a sharp maximum at the isoelectric point of the protein. Raman[375d] was quick to suggest that the effect is due to the osmotic pressure minimum of isoelectric proteins, but paid no attention to the observation made by Kraemer and Dexter that the pH dependence of light scattering was only very slight at somewhat higher temperatures. Kraemer and Dexter were correct in assuming that the sharp increase in light scattering near the isoelectric point at low temperatures is related to the gelation of gelatin solutions at higher concentrations, although the nature of this transition was not understood at the time.*

Measurements carried out by Holwerda[376] on casein solutions showed that the light scattering was highly sensitive to the addition of electrolytes or a change in pH but the author was unaware of Raman's interpretation and the result therefore seemed to be incomprehensible. Putzeys and Brosteaux,[377] however, reasoned that these disappointing results were caused by peculiarities of casein which had also been demonstrated in ultracentrifuge studies and decided to test the utility of the light-scattering method on proteins whose molecular weight had been well established by Svedberg and his collaborators. Because of experimental difficulties, Putzeys and Brosteaux did not try to obtain molecular weights from their data. However, they demonstrated convincingly that the ratio of scattering intensity to concentration extrapolated to $c = 0$ yielded, according to Rayleigh's theory, relative particle weights in good agreement with results obtained by the ultracentrifuge for proteins over a more than hundredfold range of molecular weights.

Only one other paper that dealt with light scattering of macromolecular solutions was published before Debye demonstrated the great utility of the method in 1944. In 1937 Gehman and Field[378] reported their studies of the scattering of light from rubber solutions. They cited Putzeys and Brosteaux and their use of Rayleigh's theory but concluded that this "is limited to small spherical particles"; therefore they decided to use the depolarization of scattered light for the characterization of "the length of the scattering unit." For this they estimated the enormous value of 3600 Å, and by assuming the rigid rod model for the rubber molecule they arrived at a molecular weight of 53,000.

---

*See W. F. Harrington and P. H. von Hippel, *Arch. Biochem. Biophys.*, **92**, 100 (1961); *Adv. Protein Chem.*, **16**, 1 (1961).

12. Polycondensation

The pioneering work on polymerization reactions owes its leadership to two men—Hermann Staudinger and Wallace H. Carothers. Staudinger concerned himself with addition polymerizations which take place by a chain reaction mechanism, whereas Carothers carried out an extensive study of the principles of polycondensation. Staudinger's entry into the polymer field predated that of Carothers by nine years, but since polycondensation involves types of reactions long familiar to chemists—such as the formation of ethers, esters, and amides—it is convenient to discuss it first.

The earliest study of a reaction of bifunctional reagents seems to have been that of Lourenço, who in 1859[379a] heated ethylene glycol with ethylene bromide and obtained an "intermediate ether." In the same year Wurtz[380a] showed that two or three molecules of ethylene oxide can react with one molecule of water. The products, separated by fractional distillation, were assigned the formulas $HO-(CH_2CH_2O)_2-H$ and $HO-(CH_2CH_2O)_3-H$. The first was found to be identical to Lourenço's compound. Wurtz concluded that

> 1,2,3 molecules of ethylene oxide may combine with one molecule of water forming more and more complex products which nevertheless have a simple molecular structure. If one calls glycol "ethylene alcohol" one may call the other compounds "diethylene alcohol" and "triethylene alcohol."

One year later Wurtz[380b] treated ethylene oxide with acetic anhydride and obtained the diacetates of the di-, tri-, and tetraethylene glycols. Lourenço summarized his work in this field in 1863;[379b] by that time he had extended the series up to hexaethylene glycol. For the formation of the dimer he suggested the following two-step mechanism:

$$3\ HOCH_2CH_2OH + BrCH_2CH_2Br \longrightarrow 2\ HOCH_2CH_2Br + HO-(CH_2CH_2O)_2-H + H_2O$$

$$HOCH_2CH_2Br + HOCH_2CH_2OH \longrightarrow HO-(CH_2CH_2O)_2-H + HBr\ ^*$$

*This mechanism would not be acceptable today. Presumably water liberated in the condensation of the glycol hydrolyzed the ethylene bromide and the liberated HBr catalyzed further condensations.

His results led him to formulate the following generalization:

> With bifunctional compounds, the successive condensations form a series of compounds which remain bifunctional. . . . The ability to condense in this manner belongs, therefore, to compounds which form anhydrous derivatives, such as glycol which may be transformed to ethylene oxide or lactic acid which yields lactide. . . . It is remarkable what a large number of new compounds can be expected from such reactions. . . . If we consider only the glycol derivatives of four hydrocarbons as described by Wurtz, ten alcohols may be formed when two similar or dissimilar glycols combine.

This speculation may be considered a first allusion to copolymerization. Laurenço also tried to deal with the condensation of glycerol, typical of a trifunctional reagent, and although his analysis was here less successful he pointed out that in this case the functionality increases by one unit at every step of the condensation which also increases the number of possible isomeric structures.

Another polycondensation in which it was possible to separate oligomers was studied by v. Braun and Sobecki,[381] who reacted α, ω-dibromoalkanes (containing 4,5,7 or 10 methylenes) with magnesium and isolated the dimer, trimer and tetramer of the parent hydrocarbon. Here the identification of the oligomers presented no difficulty since the properties of normal alkanes were known.

On the other hand, polycondensations in which oligomeric products could not be isolated by techniques available to the nineteenth-century chemist led frequently to fanciful interpretations, as may be exemplified by the history of salicylic acid polycondensations. In his study of acid anhydride formation, Gerhardt[382] reacted phosphorus oxychloride with sodium salicylate and obtained a product which was "extremely hard and difficult to detach from the container; on warming, it turned into a soft viscous mass which solidified after some time." Gerhardt called this material "salicylide." Kraut[383] thought that he could isolate the dimer, trimer, and tetramer which were, however, merely characterized by their "softening points"; the tetramer was supposed to condense further to the octamer, Gerhardt's "salicylide." Anschütz[384] assigned to "salicylide" the cyclic tetrameric structure and claimed that this substance, whose molecular weight was obtained by cryoscopy, formed crystals with one molecule of chloroform melting at 260–261°C. The reaction product was also supposed to contain a "polysalicylide" that was "too insoluble for molecular weight determination" but was reported to form needlelike crystals melting at 322–325°C.

An interesting argument for the formation of a large ring in a polycondensation reaction was used by Vorländer[385] in his study of the interaction of ethylene glycol with succinic acid. He reasoned that eight-membered rings form with difficulty and that the esterification therefore probably leads to a 16-membered ring with two succinate and two ethylene residues. To prove the point he tried to carry out the reaction

$$\begin{array}{c} \text{CH}_2\text{—COCH}_2\text{CH}_2\text{Cl} \\ | \\ \text{CH}_2\text{—COCH}_2\text{CH}_2\text{Cl} \end{array} + \begin{array}{c} \text{AgOC—CH}_2 \\ | \\ \text{AgOC—CH}_2 \end{array} \quad \begin{array}{c} \text{CH}_2\text{COCH}_2\text{CH}_2\text{OCCH}_2 \\ | \quad\quad\quad\quad\quad\quad | \\ \text{CH}_2\text{COCH}_2\text{CH}_2\text{OCCH}_2 \end{array} + 2\text{AgCl}$$

obtaining a product with the same melting point as that in the reaction of ethylene glycol with succinic acid. This he took as evidence that ethylene succinate has the ring structure, never considering the possibility that both reactions could have led to a crystalline open-chain polymer.

If a polymer of appreciable chain length was formed the end groups made a negligible contribution to the elemental analysis, so that the cyclic structure was in good agreement with the empirical formula. In one investigation that concerned a poly(ethylene oxide) with a molecular weight that indicated a chain of thirty monomer residues the failure to detect terminal hydroxyl groups was taken as evidence of a cyclic structure.[386] In addition, knowledge of molecular geometry was meager; therefore Bischoff and Hedenström,[387] for instance, did not hesitate to use for the products of ester interchange between diphenyl carbonate and the three dihydroxybenzenes the formulas

Thus, they ignored arguments advanced by Einhorn,[388] who pointed out that

> in contrast to catechol carbonate . . . the carbonates of resorcinol and hydroquinone are practically insoluble . . . which unfortunately precludes molecular weight determination. . . . The insolubility, the high melting point, the impossibility to distill without decomposition, point to a higher molecular weight than that of the catechol carbonate.

It is surprising that investigators of this period interpreted cryoscopic measurements in terms of the molecular weight of a single species, never considering the possibility that it might reflect the average molecular weight of a mixture. In an interesting study Blaise and Marcilly[389] dehydrated hydroxypivalic acid and found that the product was a linear polymer, as indicated by the formation of isobutene on thermal degradation:

$$\text{HO—(CH}_2\text{C(CH}_3)_2\text{COO)}_n\text{—CH}_2\text{C(CH}_3)_2\text{COOH} \rightarrow \text{HO—(CH}_2\text{C(CH}_3)_2\text{COO)}_n\text{—H} + \text{CO}_2 + \text{CH}_2\text{=C(CH}_3)_2$$

From the molecular weight they concluded that the polymer was the hexamer, although they wrote that "the portion which is ether soluble is clearly made up of similar anhydrides of lower molecular weight." Curtius[390] was even more explicit in claiming that the thermal decomposition of glycine methyl ester yields *exclusively* tetraglycine.

Other references to early work which must have resulted in polycondensation, although the scientists involved were usually unaware of this result, were compiled by Flory.[391] Clearly, there was little understanding of such processes in 1928 when Carothers was hired by the DuPont Company and "encouraged to work on problems of his own selection."[392a] According to a memorandum written in 1936[392b], he

> first became interested in what has since been called bifunctional condensations during the autumn of 1927 in connection with some work . . . on acetylene di(magnesium bromide). It seemed inevitable that such condensations would, in many cases, lead to long chains.

He planned "in a rather vague way" some experimental work while still at Harvard but after visiting DuPont he concluded that in studies of condensations of bifunctional reagents "for purposes of theoretical exploration esterification would be more suitable than Grignard reactions." Staudinger had shown little interest in the production of chain molecules by this mechanism; he believed that "polycondensations cannot lead to products of high molecular weight . . . since the reactivity of a polymer molecule decreases rapidly with increasing chain length."[393] On the other hand, Staudinger established the essential principle that scientific work on high polymers should not be concerned with the isolation of a unique molecular species—that, indeed, the study of the properties of a mixture of members of a "polymer homologous series" is a worthy aim of research.

In his first publication from the DuPont laboratory,* Carothers[394] pointed

---

*For a study of Carothers' work it is convenient to use *Collected Papers of Wallace M. Carothers on Polymerization*, H. Mark and G. S. Whitby, Eds., Interscience, New York, 1940. The book also contains Carothers' biography and a list of his patents.

out that the definition of a polymer as given by Berzelius[8] (see p. 5) is quite unsatisfactory. Thus butyric acid and hydroxycaproic acid, with empirical formulas $C_4H_8O_2$ and $C_6H_{12}O_3$, would certainly not be considered the dimer and trimer of acetaldehyde (C_2H_4O). On the other hand, the empirical formula of poly(ethylene glycol) is different from ethylene glycol, the monomer from which it can be derived. It is therefore more useful to apply the designation of polymer to substances whose "structure may be represented by —R—R—R—R—where '—R—' are bivalent radicals which, in general, are not capable of independent existence."

Carothers recognized from the outset that the production of a high polymer from a reagent x-R-y or x-R-x + y-R'-y, where x and y undergo a condensation reaction, requires that the intermolecular process be highly favored over the intramolecular ring formation. It had been known for a long time that the formation of larger rings is very difficult but the cause of this behavior was poorly understood. Baeyer[395] in a theory formulated in 1885 considered ring strain to be due to a distortion of the tetrahedral bond angle but assumed, rather illogically, that rings were planar. This restriction was removed by Sachse,[396] who first proposed the chair and boat models for cyclohexane but assumed that conformational transitions could not take place.* His theory was therefore generally disregarded since it required the existence of two isomers for monosubstituted cyclohexane. Only in 1918 did Mohr[397] conclude that ring inversion in cyclohexane should be easy at ambient temperatures, and this led to the rapid acceptance of Sachse's model. However, the great difficulty with which larger rings are formed remained mysterious. For instance, as Mohr pointed out, it is possible to trace out a ring of 10 bonds in the diamond lattice in which all bond angles are tetrahedral—why would it then be nearly impossible to transform a linear chain into such a ring? The solution to this problem was provided in 1930 by Stoll and Stoll-Comte.[398] (Stoll, who had worked with Ruzicka on large ring compounds, was puzzled by the fact that the densities of cycloalkanes pass through a maximum for rings of 13 members, whereas densities of the series of linear alkanes increase smoothly to a limiting value as the chain length is extended.) They showed that the effect can be explained by the steric interference of methylene hydrogens in rings of intermediate size and that the density maximum occurs in rings of a size which is close to that at which the formation of cyclic ketones from dicarboxylic acids is most difficult. For rings of more than 17 members the steric constraint is no longer important, but the rate of ring closure is limited by the low statistical probability that the two reactive chain ends will lie close to one another so that the reaction can take place. Kuhn[370] showed in 1934 that for his model of a randomly kinked chain this probability decays as the 3/2 power of the chain length but Carothers seems to have been unaware of this result. Nevertheless, from qualitative arguments he arrived at the important conclusion that whenever a condensa-

---

*Sachse cited van't Hoff as having stated that molecules will assume their "favored configuration" (cf. footnote p. 13).

tion does not lead to relatively small rings the possibility that large polymeric rings might form can be entirely disregarded.

A striking demonstration that the probability of ring formation passes through a minimum as the ring size is increased was provided by the finding that the thermal decomposition of poly(tetramethylene carbonate)[399] and poly-(trimethylene oxalate)[400] yielded the 14-membered cyclic dimer rather than the seven-membered cyclic monomer. Eventually, detailed data on the ease with which rings of various sizes are formed were obtained by Stoll and Rouvé[401] from the yield of lactones in the dehydration of hydroxycarboxylic acids and by Spanagel and Carothers[402] from the yields of cyclic esters and anhydrides in the thermal decomposition of polyesters and polyanhydrides, respectively. The range of intermediate-sized rings with low stability was found to depend only to a minor degree on their chemical nature.

In a series of papers dealing with polyesterification* Carothers and his collaborators[399,400,403,406] showed that a reaction of an excess of glycol with dicarboxylic acids, their anhydrides and their diesters, or with diethyl carbonate leads to polyesters with molecular weights in the range of 2000–5000 whenever there is no possibility of forming a five or six-membered ring. The polyesters were found to be crystalline, and Carothers and Arvin[403] concluded, incorrectly, that this "indicates that the varieties of molecular species present in a given sample probably do not include a very wide range." Apparently Carothers was unaware of the work of Hengstenberg and Staudinger.[253,254] In a review that Carothers[313] wrote later on polymerization, he presented a schematic picture of a polymer crystal with chain ends distributed at random in the lattice, but again he seemed unaware of Staudinger's earlier suggestion of this model.

Carothers speculated about the reason why the polyesterification should not have proceeded beyond a certain point. Staudinger[404] had written

> It is well known that the stability of paraffins decreases with an increasing number of links. . . . Those with 30–40 carbon atoms degrade at 400°–500° . . . with still higher molecular weight one may expect degradation at 100°–200°. . . . For every temperature there is a limit for the size of molecules which can exist.

It was therefore important to show whether the chain length of the molecules was indeed thermodynamically determined. Carothers et al.[405] showed that heptacontane (obtained by the action of sodium on 1,10-dibromodecane) remained unchanged after 5 minutes at 400°C and concluded that "it would be somewhat unsafe to infer that a paraffin hydrocarbon of molecular weight 200,000 or even greater might not persist at room temperature." He assumed, at first,[406] that "there can be no question that the reactivity of functional

---

*One of Carothers' early difficulties involved the supply of the required reagents. In February 1928 Marvel wrote to Carothers that Eastman Kodak charges six cents per gram of adipic acid. During the following month he offered him the uncrystallized acid for $30/kg.

groups diminishes with the size of the molecules"* and this seemed to account for the difficulty of carrying polyesterifications beyond a certain point. The crucial break in this program occurred with the introduction of a "molecular still," an arrangement in which the heated polyester was kept under a high vacuum (less than 10^{-5} Torr), with the condensing surface at a distance less than the mean free path. This, it was hoped, would allow the small molecules to be eliminated much more efficiently, so that the polycondensation could be carried to higher conversion. The experiment was started by J. W. Hill on April 17, 1930. Later Carothers[392b] recollected "I do not think we had in mind to make a fiber, but we did want to make a molecule as large as we could get." When Hill tried to remove the molten polymer from the still he found, to his surprise, that it could be drawn into a fiber and the fiber could be extended severalfold by cold drawing.[392b,407] The extension occurred at a sharply defined boundary and the observation of this unexpected "necking" phenomenon must have amazed the investigators. The material, which had a molecular weight above 10,000, was transformed from a brittle opaque substance to a transparent, lustrous, highly elastic fiber "sufficiently tough and flexible to be tied into hard knots" and with a tensile strength that "compares well with those of cotton fibers and silk."† The enthusiasm generated by this result led to the hope that "the entire class of synthetic linear condensation polymers having molecular weights above 10,000, as well as a synthetic polymer in the form of a permanently oriented fiber capable of being tied into knots,"[392b] could be covered by a patent.

In spite of this exciting demonstration that polycondensation could lead to the preparation of fully synthetic fibers with outstanding physical properties, the low melting points, and the solubility of the "superpolyesters" made "the possibility of laying the foundation for a new commercial fiber appear remote" so that research was discontinued from June 1933 to early in 1934[392] when Carothers decided to "make one more effort." This time he planned to concentrate on polyamides. A "superpolymer" of ε-aminocaproic acid had been a

---

*In this Carothers agreed with Staudinger.[393] As we shall see later this fallacy was also widespread among those working on addition polymerizations of vinyl compounds.

†Carothers recognized that "the ability to crystallize appears to require a high degree of linear symmetry in the structural unit"[313] and he understood that this was a reason why polyesters derived from phthalic anhydride[398a] failed to crystallize. It is strange that he failed to draw the logical conclusion that terephthalic acid polyesters would be likely to lead to valuable fiber-forming polymers and missed the opportunity to make another important invention. The "superpolymer fibers" were reported by Carothers at the 82nd meeting of the American Chemical Society in Buffalo in September 1931. Under the caption "Castor oil silk," TIME wrote on September 14, 1931: "By heating castor oil with an alkali and mixing the result with the motor antifreeze compound . . . DuPont chemists produced an artificial silk fiber. . . . It is too expensive to manufacture commercially and is mainly a demonstration of chemical knowledge and skill." It may be noted that an "artificial silk" based on a polyester was claimed one year earlier in the British Patent 303 867. This patent, however, did not recognize the need to use only bifunctional reagents in the polycondensation and cited specifically the condensation products of phthalic acid and glycerol. Clearly, this could not have led to a fiber-forming polymer.

disappointment and at the time it was investigated[408] had failed to yield a fiber.* Carothers "suspected that the viscosity of amides was so high that reaction stopped prematurely"[392b] and the high melting point made it difficult to handle the polymer melt without chemical decomposition. Attempts to reduce the melting point by copolymerization of aminocaproic acid with hydroxycarboxylic acids led to weak fibers.[408] This time Carothers planned to use for monomers the higher homologs of aminocarboxylic acids, assuming that in this way he could "reduce the melting point to any desired degree."[392b] In fact, he found that the polymers of 9-aminononanoic acid† yielded a polymer melting at 195°C. On February 25, 1935, his laboratory produced poly-(hexamethylene adipate), and DuPont concentrated its effort on developing the technology for the spinning this material. No research papers were published on this development; patents covering the purification of the hexamethylene diamine adipate crystals and their polycondensation were granted in 1938,[409] a year after Carothers committed suicide.‡

In 1934 P. J. Flory joined the DuPont laboratory headed by Carothers and it may well be considered one of Carothers' major achievements that he should have fired the twenty-four year old Flory with the enthusiasm which led him to devote his life as a scientist to the polymer field.• Already in 1935, at a Faraday Society meeting devoted to polymers, Carothers[411] presented results of Flory's derivation[412] of the molecular weight distribution to be expected in polycondensations, assuming that the reaction rate constants that characterize the functional groups are independent of the length of the interacting chains. At the time many found this assumption implausible,[413] but the point was convincingly substantiated by Flory's detailed studies of the kinetics of polyesterification.[414] When the assumption was made that the reactivity of functional groups was independent of the length of the chains to which they were attached the kinetic analysis was greatly simplified, since it could be treated in terms of the concentration of these groups rather than in terms of the

---

*According to an interview conducted by D.O. Hummel with P. Schlack on December 22, 1982, Schlack attempted the polycondensation of aminocaproic acid at the same time as Carothers, unaware of his work. He thought that the initial failure was due to an insufficient polymer chain length. Later Schlack overcame this problem and became the inventor of Nylon 6[408a].

†This compound was obtained as follows: the polymeric anhydride of sebacic acid $HO-[C(CH_2)_8CO]_n-H$ was reacted with ammonia to yield 50% of the monoamide of sebacic acid. Hofmann rearrangement produced 9-aminononanoic acid, $H_2N(CH_2)_8COOH$.[229a]

‡News of the new fiber was carried in *The New York Times* on September 22, 1938, under the headline "New Silk Made on Chemical Base Rivals Quality of Natural Product." On September 26 an editorial reported that "out of castor oil and coal... a new kind of rayon" had been produced. Because of its impact on the silk trade "Japan has reason to worry." The official announcement by DuPont with the designation of the new fiber as Nylon appeared on October 28.

•In an interview Flory[410] referred to Carothers' death as "one of the most profoundly shocking events of my life.... It just pulled the rug out from under my hopes.... I realized how much a shield he had been and how much of an influence... when he was gone."

concentrations of the various polymeric species. Also, whereas in his first paper Flory[412] assumed that "the increase in viscosity as the reaction progresses might slow down the reaction," he now showed that for reactions with an appreciable activation energy the reaction rate constants should be independent of viscosity. This is so because an increased viscosity will not only reduce the rate at which functional groups diffuse toward one another but also the rate at which they diffuse apart so that the total number of their encounters is viscosity independent. No data were available at the time to confirm this assumption and Flory felt that polyesterification reactions would be particularly suitable to prove the point.

When a dicarboxylic acid reacted with an equivalent amount of glycol without addition of a strong acid catalyst the reaction followed third-order kinetics because the carboxyl groups acted as catalysts as well as reagents. In the presence of a sulfonic acid catalyst second-order kinetics were observed. If p is the fraction of the reactive groups which have reacted at any time a linear plot of $1/(1-p)^2$ against time with no catalyst added a linear plot of $1/(1-p)$ against time in systems that contain added sulfonic acid would prove that the reaction rate constant is independent of both chain length and viscosity. These plots were, in fact, linear over an enormous range of viscosities except for a slight initial curvature attributed to a change in the reaction medium.

Although Carothers' main interest was directed to polycondensation reactions involving only bifunctional reagents, he also displayed some interest in reactions of reagents with a higher functionality. In a review published in 1931[313] he discussed the condensation of glycerol with phthalic acid and stressed that "analysis of the resin before the infusible stage is reached shows that glycerol and phthalic acid are far from having completely reacted with each other." He also pointed out that phenol behaves like a trifunctional reagent in condensation with formaldehyde because reaction can take place at the two ortho positions and the para position; therefore this polycondensation may lead to crosslinking, whereas the condensation of p-cresol with formaldehyde leads only to linear chain products. In the paper presented at a meeting of the Faraday Society in 1935[411] he attempted to formulate a theory of the gel point, and the discussion which followed his presentation touched on the most vexing problem of gelation theory; the fact that multifunctional branched structures can take part both in intermolecular and intramolecular reactions.

A theory of gelation in polycondensations involving reagents with more than two functional groups was formulated in 1941* by Flory.[414a] He defined the "gel point" as the stage in the reaction in which a network of covalent bonds first extends throughout the system, although most of the polymer consists of relatively small molecules. When following along the chain an infinite structure

---

*To appreciate early difficulties in dealing with polycondensations that lead to gelation it is instructive to read the paper by Kienle and Hovey[415] on the reaction of glycerol with phthalic anhydride. Although they asserted that "it is probably safe to say that a resinous product results whenever a polyhydric alcohol and a polybasic acid are heated together", they made no distinction between divalent reagents and reagents of higher functionality.

will form, according to the theory, if the probability of reaching a trifunctional branch point is larger than the probability of reaching a chain end. In the simplest case, that of a mixture of bifunctional and trifunctional reagents which carry the same functional groups, the theory predicts that when ρ is the fraction of the reactive groups contained in the trifunctional reagent the gel point occurs when $p\rho/(p\rho + 1 - p) > 1/2$. Somewhat later, Stockmayer[415a] obtained a similar expression for the gel point by arguments which revealed the analogy between gelation and the condensation of a gas resulting from a progressive molecular cluster formation. Experimentally, gelation was found to occur somewhat later than predicted by these models, a result that was not unexpected in view of the probability of some intramolecular condensations.

Another striking discovery of a class of condensation polymers, the polysiloxanes, was based on work carried out shortly before the entry of the United States into the second world war so that much of the pertinent information could be published only after the end of hostilities. The development of organo-silicon chemistry has a long history. The first compound with silicon-carbon bonds was made by Friedel and Crafts as early as 1863,[415b] but a sustained effort to study silicon-organic compounds was the life work of F. S. Kipping. This research, which extended over almost four decades, was started in 1899 with the initial objective of preparing an optically active compound in which a silicon atom is the center of asymmetry but led eventually to a monumental exploration of this field. Kipping felt that he had left little to be done, and his summary, contained in the Bakerian Lecture he delivered in 1936,[415c] ends with the unduly pessimistic conclusion that "the prospect of any immediate and important advance in this section of organic chemistry does not seem to be very hopeful."

Kipping could not have been more mistaken in his skeptical assessment of the chances of an "important advance" since within five years his work provided much of the basis of a substantial industrial development. He had amply demonstrated "fundamental differences in the properties of the atoms of silicon and carbon" by showing, for instance, that $C_2H_5SiCl_3$ and $C_6H_5SiCl_3$ do not hydrolyze, as previously reported, to "silicopropionic acid" and "silicobenzoic acid" (C_2H_5SiOOH and C_6H_5SiOOH) but lead on hydrolysis to poorly defined condensation products.[415d] He also found that diphenyl dichlorosilane "when warmed with solvents was converted to viscid glue-like products"[415e] and concluded that "it seemed worth-while to attempt the otherwise uninviting task of isolating the different components of the oils and glues."[415f] This led to the isolation of the linear dimer and trimer as well as the cyclic trimer and tetramer (characterized by cryoscopy). Eventually, chain molecules with a molecular weight of 3900 were obtained, thus eliciting the comment that "it would seem that silicanediols, like methylene glycol, may give rise to a very extensive series of condensation products."[415g]

At the time of Kipping's Bakerian Lecture the Corning Glass Works were busy developing woven glass fiber tapes for electrical insulation. This called for a resin stable at high temperatures with which the tape could be impregnated and J. F. Hyde tried to use polydiphenylsiloxanes for this purpose. Samples were shown to representatives of the General Electric Company research laboratory, where E. G. Rochow was encouraged to follow up on the Corning approach.* He reasoned that "high heat stability would call for a polymer having a minimum of organic content" and decided therefore to work on the hitherto unknown methyl-substituted chlorosilanes, which condensed after hydrolysis to resins with desirable properties.[415i] Rochow's patent states that "the polymeric bodies of this invention are unique in that they have no carbon-to-carbon bonds and are therefore free of the thermal decomposition which initiates in the rupture of a carbon-to-carbon bond." Although stability at 200°C, observed with the early products, was far exceeded in polymers of a later era, it seemed almost miraculous at the time.

Before this new class of polymers could be considered for industrial exploitation Kipping's procedure for monomer synthesis from $SiCl_4$ and Grignard reagents had to be replaced by an economically more feasible method. Here good fortune played an important part: Combes[415j] had in 1896 used an alloy of silicon and copper for the synthesis of $HSiCl_3$ because, unlike pure silicon, it was industrially available. Rochow[415k] found that the attack of methyl chloride on a similar alloy leads to $(CH_3)_2SiCl_2$ and later work showed that copper is unique in its catalysis of this reaction.[415h]

During the period we are considering the first information also became available on the mechanism by which living organisms synthesize condensation polymers. An enzyme-catalyzed polycondensation in which a simple sugar is converted to a polysaccharide seems to have been observed first by Hibbert and his associates,[415ℓ] who in 1930–1931 found that a sterile filtrate of a microorganism produced levan, an anhydrofructose polymer, from sucrose. Several other enzyme-catalyzed polymerizations were discovered in the next ten years. Cori and Cori[415m] found that glucose-1-phosphate ("Cori ester") is converted to glycogen and inorganic phosphate in the presence of mammalian phosphorylase with the reaction reaching an equilibrium at a phosphate/Cori ester ratio which depends on pH; Hanes[415n] discovered a similar conversion of Cori ester to amylose in the presence of a plant phosphorylase. Both reactions depended on the presence of a "primer," a trace of glycogen in the system studied by the Coris, starch or maltose in Hanes's experiments. Since the enzyme-catalyzed process requires the addition of glucose units to the end of the primer molecules so that a fixed number of chain molecules propagate, the kinetics of the reaction would be expected to be similar to the polymerization of ethylene oxide (see p. 138) leading to a narrow distribution of chain lengths. Such an effect was actually observed by Husemann et al.[415o] in the enzymatic synthesis of amylose.

---

*An excellent history of the silicone polymer work at General Electric, which contains laboratory notes and personal recollections of the large number of men engaged in its scientific and engineering development, was written by Liebhafsky.[415h]

13. Addition Polymerization

In his first paper, "On Polymerization," published in 1920,[299] Staudinger described the formation of long-chain molecules as involving the "assembling" of monomer molecules until equilibrium is attained. At this stage the manner in which the monomers "assembled" to form a polymer chain was undefined. The notion that the chain length is thermodynamically determined, that is, the chain will grow until it attains the maximum length stable at a given temperature, was elaborated by Staudinger in a review lecture on his macromolecular concept delivered at the Düsseldorf 1926 meeting.[404] In 1928 Whitby and Katz published their work on the polymerization of indene[416] in which they assumed that the chain growth of vinyl derivatives involves hydrogen migration:

$$2CH_2 = CHR \rightarrow CH_3—CHR—CH=CHR$$
$$CH_3—CHR—CH=CHR + CH_2=CHR \rightarrow CH_3—CHR—CH_2—CHR—CH=CHR, \text{ etc.}$$

Neither they, nor anyone else writing at that time, were aware that such a mechanism of vinyl polymerization had been proposed by Berthelot more than sixty years earlier[417] (see also p.18). However, in an extensive 1929 paper by Staudinger and his collaborators[418] dealing with the polymerization of styrene, it was pointed out that the linear styrene dimer is quite unreactive so that it cannot be an intermediate in the formation of the high polymer. They concluded that polymerization must involve an activated monomer to which monomers can be added preserving the activated state until the process is terminated by ring formation. This conclusion was reinforced when Staudinger and Kohlschütter[419] found that the polymerization of acrylic acid led to a product that contained no detectable amounts of oligomers. They reasoned therefore that "polymerization occurs first by the activation of a molecule followed by a fast addition of more molecules in a chain reaction . . . the chain ends have free valences, they are unstable and continue to grow until the end groups are inactivated in an as yet unknown manner."

The first attempt to study the kinetics of a vinyl polymerization was made in

1909 by Stobbe and Posnjak,[419a] who studied the conversion of styrene to "metastyrol" by following the increase in viscosity after heating for varying times at 200°C. They found that freshly distilled styrene polymerized only after an induction period of seven hours, while no induction period was observed if the styrene was allowed to age for two weeks at room temperature before polymerization. This result was confirmed later[419b] in experiments in which the change in the refractive index was used to follow the reaction. Stobbe and Posnjak concluded that in the aged styrene "a polymerization nucleus must have been formed which leads to an extraordinary acceleration of the polymerization." Lautenschläger,[298] in his 1913 thesis, wrote that the observation of Stobbe and Posnjak "should be explicable by the formation of small quantities of peroxides" because added peroxides acted as polymerization catalysts.

Valuable accounts of early studies aimed at the production of technically useful polymers of monovinyl derivatives are contained in collections of documents from the archives of Farbwerke Hoechst AG (*Dokumente aus Hoechster Archiven*, No. 10, 17, and 23) and in the memoirs of W. O. Herrmann.[419c] As early as 1913 a patent based on the work of F. Klatte (Ger. Pat. 281877) claimed "the preparation of horn substitute, artificial fibers, lacquers, etc." from poly(vinyl chloride). The patent stated that the polymer previously described by Baumann[56] was "an opaque porous chalky mass . . . while this invention led to the surprising realization that this useless substance can be transformed to technically useful materials by dissolution or plasticization and reconstitution of the solid form." Yet Klatte's patent was never used. The IG Farbenindustrie laboratories resumed work on PVC in 1927. In early work polymerization was effected by exposure to sunlight and only in 1929 was a peroxide-initiated thermal polymerization reported by E. D. Dickhaeuser. Yet in a report of January 31, 1930, Klatte writes pessimistically that "no great hopes are being entertained as to the technical utility of poly(vinyl chloride)." It should also be noted that some attempts were made to convert PVC to poly(vinyl alcohol).

The first vinyl polymer to be produced on an industrial scale was poly(vinyl chloroacetate), which was used in Germany as a lacquer during the first world war. Work on poly(vinyl acetate) was started in 1924 and in 1926 German Patent 471,278 claimed "an adhesive consisting of poly(vinyl acetate)." Again, polymerization was effected by sunlight and as late as 1929 Klatte suggested that the poor light intensity during the winter months "may be taken care of by increased production from March to October." In 1924 Haehnel and Herrmann hydrolyzed poly(vinyl acetate) and found to their surprise "that the extremely labile monomeric vinyl alcohol can exist in polymeric form." They also found it "surprising that the poly(vinyl alcohols) were distinguished in many ways as to molecular weight or solubility just as the parent substances." These observations were in strong support of Staudinger's contention that polymers consist of covalently bonded chain molecules and the report by Herrmann and Haehnel[419d] was published next to a similar one from Staudinger's laboratory.[419e] The cost of poly(vinyl alcohol) restricted its early use to textile sizing. The possibility of dehydrating poly(vinyl alcohol) to a rubber was also explored.*

About the same time that Klatte obtained his first patent on poly(vinyl chloride) Otto Röhm began his painstaking work on polyacrylates.† Röhm's 1901 doctoral dissertation at Tübingen dealt mainly with the dimerization of acrylic esters by sodium ethoxide, but it also described "a solid modification of acrylic esters" which was "transparent, highly elastic with only a faint odor of the liquid ester." Regarding the nature of this substance Röhm firmly excluded the possibility of a covalent bonding of the monomer residues and concluded that the recovery of the acrylic ester when the solid was heated proved that it was a "pseudopolymer." Plans for an extended study of the "solid modification"[419f] were not realized because of the death of v. Pechmann, Röhm's research director. Later, under the influence of F. Hofmann's work on synthetic rubber, Röhm initiated studies on methyl acrylate polymers in the hope that they would lead to a rubbery product. Although these experiments led to few useful results,‡ Röhm spent fifteen years searching for a cheap synthesis of methyl acrylate. By that time Röhm had founded the Röhm and Haas Company to exploit his discovery of a revolutionary tanning process and the profits from that invention provided the funds for his studies of acrylates. The first important result was obtained by accident in 1927. An attempt to prepare a sheet of poly(methyl acrylate) by pressing it between two glass plates led to the discovery of "plexiglass," which was far superior to older safety glasses using cellulose nitrate as the interlayer.

Starting with 1935, important advances in the understanding of the mechanism of vinyl polymerization were achieved by increasingly careful studies of the kinetics of the process in combination with estimations of the chain lengths of products obtained at various monomer conversions. Staudinger and Frost[420] found that the degree of polymerization of polystyrene remained virtually unchanged with the degree of monomer conversion. Addition of stannic chloride catalyst led to a decrease in chain length and this was believed to indicate that the catalyst accelerates both chain initiation and termination. They wrote:

> This connection is not necessary since there might exist catalysts which speed up initiation but not termination. Such catalysts would produce high polymerization rates and high molecular weight products.

Staudinger and Frost also observed that high rates could be obtained by polymerizing styrene in emulsion and speculated that "on the surface of the colloid particles numerous active sites may be formed."

*Report by F. Hofmann, January 29, 1931, Hoechst Archives.
†A fascinating account of Röhm's career is to be found in Tromsdorff's *Dr. Otto Röhm, Chemiker und Unternehmer* Econ Verlag, Düsseldorf, 1976.
‡In 1915 Röhm patented such polymers as "substitutes for drying oils." *Ger. Pat.* 295,340.

Schulz[421] first interpreted the findings of Staudinger and Frost that the polystyrene chain length remains constant with increasing monomer conversion as indicating that chain propagation and termination involve an interaction of the same two species; that is, the "activated" chain end and a monomer molecule. Thus the chain length would depend only on the ratio k_p/k_t, characterizing these two processes, but would be independent of the initiation rate. In this paper Schulz made the important comment that the distribution of molecular weights of a polymer should illuminate the mechanism of its formation and he derived the well-known distribution function obtained when the relative probability of propagation and termination is independent of chain length.* A distribution obtained by fractionating polyisobutene agreed with the theoretical prediction.†

In the same year in which Schulz published his first study of styrene polymerization the problem was also investigated by Dostal and Mark.[422] They understood that the process involves a chain reaction but assumed no chain termination. A later study by Mark and Raff[423] used the change in the refractive index for the continuous monitoring of the polymerization process. They reported an initial accelerating phase in accord with their assumed mechanism. The paper read by Dostal and Mark[424] at the 1935 Faraday Society meeting on polymers introduced a chain termination mechanism involving an intramolecular hydrogen transfer in a chain biradical:

$$\cdot CH_2CH-(CH_2CH)_n-CH_2CH \longrightarrow CH_3CH-(CH_2CH)_n-CH=CH$$
$$\quad\; | \qquad\quad | \qquad\qquad | \qquad\qquad\; | \qquad\quad | \qquad\qquad |$$
$$\quad\; R \qquad\quad R \qquad\qquad R \qquad\qquad R \qquad\quad R \qquad\qquad R$$

In other papers presented at that meeting Gee and Rideal[425], Gee,[426] and Bawn[427] assumed that the length of chains produced in vinyl polymerization was limited by a gradual decrease in the reactivity of the active chain end due to an increasing steric constraint. To quote Bawn: "As the chain length increases, reaction becomes more difficult and will finally become immeasurably slow."‡

---

*In a later paper[430] Schulz and Husemann referred to a paper of Carothers[313] in which he announced the chain-length distribution function derived by Flory for polycondensation reactions. They were clearly puzzled by the identity of this function with the one obtained by Schulz for vinyl polymerizations involving chain propagation and termination. It seems that when Flory wrote his paper on chain-length distributions in condensation polymers[412] he also did not realize that vinyl polymers obtained by a chain reaction may be characterized by a similar distribution. In the later literature this is commonly referred to as the "normal distribution."

†In this work Schulz used the Staudinger "viscosity law" to estimate molecular weights.

‡It should be noted that in his 1935 paper Schulz[421] stated in a footnote that "it is assumed that the reaction rate of X_nX is independent of n. A dependence is only to be expected for the lowest degree of polymerization." This first statement that free radical polymerization is characterized by a propagation rate constant independent of chain length was apparently unknown to Gee and Bawn and would not have been widely believed [e.g., H. Dostal, *Monatsh. Chem.*, **67**, 63 (1936)] before Flory's demonstration[111] that the reactivity of functional groups is indeed independent of chain length in polycondensations.

Important progress resulted from the suggestion by Staudinger and Steinhofer[428] that the termination of the polymer chains involves the interaction of *two* chain radicals in which a hydrogen atom is transferred (referred to later as a "disproportionation"):

$$2\text{\small\textasciitilde\textasciitilde}-CH_2\dot{C}H \longrightarrow \text{\small\textasciitilde\textasciitilde}-CH=CH + \text{\small\textasciitilde\textasciitilde}-CH_2-CH_2$$
$$\qquad\quad |\qquad\qquad\qquad |\qquad\qquad\quad |$$
$$\qquad\quad R\qquad\qquad\qquad R\qquad\qquad\quad R$$

They were somewhat troubled by the fact that Signer and Weiler[429] were unable to detect a prominent line characteristic of carbon-carbon double bonds in the Raman spectrum of low molecular weight polystyrene.

In spite of this negative spectroscopic evidence Schulz and Husemann[430] accepted the principle that chain termination in styrene polymerization involves an interaction of two growing chains. They argued that such a mechanism is strongly supported by the general observation that changes in reaction conditions which lead to an increase in the reaction rate also produce a decrease in the chain length of the product. Using the steady-state approximation* for the concentration of free radicals, they showed[431] that the polymerization rate should be proportional to the square root of the initiation rate, $R_i^{1/2}$, whereas the molecular weight of the polystyrene should be proportional to $1/R_i^{1/2}$. They also speculated that the polymer formed might reduce the rate of bimolecular interactions of chain radicals.

Some years before the appearance of these publications, work in the United States, which had apparently escaped the notice of Europeans studying vinyl polymerization, adduced clear evidence for the role of chain reactions in these processes. Taylor and Jones[432] observed in 1930 that the thermal decomposition of diethyl mercury in the presence of ethylene leads to the formation of liquid ethylene polymers. Starkweather and Taylor[433] studied the benzoyl peroxide-initiated polymerization of vinyl acetate following the reaction by a continuous observation of the contraction of the system and established dilatometry as a method which later became the preferred technique in the study of polymerization kinetics. They also found that the reaction followed first-order kinetics and reached two important conclusions:

> Benzoyl peroxide acts as a trigger catalyst in starting reaction chains and is destroyed in the process. It seems likely that we are dealing with a chain reaction of some kind. The reaction involves the combination of many molecules to form a polymer and yet it follows a unimolecular reaction rate law.

---

*The steady-state approximation for the concentration of a reaction intermediate in a chain reaction seems to have been used first by Berthoud and Ballenot[558] for the iodine atom concentration in the photooxidation of oxalate in the presence of iodine. It was later applied by Rice and Herzfeld [*J. Am. Chem. Soc.*, **56**, 284 (1934)] to the radical concentration in gas-phase chain reactions.

A little later Taylor and Vernon[434] published their study of the photopolymerization of vinyl acetate and styrene. They suspected the existence of chain reactions, since Moureu and Dufraisse,[435] the discoverers of antioxidants, had found (as early as 1924) that hydroquinone, which protects acrolein against photooxidation (known to be a chain reaction), also inhibits the conversion of acrolein into a resin. Moureu and Dufraisse observed also that although oxygen induces acrolein to form resin in the dark it inhibits formation of the resin when acrolein is exposed to light. They concluded therefore that the oxidation and formation of the resin compete for the same active species.* Taylor and Vernon found that the rate of the photochemical reaction increased by about 30% for a temperature rise of 10°C and that the "aggregates of styrene" increase in size as the temperature is increased, in contrast to the decrease in the chain length of the product with an increasing polymerization temperature in the thermal reaction observed by Staudinger. They obtained a quantum yield of 935 for the polymerization of vinyl acetate, which clearly established a chain mechanism which they described as follows:

> The initially activated molecule initiates the reaction by combining with another molecule. This process is exothermic and as a result these two are energy-rich and can add a third to their aggregate. This addition process continues until the energy is insufficient to cause the addition of another member. This loss of energy may be caused by collision with foreign molecules of the solvent, by collision with the walls or by distribution over the whole aggregate to such an extent that further reaction is impossible.†

It is striking that the authors could not bring themselves to use the term "polymer" but instead wrote vaguely about "aggregates." Carothers and his collaborators[229] wrote, almost at the same time, that

> the transformation of chloroprene into the polymer is evidently a chain reaction. It is enormously susceptible to catalytic and anticatalytic effects, it is accelerated by light and although a large number of molecules is combined to form a single large molecule, the apparent order of the reaction is low,

a beautiful summary of the evidence and no hesitation to speak of "polymers."

---

*The opinion that "substances which autoxidize exhibit frequently polymerization" was stated as early as 1904 by C. Engler and J. Weissberg (*Kritische Studien über die Vorgänge der Autoxydation*, F. Viewig, Braunschweig, p. 179). Engler later sponsored Lautenschläger's thesis "Autoxidation and polymerization of unsaturated hydrocarbons,"[298] which provided Staudinger with his first contact with polymerizations.

†In the mechanism postulated by Taylor and Vernon inhibitors were supposed to act as energy traps without undergoing chemical change. This concept was used by Jeu and Alyea[436a] to predict that the inverse of the initial polymerization rate and the inverse of the polymer chain length should be linear in the inhibitor concentration. Chalmers[436b] pointed out that this mechanism cannot be correct since it predicts, in disagreement with experimental results, that with high inhibitor concentrations high oligomer yields should be obtained.

Conant and Peterson[436d] noted that the polymerization of isoprene under pressure is accelerated if the monomer is first exposed to atmospheric oxygen and concluded that the formation of an isoprene peroxide must be responsible for this effect. This assumption was confirmed by the catalytic action of added peroxides.* Conant and Peterson wrote that the data suggest

> a series of chain reactions initiated by the spontaneous decomposition of peroxide. The accelerating effect of great pressures is due, according to this theory, to the orientation of the molecules of isoprene into a more compact bundle in which longer reaction chains would be propagated by the decomposition of a single peroxide molecule.†

They never mention free radicals explicitly, although their involvement is implied by the comment that "the efficiency of a peroxide as a catalyst will depend on its instability." Their hesitation may have been due to the dogmatic statement of the prestigious Wieland:[437]

> ... in the thermal decomposition of diacyl peroxides *in solution* free radicals are not formed even as transitory species but one has to view the reaction as a consequence of an interaction of the solvent and the peroxide.

A review published in 1937[438] still shows that the generation of free radicals from peroxides was considered controversial at that time. As a result, we find for some years in the literature rather vague references to the "activation of monomers" by polymerization catalysts. (This is so even in Schulz's 1935 paper,[421] although Staudinger and Kohlschütter[419] had written about free valences at the end of propagating chains four years earlier.)

When discussing vinyl polymerization in 1937, Flory[439] had no doubt that free radicals at the end of growing chains were responsible for chain propagation. He pointed out that these radicals can disappear only in pairs (by disproportionation or recombination), and by introducing the concept of "chain transfer" he explained that the kinetic chain length is not necessarily related to the length of the polymer chain molecules formed. (Schulz and Husemann,[431] in a discussion of Flory's proposal of a chain transfer process, concluded rather rashly "It is possible that such reactions occasionally occur ... but it is unlikely that they have a significant effect on the course of the reaction.")‡ If

---

*Conant and Peterson were unaware of the fact that similar observations had been made by Lautenschläger[298] and Staudinger[224] many years earlier.
†Conant and Tongberg[436c] claimed that isoprene polymerized a hundred times faster when the pressure was raised from 2000 to 12,000 atmospheres and estimated that the rate under 2000 atmospheres was at least a hundred times as fast as in the absence of applied pressure. It is difficult to believe that the pressure dependence of the reaction rate could be so large.
‡On the other hand, Flory exaggerated the importance of chain transfer when he generalized, on the basis of limited data, that "the production of each active center will, on the average, bring about the formation of many polymer molecules."

chain transfer occurs by the abstraction of an atom from a polymer molecule the radical formed may lead to chain branching. Flory also noted that before a steady-state radical concentration is attained the polymerization should exhibit an accelerating phase; he thought that this explained the initial acceleration reported by Dostal and Mark. Although this conclusion was incorrect (the nonsteady state is too short-lived to have been observed in these thermal polymerization experiments), the principle became important later in studies in which the polymerization was initiated photochemically.

It is rather curious that in the first publications dealing with termination involving two propagating polymers Staudinger and Frost[420] and Schulz and Husemann[430] considered only hydrogen transfer between two polymer chain radicals and not a radical recombination reaction that might be thought of as a more obvious process. This is particularly surprising since these investigators stressed that spectroscopy revealed no evidence of carbon-carbon double bonds in low molecular weight polystyrene, which should have been detected if hydrogen transfer had been the mechanism of chain termination. However, two years after publication of Flory's paper Schulz[440] pointed out that chain termination by radical recombination should lead to a narrower distribution of chain lengths than termination by hydrogen transfer and the distribution obtained from a careful fractionation of polystyrene (using the Staudinger relation between solution viscosity and molecular weight) seemed to agree with that predicted for the radical recombination mechanism.[441]

An acceleration of the reaction in the later stages of polymerization was first reported by Norrish and Brookman,[442] who concluded that it resulted from a breakdown of thermostatic control when the system became highly viscous.* Schulz and Blaschke[443] made similar observations and thought that the acceleration was due to a branching chain reaction that resulted in an increase in the concentration of actively growing chain ends. The real cause of the phenomenon was pinpointed in 1942 by Norrish and Smith,[444] who carried out a very slow polymerization of methyl methacrylate to keep the temperature constant and followed the process continuously by dilatometry. They found that even under these conditions the accelerating phase of the reaction was observed. They also found that during this phase the chain length of the polymer increased and so they arrived at the correct conclusion that the effect was caused by a decrease in the rate at which the active chain ends could diffuse to each other through the viscous medium to terminate chain growth. The acceleration was, as expected, more pronounced in poor solvent media. It is of some interest that Norrish and Smith, writing at the height of World War II, were aware of Schulz and Blaschke's publication in a German journal. Five years later Schulz and Harboth,[445] as well as Trommsdorff et al.,[445a] elaborated on the phe-

---

*It is rather surprising that the authors were unfamiliar with Flory's paper[439] published two years earlier. This explains their contention that in view of the "bifunctionality of vinyl monomers" in their polymerization "only straight chains can be produced." The branching of such chains had already been postulated by Schulz and Husemann[430] to explain lower solution viscosities for any given molecular weight when the polymerization temperature was increased.

nomenon. It is hard to understand why it should have become known as the "Trommsdorff effect," although both German publications referred properly to its discovery by Norrish and Smith.

The story of the discovery of the high pressure process for ethylene polymerization has been recounted by Perrin[445b] in fascinating detail. In some respects it parallels strikingly the discovery of Nylon. Just as DuPont decided in 1928 to hire Carothers to "work on problems of his own selections," so at about the same time Imperial Chemical Industries (ICI) sent two of their research scientists to Amsterdam to gain experience in A. Michels' laboratory in the study of reactions under high pressure. A research group was then formed at ICI to explore this field. A number of basic studies in that area published by R. V. Gibson, M. W. Perrin, and E. G. Williams suggested no industrial utility but "established active cooperation with several of the leading schools of physical chemistry." A report written in 1932 by Perrin and Swallow seems to have been influenced by Conant's finding[436c,d] that high pressures may produce a strong acceleration of polymerizations and in particular by the fact that he had obtained under high pressure a hitherto unknown polymer of butyraldehyde. Perrin and Swallow concluded that high pressure studies would "probably" yield "results of chemical interest." An early result of the experimental work started at the time this report was written was a claim (probably incorrect) that polyacrolein was produced from ethylene and carbon monoxide. Then in March 1933 an attempt was made to react ethylene with benzaldehyde at 170°C under a pressure of 1400 atmospheres. The benzaldehyde was found to remain unchanged but "a waxy solid" was recovered and this was identified as a polymer of ethylene. An attempt to improve the yield by increasing the pressure to 2000 atmospheres led to an explosion with decomposition of the reagents and work on the reaction was discontinued. A 1934 publication by Fawcett and Gibson[445c] lists a large number of reactions studied under high pressure but makes no mention of polyacrolein or polyethylene.*

By 1935 great progress had been achieved in the experimental technique of conducting reactions under high pressure. The 1933 result now appeared in a new light and the polymerization of ethylene was attempted in December of that year (this time without benzaldehyde). Eight grams of polyethylene were recovered. During the following year the importance of a small concentration of oxygen to initiate the reaction and the removal of the heat liberated in the polymerization were recognized. A patent[445e] was granted in 1939.‡ At first

---

*In the discussion following the reading of Staudinger's paper at the meeting of the Faraday Society, held in September 1935, Fawcett[445d] reported the initial results of ethylene polymerization to a product with a cryoscopic molecular weight of 4000 but his disclosure elicited no reaction. Nevertheless, this indiscretion led to Fawcett's discharge by ICI.

‡The first attempt to polymerize ethylene by applying pressure at elevated temperatures was reported by V. Ipatiev as early as 1911 (*Ber.*, **44**, 2978). Under a pressure of 70 atmospheres at 400°C he obtained liquid products with 21% boiling above 280°C. The ICI patent specified that solid polymers could be obtained only at pressures exceeding 1000 atmospheres using small concentrations of oxygen to catalyze the reaction and removing as effectively as possible the heat of polymerization. The molecular weight of the product increased with increasing pressure and with decreasing temperature and oxygen content.

no one could think of an application for this waxy polymer. However, B. J. Habgood, who had come to ICI from the cable industry, pointed out that polyethylene had similar (and superior) characteristics when compared with gutta percha, used for submarine cable insulation, and this became the first application of polyethylene. One hundred tons of polyethylene were produced in 1939. This development proved to be of crucial importance in the war, which broke out in that year, since the extremely low dielectric loss of polyethylene made it an ideal material for radar cable insulation.*

The outbreak of the second world war also stimulated the first use of a synthetic polymer for a medical application. As related by W. Reppe,[445f] H. Weese of the IG plant at Elberfeld visited the Ludwigshafen laboratory in February 1940 and inquired about polymers suitable as blood plasma expanders to prevent heart failure after heavy loss of blood. Reppe suggested the use of polyvinylpyrrolidone. Three years later, after the Allied landing in Africa, military authorities were puzzled by the purpose of the polymeric powder found in captured German stores.

The first study of copolymerization appears to have been undertaken by Herrmann, who wrote in his memoirs (Ref. 419c, pp. 96–97):

> We considered it a particular success that it was not only favorable for their application to blend polymers but that one obtained novel substances when one copolymerized chemically different monomers. Thus, we copolymerized in 1928 vinyl esters and acrylic esters obtaining polymers useful for applications for which the homopolymers were not satisfactory. These studies led to the surprising finding that copolymerizations of vinyl esters succeed even with unsaturated halogenated hydrocarbons which polymerize by themselves with difficulty so that they must receive an impulse for their polymerization. Thus we obtained in 1928 copolymers of vinyl acetate with trichloroethylene and dichloroethylene.

Yet an understanding of copolymerization progressed so slowly that two years later G. Kränzlein, chief proponent of polymer research at the IG laboratory in Hoechst, reported as a novel finding† that "copolymers are quite different from blends. Each monomer acts as a regulator on the other and they polymerize into each other."

The first academic study of copolymerization appears to have been undertaken in 1930 by Wagner-Jauregg,[446] who found that maleic anhydride, which would not polymerize by itself, formed copolymers with stilbene and styrene. Most remarkably, the stilbene copolymer contained always equal concentrations of maleic anhydride and styrene residues, no matter what the ratio was in which the comonomers were mixed. (Surprisingly, this result seems to have attracted no notice before Mayo and Lewis embarked on their study of copolymerization fourteen years later.) In another early study Staudinger and

---

*When I was hired by the Bakelite Co., a subsidiary of Union Carbide Co., in 1945 I was told that it was important to find new uses for polyethylene, "now that the war is over." U.S. production was then 10,000 tons a year; by 1982 it had increased four hundred-fold.

†Meeting the "Kuko" *(Kunstoffkommission)* on July 21, 1930. Hoechst archives.

Husemann[446a] initiated a study of the copolymerization of styrene and divinylbenzene, arguing that "there must be a relation between the length of polystyrene molecules and the amount of divinylbenzene required for the crosslinking to a polystyrene of limited swelling"; their data confirmed this expectation. They also found that swelling decreased with increasing crosslink density and was independent of the chain length which the polystyrene would have attainedin the absence of the divinyl benzene. Rather remarkably, although they observed swelling ratios up to 200, they insisted that the results were compatible with the assumption of rigid polystyrene chains. Somewhat later Norrish and Brookman[446b] demonstrated the formation of crosslinked products from the polymerization of styrene or methyl methacrylate in the presence of a variety of divinyl compounds. Although these studies could be interpreted only by assuming a copolymerization, Hill et al.[446c] believed that it was important to prove chemically the presence of comonomers in the same polymer chain. They copolymerized an equimolar mixture of methyl methacrylate and butadiene and isolated from the product after ozonolysis and hydrolysis 2-methylbutane tricarboxylic acid:

$$—CH_2CH = CHCH_2\underset{COOCH_3}{\underset{|}{CCH_2}}CH_2CH = CHCH_2— \longrightarrow HOOCCH_2\underset{COOH}{\underset{|}{CCH_2}}CH_2COOH$$

with CH$_3$ groups on the quaternary carbons.

Industrial experience soon showed that differences in the reactivities of comonomers leads to a heterogeneity in the copolymers produced. A 1937 patent by Fikentscher and Hengstenberg[446d] suggested that this difficulty could be avoided by continuous addition of the more reactive to the less reactive comonomer.* Staudinger and Schneiders[446e] fractionated an equimolar vinyl chloride-vinyl acetate copolymer and found that the chlorine content increased with decreasing solubility, and copolymerization studies by Marvel et. al.[446f] showed that the vinyl chloride content in copolymers decreased with increasing monomer conversion.

An attempt to derive the relation between the composition of a binary monomer mixture and a composition of the copolymer formed from it was made in 1936 by Dostal.[446g] He was the first to suggest that the reactivity of a propagating chain should depend on its terminal monomer residue so that four rate constants have to be specified to characterize the reactions of the two monomers with the two kinds of reactive chain ends. Dostal, however, arrived at the wrong result because he assumed that the ratio of the concentration of

---

*This technique to avoid the cloudy appearance of vinyl chloride copolymers in which the polymerization was carried to high conversion was suggested in July 1932 (J. Hengstenberg, report to the IG laboratory at Ludwigshafen, April 18, 1934). Yet as late as 1938 the German Patent 655,570 (to Röhm and Haas) claimed that catalysts accelerating copolymerization can prevent the formation of cloudy products, thus showing a complete misunderstanding of the underlying principles.

the two kinds of reactive chain ends is equal to the ratio of the two monomer residues in the copolymer. Somewhat later Norrish and Brookman[442] studied the rates of copolymerization of styrene and methyl methacrylate and also used four propagation rate constants in the analysis of their data, although they were unaware of Dostal's paper. They thought that such a study would allow them to distinguish between different polymerization mechanisms and reasoned that

> if the rate of activation of monomer is the limiting factor, the speed of polymerization should be determined by the more readily polymerizable substance . . . if the rate of chain growth were the limiting factor, the rate . . . should be sharply reduced by comparatively small additions of the less polymerizable substance.

They made the unjustified assumption that the concentration of propagating chains is independent of the composition of the monomer mixture and that the propagation rate constants characterizing a monomer in copolymerization are related to the propagation rate in the formation of the homopolymer. Because a plot of the polymerization rate against the composition of the monomer mixture was strongly convex toward the abscissa, they concluded that "without doubt the rate of propagation is the limiting factor." Two years later Wall[446h] tried to develop a theory for the drift of copolymer composition for two monomers of different reactivity but went wrong in assuming that each co-monomer could be characterized by a single propagation rate constant.

Although some reports of a peroxide-catalyzed copolymerization of sulfur dioxide with olefins had appeared in the older literature, careful studies of these reactions are due to simultaneous work by Staudinger and Ritzenthaler[446i] and by Marvel and his students.[446j] Staudinger and Ritzenthaler found that sulfur dioxide reacted with olefins to form polysulfones, with the two monomer residues alternating in the chain molecule, but that no copolymers were obtained with vinyl chloride or acrylic esters. They arrived at the important conclusion that "there is no direct relation between the ability of unsaturated compounds to form a normal polymerization product and to copolymerize with sulfur dioxide." Marvel proved the structure of the polysulfones by chemical degradation studies.

While foundations were being laid for the understanding of free radical polymerization Ziegler and his associates published a series of remarkable papers on anionic polymerization.[447–451] These studies were stimulated by the production of "sodium rubbers which have attained temporarily some technical importance" in the German industry during the war of 1914–1918.* Ziegler was particularly intrigued by the difference in the properties of *Hevea* rubber and synthetic polyisoprene and anxious to define the chemical cause of this difference. In the early studies cumyl potassium was used to effect the polymerization of butadiene which was formulated as a successive 1,4 addition of the metal-organic compound to the double bond of the monomer

---

*Chalmers[436b] noted in 1934 that styrene and the nitriles and esters of acrylic and methacrylic acid can also be polymerized by sodium.

$$\text{RK} + \text{CH}_2=\text{CHCH}=\text{CH}_2 \longrightarrow \text{RCH}_2\text{CH}=\text{CHCH}_2\text{K}$$

$$\text{RCH}_2\text{CH}=\text{CHCH}_2\text{K} + \text{CH}_2=\text{CHCH}=\text{CH}_2 \longrightarrow \text{RCH}_2\text{CH}=\text{CHCH}_2\text{CH}_2\text{CH}=\text{CHCH}_2\text{K, etc.}$$

As early as 1929 Ziegler and Kleiner[448] pointed out that the reaction product would depend on the relative rates of the first and subsequent steps of the process: If the first step is very fast the polymer chain lengths will depend on the relative concentrations of the metal-organic compound and monomer, whereas a slow initial addition of the metal-organic reagent to the butadiene would lead to the production of long chains in the presence of unreacted monomer. (We should note that this principle was stated here long before it was formulated by those studing free radical polymerization.) Eventually, Ziegler et al.[449] employed butyl lithium as polymerization initiator, so that the hydrogenated products could be compared to corresponding normal alkanes. It was concluded that the first addition of butadiene was exclusively 1,4 but that 1,2 addition has a significant probability in subsequent monomer additions. This was surmised, in an age long before chromatographic methods were available, from arguments such as the low melting points of a fraction boiling at some given temperature, suggesting that products of a given molecular weight consisted of a mixture of isomeric alkanes.

In some cases, for instance, in the polymerization of butadiene induced by triphenylmethyl sodium, the metal-organic compound apparently did not disappear and Ziegler and Jakob[450] wondered whether this implied a "true catalytic action." They reasoned, however, that chain propagation might be much faster than the addition of the first butadiene and carried out the reaction in the presence of chain terminators. This allowed them to isolate triphenylpentane after hydrogenation, showing that even in this case the metal-organic compound adds to the monomer molecule. Finally, it remained to be demonstrated that elemental sodium reacts with a diene to form the alkali compound that induces the polymerization. This was proved[451] by producing the highly colored triphenylmethyl sodium from sodium and triphenylmethane in the presence of butadiene. This result was interpreted by the mechanism

$$2\text{Na} + \text{CH}_2=\text{CHCH}=\text{CH}_2 \longrightarrow \text{NaCH}_2\text{CH}=\text{CHCH}_2\text{Na}$$

$$\text{NaCH}_2\text{CH}=\text{CHCH}_2\text{Na} + 2(\text{C}_6\text{H}_5)_3\text{CH} \longrightarrow \text{CH}_3-\text{CH}=\text{CH}-\text{CH}_3 + 2(\text{C}_6\text{H}_5)_3\text{CNa}$$

A review of this early work by Karl Ziegler is not only of great intrinsic importance but also of particular interest in that it shows that his involvement with polymerizations had a long history before it resulted in the spectacular discovery of the low-pressure polymerization of ethylene a quarter of a century later.

Addition polymerization of monomers other than vinyl derivatives was studied only to a limited extent in the period we are considering here. The polymerization of formaldehyde* to materials melting in the range of 163–170°C had been described in 1907 by Auerbach and Barschall.[452] Staudinger[453] first assumed that

> formaldehyde molecules are deposited on crystal nuclei in such a manner that on addition to the crystal lattices a covalent bond is formed. Thus, polymerization proceeds concurrently with the crystal growth.

Later he postulated that the polymerization of pure formaldehyde at −80°C is a chain reaction initiated by the activation of a formaldehyde molecule to a species —CH_2O— to which formaldehyde molecules add until "the reaction is eventually terminated by another process."[454] A similar mechanism was assumed by Staudinger[455] for the polymerization of ethylene oxide in the presence of acidic or basic catalysts. In this case the chain length of the polymer obtained was independent of the concentration of $ZnCl_2$ or triethylamine catalysts but decreased with an increasing concentration of NaOH catalyst, "which is understandable since the chain ends are saturated by water which terminates them by hydroxyl groups."

In 1906 Leuchs[456a] found that when glycine N-carboxyanhydride was treated with a small amount of water it lost carbon dioxide and was converted to a water-insoluble product. Later[456b] he formulated the reaction by

$$x \begin{array}{c} NHCH_2\,CO \\ | \quad\quad | \\ OC \!-\!\!\!-\!\! O \end{array} \longrightarrow \left[CH_2 \begin{array}{c} NH \\ / \;\;\; | \\ \;\;\;\;\;\; | \\ \backslash \;\;\; \\ CO \end{array} \right]_x + xCO_2$$

being quite vague, as was customary at the time, concerning the manner in which the "glycine anhydrides" polymerize. As late as 1926 Wessely and Sigmund[457] could not decide whether the polymer obtained from sarcosine N-carboxyanhydride was to be described by the formula

$$\left[CH_3N \begin{array}{c} CO \\ / \;\;\; | \\ \;\;\;\;\;\; | \\ \backslash \;\;\; \\ CH_2 \end{array} \right]_x$$

---

*An aldehyde polymerization, that of chloral, had been noted by Liebig as early as 1832 (*Ann.,* **1**, 194, 209) and this is probably the first literature reference to the formation of a polymer, although the nature of the product was, of course, not understood.

or as a polypeptide. The polypeptide structure of the product obtained from glycine N-carboxyanhydride was proved only in a crystallographic study by Meyer and Go.[458]

In 1940 Flory[459] analyzed the consequences of a polymerization process, such as that of ethylene oxide, in which a fixed number of chains is initiated simultaneously, chain propagation involves addition of one monomer molecule at a time to the active chains but in which there is no chain termination. He showed that such a polymerization would lead to a Poisson distribution of chain lengths, with the breadth of the distribution becoming narrower relative to the mean chain length as the polymerization continues.* This principle was to lead in a later era (see p.186) to the synthesis of polymers with remarkably narrow molecular weight distributions.

---

*The ratio of the weight and number averages of the degree of polymerization for a Poisson distribution is given by $\bar{P}_w/\bar{P}_n = 1 + \bar{P}_n/(\bar{P}_n + 1)^2$.

14. Advances in the Understanding of Proteins

During the 1920s a number of respected investigators expressed views concerning the structure of proteins, which represented a step back from the early insights of Emil Fischer. Abderhalden[460] asserted that proteins contain diketopiperazine rings held together by "secondary valences," and similar views were expressed by Bergmann,[461] who wrote that "the high molecular state of proteins is of a different type . . . than the polypeptides prepared in the belief that they are models of natural substances."

The notion that proteins are association colloids made up of relatively small "building blocks" was reinforced by Brill's[243] interpretation of the crystal structure of silk fibroin (see pp. 72–73) and by various reports that cryoscopic measurements of proteins in phenol yielded low molecular weights. (Cohn and Conant[462] showed that when anhydrous calcium chloride was added to buffer the activity of any water present in the system the addition of various proteins resulted in a very small melting-point depression indicating a molecular weight of at least 10,000.)*

Svedberg's deonstration by equilibrium ultracentrifugation that various proteins consist of large particles† would, by itself, not have distinguished between large covalently bonded macromolecules and association colloids. The decisive point was the finding that the distribution of the protein in the ultracentrifuge cell showed that *all particles have the same weight*, a result clearly incompatible with the concept of an association colloid. Svedberg was quite unprepared for this result. Reminiscing, he wrote:[463]

---

*Cohn and Conant also made the interesting point that the amino acid composition of well defined proteins could be reconciled with a low molecular weight only by the implausible assumption that a number of *different* small molecules form an association complex.

†Svedberg told B. G. Rånby that he was entirely unaware of Staudinger's studies of macromolecules when he embarked on his protein investigations. (Rånby, personal communication.) In fact, there has been in general little contact between those studying synthetic and biological macromolecules.

> When the author, after measurements on gold sols, took up the task of studying the polydispersity of proteins, he was firmly convinced that they would be as polydisperse as the gold sol particles. . . . The calculation yielded, quite contrary to our expectation, a system which was monodisperse.

Thus the ultracentrifuge data established proteins as covalently bonded macromolecules or, in some cases, association products of a small discrete number of macromolecular subunits which might separate from one another in some solvent media.[464] Moreover, ultracentrifuge sedimentation provided a sensitive technique for establishing the purity of a protein species.

Early in his studies of proteins Svedberg[465] concluded that molecular weights of proteins are multiples of the "fundamental" weight 34,500. Later he considered half of this value the basic unit. The significance that he attached to this presumed regularity was expressed by the following passage in a lecture delivered at a meeting of the Royal Society in 1938:[466]

> If we choose 17,600 as the basic unit, the majority of proteins may be divided into eleven classes with molecular weights which are multiples of this unit by factors containing powers of 2 and 3. The rule is only approximate, indicating that the underlying principle, which probably means a similarity in the architecture of proteins, is obscured by some secondary factor.

These views of Svedberg's also exerted a strong influence on concepts concerning the behavior of proteins in solution. Sørensen, an influential student of proteins, published in 1930[467] a long paper under the title "The structure of soluble proteins as systems of reversibly dissociable components." He claimed that the solubility of proteins is not well defined (a result undoubtedly due to the difficulty of isolating pure protein species) and tried to explain his result as due to an association equilibrium. The following passages provide a flavor of his thinking:

> A complete fractionation we have not achieved and not even aimed at . . . the task is similar to Staudinger's attempt to fractionate polyoxymethylene diacetates. The component, probably a polypeptide chain of very considerable dimensions and with strong residual valences would presumably combine to a system of components similar to the original one.

Sørensen cited an interesting example for his argument: As we have seen, Brill's finding[243] that the unit cell dimensions of the silk fibroin crystallites cannot accommodate amino acids with bulky side chains known to be present was interpreted by assuming that silk fibroin contains, in addition to a crystalline protein, one or more amorphous proteins. Yet, attempts to separate these components failed[468] and this was interpreted by Sørensen as showing that the association between the protein components is so strong that it resists all attempts at separation. (He could not conceive of the same polypeptide chain running through crystalline and amorphous regions.)

Whereas Svedberg's ultracentrifuge had established beyond doubt the macromolecular nature of proteins, a second revolutionary development led to previously inconceivable possibilities to analyze complex protein mixtures. In his doctoral dissertation Arne Tiselius showed that by conducting electrophoresis at 4°C, at which the temperature coefficient of the density of water vanishes, the elimination of convection currents leads to stable concentration boundaries that can be visualized by taking advantage of the bending of light when it passes through refractive index gradients (see p. 102). Using his instrument, Tiselius[469] demonstrated that blood serum globulin, which had previously been considered a single species, consists of three electrophoretically distinct proteins, designated as α, β, and γ globulin. The importance of this separation was emphasized by the finding[470] that the immunization of rabbits with ovalbumin leads to a sharp increase in the concentration of the γ-globulin and that this fraction contains all the antibodies. The resolving power of the electrophoretic method was strikingly demonstrated when Longsworth et al.[471] showed that repeatedly recrystallized ovalbumin contains two species with different electrophoretic mobilities. Thus crystallization was no longer to be depended on to ensure the isolation of a single protein species.

With the techniques available in the 1920s and 1930s, it was difficult to advance the understanding of the structure of protein molecules. Emil Fischer had been puzzled by the inability of pepsin to catalyze the degradation of his polypeptides and this continued for some time to support the assumption that not all the bonds linking the amino acids of proteins can be peptide bonds. Vickery and Osborne[472] speculated, in considering the reason for the compact shape of globular protein molecules, that "the explanation will almost certainly be found when the nature of the bonds attacked by pepsin is discovered." Improved analytical techniques eventually made it possible to prove that equivalent amounts of carboxyl and amine groups were liberated by the action of pepsin on proteins,[473] thus eliminating the need to postulate other than peptide bonds.*

Various experimental data concerning the shape of dissolved protein molecules also became available. Flow birefringence measurements on proteins, introduced by v. Muralt and Edsall,[474] showed that the myosin molecule† behaved like a highly asymmetric particle easily oriented in a shear gradient but that this characteristic was lost when myosin was denatured. By contrast, Boehm and Signer[475] found no flow birefingence in solutions of ovalbumin and concluded that molecules of this protein are spherical. Data obtained by Svedberg by equilibrium centrifugation and by ultracentrifuge sedimentation allowed him to calculate the frictional coefficients of protein molecules which in many cases also suggested a compact spherical shape. This induced a great

---

*Linderstrøm-Lang, one of the great protein chemists, remained unconvinced. He argued that peptide bonds may be formed during protein denaturation, before attack by proteolytic enzymes [*Proc. Roy. Soc London A*, **170**, 56 (1939)].
†The "myosin" probably contained actin leading to much longer rodlike aggregates [J. T. Edsall, and A. v. Muralt, *Trends Biol. Sci*, **5** 228 (1980)].

deal of speculation concerning the forces responsible for such a close packing of the polypeptide chains. A rather fanciful suggestion was made by Wrinch[476] who postulated that units of type

$$\begin{array}{c} \diagdown \phantom{\text{HO}-\text{C}-\text{N}} \diagup \\ \text{HO}-\text{C}-\text{N} \\ \text{H} \diagdown \phantom{\text{C}} \diagup \phantom{\text{C}} \diagdown \phantom{\text{C}} \text{H} \\ \phantom{\text{H}} \text{C} \text{C} \\ \text{R} \diagup \phantom{\text{C}} \diagdown \phantom{\text{C}} \diagup \phantom{\text{C}} \text{R} \\ \text{N}-\text{C}-\text{OH} \\ \diagup \phantom{\text{N}-\text{C}-\text{OH}} \diagdown \end{array}$$

are built into three-dimensional honeycomblike structures.* She predicted that these structures would contain $72n^2$ residues (with n a small integer), presumably explaining the regularity in the distribution of the molecular weights of proteins suggested by Svedberg.

A number of observations indicated that protein denaturation may be reversible. The first such report came as early as 1926 from Spiegel-Adolf,[477] who found that heat-coagulated serum albumin or ovalbumin could be dissolved by careful treatment with dilute alkali to yield, after neutralization and electrodialysis, a protein solution with the same optical activity, heat denaturation, and immunological properties as the original protein. In the case of hemoglobin the recovery, after heat denaturation, of a substance with the characteristic spectrum of the native protein was considered by Mirsky and Anson[478] an additional indication that denaturation had been reversed. In the case of serum albumin they[479] showed that the renatured material could be crystallized. Yet the possibility that denaturation may be reversed met at the time with widespread skepticism. This is hardly surprising, since proof that the native state had been recovered rested on the demonstration that no property of the "renatured" protein was different from that of the native molecule. Obviously, no matter how many properties satisfied this test, there was always the possibility that others might not. Thus Lewis[480] wrote, after citing the results of Anson and Mirsky, that

> by the setting up of a new configuration within the original structure involving new spatial arrangements . . . the probability of reproducing all the original conditions simultaneously is so small as to be negligible.

In 1931 Wu[481] published an impressive paper on "A theory of denaturation," which was far ahead of the conventional wisdom of the time but remained unknown for many years because it was published in China and apparently failed to come to the attention of western investigators. "In the language of

---

*Wrinch's "cyclol" theory was at first taken rather seriously, partly because of its endorsement by I. Langmuir [see *Proc. Phys. Soc.*, **51**, 592 (1939)]. A detailed account of exchanges relating to this proposal was given by J. T. Edsall and D. Bearman [*Proc. Am. Phil. Soc.*, **123**, 279 (1979)].

colloid chemistry," he wrote, "denaturation is a change of the protein from the lyophile to the lyophobe state. But what is the cause of this change?" He then considered the large number of conformations ("configurations" in the nomenclature of the period) accessible to a long polypeptide chain and the restriction that might be imposed on this number by secondary valence bonds. As examples of such bonds he cited the interaction between a pair of carboxyls, between carboxyl and amine groups, and the forces between the chains of silk fibroin, revealed by the crystallographic work of Meyer and Mark.[259] He suggested then that

> although the individual secondary valence bonds are not strong enough to withstand the force of collision, more or less stable configurations can be formed if a large number of secondary valence bonds are arranged "in parallel." This is evidently possible with large molecules . . . such as proteins.

Among the changes in protein properties that accompany denaturation, Wu was particularly impressed with the loss of crystallizability, the loss of biological specificity, and decreased water solubility. He felt that "it is remarkable that proteins crystallize" and that this would be virtually impossible unless "the atoms in each molecule occupy fixed positions", so that "it will be necessary only to orient the molecule as a whole" on adding it to a growing crystal.* The weakening or destruction of the antigenicity when a protein is denatured "must be due to a certain structure which is unstable . . . and it is to certain parts on the surface of this structure that the antigenic power of the protein must be ascribed." As for the characteristic change in solubility, "it is conceivable that in the natural protein molecule the majority of polar groups are directed outward, while in the denatured protein molecule the polar groups are surrounded by the non-polar groups." In this passage Wu anticipated a principle which was recognized theoretically and substantiated by crystallography only a generation later. Six years after Wu the principles of protein denaturation were considered by Mirsky and Pauling,[482] who used arguments similar to those advanced by Wu.† Rather than the vague term "secondary valence bonds," they assigned the stability of the native protein molecule to hydrogen bonding and included in their argumentation data on the enthalpy and entropy of denaturation, which had in the meantime become available. Their paper, easily available to the scientific community, had an impact unjustly denied to Wu's publication.

As I described earlier (see p. 45), the question of the chemical nature of enzymes had been the subject of a lively controversy for a long time. During the 1920s the notion that enzymes might be proteins fell into general disrepute,

---

*Note that this conclusion was reached by Wu three years before Bernal and Crowfoot[275] arrived at it on the basis of crystallographic evidence (see p. 85).

†Their assumption that "this chain is folded into a unique configuration" must be considered an inspired guess. Strangely, the paper makes no reference to the X-ray diffraction obtained with a pepsin crystal by Bernal and Crowfoot,[275] which could be understood only if the internal structure of the protein molecule was closely defined.

partly because of the prestige of Willstätter, who was, most emphatic in opposing such a concept. Thus Sumner, who was convinced that enzymes were proteins, failed to obtain a fellowship in 1921 since his intended sponsor "thought that Sumner's idea of isolating urease was ridiculous."[483] Five years later he *did* isolate urease,[484] proving to his own satisfaction that the crystals he observed were pure protein and that they carried all the characteristic catalytic activity of the enzyme. Wilstätter, however, was unconvinced; during the following year[485] he wrote that "observations on preparations to which proteins adhere tenaciously still threaten to lead to the conclusion that enzymes are of the nature of proteins." The skeptical remarks in his lecture at Cornell, where Sumner taught, may have been responsible for the poor appreciation of Sumner's epochal discovery; his promotion to a full professorship came only three years later at the age of 42. In his Nobel Prize lecture in 1946 Sumner remarked with some justice: "It does not seem difficult to isolate and crystallize enzymes now, but it was difficult twenty years ago." Surely among the greatest difficulties was the need to overcome the widespread conviction that it could not be done.

Sumner's publication on urease stimulated Northrop to attempt the crystallization of proteolytic enzymes and in 1930 he succeeded in crystallizing pepsin. His most impressive report[486] contains a long series of tests by which it was established beyond doubt that the catalytic activity was due to the protein and not to a substance adsorbed to it, as postulated by the Willstätter school. For instance, the protein nitrogen and the catalytic activity in the supernatant in equilibrium with pepsin crystals were independent of the amount of the crystals. An additional convincing argument was provided by the demonstration that in pepsin denaturation the fraction of denatured protein (characterized by a changed solubility) is identical to the fractional inactivation of the catalytic power and that renaturation restores the catalytic activity.[487] Similar results were obtained with trypsin. In this case the native and denatured enzyme were in equilibrium with each other; at 42, 44 and 50°C in N/100 HCl the equilibrium corresponded to 32.8, 50 and 87.8% denaturation.[488]* This result was of great interest for the understanding of proteins in general since it demonstrated that denaturation is driven by a very large increase in entropy (210 eu in this case), as expected for the transition of a highly ordered native, compact, globular form to a disordered denatured state.

Once a number of proteolytic enzymes became available in the pure crystalline form the clarification of their mode of action also made important progress. The stereospecificity of enzyme action suggested that the enzyme must interact with three groupings of its substrate, and this insight stimulated the synthesis of a variety of peptides with a study of their susceptibility to catalytic attack by various proteolytic enzymes. Fruton[488a] has written a beautiful

---

*Northrop also relied heavily on a well defined solubility as a criterion of purity of his crystalline enzymes. Later it was found that this criterion is not reliable since mixed crystals of related protein species exhibit a similar phase equilibrium.

account of the history of these investigations in which he played a crucial role. They led not only to the long-sought discovery of peptide substrates for pepsin but also to the definition of the specificities of other proteases.[489] It led Bergmann to a reversal of his long-standing conviction when he proclaimed[490] "We can now be certain that proteins have the structure of peptide chains."

Long before Svedberg's ultracentrifuge and Tiselius' electrophoretic techniques were developed serological tests provided a powerful method of differentiating between closely related proteins. These tests are based on the ability of animals to synthesize, after injection of a foreign protein (the "antigen"), an antibody which interacts with high specificity with that particular species. The most common method of studying this interaction was based during the development of immunochemistry on the formation of a precipitate when solutions of an antigen and an antibody are mixed. Landsteiner, who contributed more than anyone to the development of this field, published in 1933 a beautiful monograph on "The Specificity of Serological Reactions" which is also available in expanded later editions.[491] Early experiments were carried out with biological fluids that contained many protein species. For instance, a rabbit injected with ox-blood serum produced antibodies that interacted much more strongly with the blood serum of an ox than with sera of sheep, goat or pig, and the rabbit antibody to the blood serum of a hare precipitated hare serum but not rabbit serum. In general, the strength of "cross reactions" (i.e., the interaction between antibodies to a given protein of one animal species with a corresponding antigen from another animal species) was considered a measure of the relatedness of the animals. Serological specificity was thus an early forerunner of techniques used by molecular biology for the study of evolution.

In 1918 Lansteiner and Lampl[492] introduced a technique that was to have a profound effect on the study of antigen-antibody reactions. They coupled a variety of diazotized aniline derivatives to proteins and showed that antibodies to such "conjugated antigens" interacted with high specificity with these protein substituents ("haptens"). For instance, when the three isomeric aminocinnamic acids were used as haptens the antibody reacted only with the antigen used in its production. Also, low molecular weight derivatives of the haptens which were not antigenic could block the antigen-antibody reaction showing that they competed with the antigen for antibody binding sites. Such binding was later demonstrated by Marrack and Smith[493] and by Haurowitz and Breinl[494] using dialysis equilibrium measurements of haptens with their antibodies.

The antigenicity of cells of pathogenic organisms, which is responsible for the immunity acquired after survival of microbial infections, is due to polysaccharides of the microbe cell surface and Heidelberger[495] discovered that soluble polysaccharides also function as antigens. This finding was important since it allowed an easy determination of the composition of precipitates containing polysaccharide antigens and protein antibodies. Antigens on the surface of the red blood cells are also responsible for the fact that the blood serum of some individuals may precipitate or destroy red blood cells of other individuals *of the same species*. A study of this phenomenon led Landsteiner to the recognition of blood groups and rules specifying which individuals may serve as blood donors in transfusions to a given patient. Landsteiner detailed these studies in his Nobel Prize address,[496] in which he also showed that the blood group antigens are inherited by Mendelian genetics.

The presence of several "determinant sites" that interact with antibodies on the surface of an antigen molecule was demonstrated by Haurowitz[497] by the use of a rabbit antibody to horse globulin, coupled with arsanilic acid. Some of this antibody was precipitated by rabbit globulin, which carried the arsanilic acid hapten, some by horse globulin free of the hapten and some only by horse globulin coupled with the hapten. While the "polyfunctionality" of antigens was thus established (as was the heterogeneity of antibodies to a given antigen, which might contain molecular binding to different "determinants"), the number of binding sites on an antibody molecule remained controversial for some years. Heidelberger and Kendall[498] believed that both antigens and antibodies are polyfunctional, so that formation of a crosslinked network is the cause of precipitation when they are mixed, whereas Boyd and Hooker[499] insisted that antibodies are monovalent and that precipitation is due to a nonspecific aggregation of antibody-covered antigens. In any case, the relative ease with which the antigen-antibody complex could be dissociated made it clear that no covalent bonds were formed but that the association depended on steric complementarity.

Because Landsteiner's work on conjugated antigens seemed to have established that antibodies of an almost infinite variety may be produced by an organism, the question of the mechanism by which they are formed intrigued many investigators. Breinl and Haurowitz[500] wrote in 1930, when considering the conditions under which antibodies are synthesized,

> It must be assumed that antigens have weak residual valences which cause the amino acids to assume a certain position as they combine to the globulin. . . . The difference (between the antibody and a normal globulin) can lie only in the grouping of the amino acids, their spatial location in the molecule.

They believed that the amino acid composition of different antibodies is "identical or near-identical" but their formulation does not make it quite clear whether they thought that the presence of the antigen modifies the sequence of the amino acids or only the shape of the polypeptide chain. In 1940 Pauling[501] dealt with the same problem in a more definite manner. In considering the antibody functionality, he stated that "the rule of parsimony suggests" that antibodies are bivalent. He postulated then that the amino acid sequence is identical in all antibodies. The central part of the chain was always supposed to be folded in the same manner but the two chain ends were assumed to fold to a structure spatially complementary to antigen determinants. On the basis of this theory Pauling assumed that antibodies could be prepared artificially by denaturing "normal globulin" and removing the denaturation reagent in the presence of an antigen. He claimed success in such experiments.[501a] A quarter century had to pass before it was established that the specificity of antibodies depends on specific sequences of amino acid residues so that Pauling's procedure could not have led to the desired result.

15. The Elasticity of Flexible Chain Polymers

Among the properties that distinguished high polymers from substances composed of small molecules none seemed more spectacular than rubber elasticity. I have already quoted (p. 35) the rhapsodic passage in Goodyear's memoirs[103a] which referred to this phenomenon. Naturally, as the understanding of macromolecular structure advanced a wish developed to explain this elasticity in molecular terms.

Proponents of the "colloid school" tended to favor models describing rubber as a two-phase system. Freundlich and Hauser[502] claimed to have proved that the particles of *Hevea* rubber latex contain a solid skin surrounding a fluid "of honey-like viscosity." Observing the drying of latex under a microscope they reported that the particle structure remained distinct and speculated that "the complex structure of the *Hevea* latex particle determines the quality of the rubber with the external skin responsible for the firm structure." Ostwald[503] modified this theory by suggesting that the interiors of the latex particles be thought of as having a network structure. This was also supposed to account for the appearance of crystallinity when rubber was stretched. Ostwald felt that Katz's failure to observe crystallization with synthetic rubbers was in striking support of his model since these rubbers had, as he assumed, been prepared by bulk polymerization and could therefore not be expected to exhibit properties dependent on the structure of latex particles. Hauser[504] suggested later that the crystallization of rubber on extension is the result of the expulsion of a fluid "α-component" from the crystallizable "β-component."

In 1924 Hock[505a] reported that when a stretched rubber band was kept extended until the heat generated by the extension was dissipated it no longer retracted to its original dimensions as long as the temperature was kept below 18°C. However, when such a specimen was heated, it retracted and could even lift a substantial load. He concluded that "this temperature-sensitive hysteresis is due to an orientation of particles—as also suggested by the bifringence. The mobility of the oriented particles is, it seems, insufficient to respond to the retractive elastic forces" unless the temperature is raised. A year later Hock[505b] made the spectacular discovery that fibrous particles are obtained when oriented rubber is cooled to −195°C and shattered. Feuchter[506a] introduced the term "racking" for the production of such anisotropic rubber specimens; he was impressed by the fact that they would recover their original shape when heated after prolonged storage at a low temperature and wrote[506b] that "we must assume an extraordinary concentration of forces of such long range that . . . two particles separated a hundredfold over their original state still have a "memory" of their original position." A number of investigators also found that the tensile strength of racked rubber at −195°C may be up to six times as high as that of an isotropic sample.[507]

The contraction observed when tendon is heated seemed to Wöhlisch[508] to be analogous to the behavior of rubber (as observed by Gough[104] and Joule[106]; see pp. 35–36). He reasoned that "the fibers consist of extended molecules or micelles oriented parallel to the fiber axis. . . . Any temperature increase must oppose the orientation and tend towards its randomization." To this view that the retractive force of an extended rubber specimen is produced by a tendency of rigid rodlike particles to lose an imposed orientation because of thermal motion Meyer[509] opposed a convincing argument: A system of disoriented rods would necessarily occupy a much larger volume than a system in which the rods are packed in parallel, yet no such drastic expansion is observed when rubber or tendon retract. The fibers obtained when oriented specimens are shattered at liquid air temperature indicate that such samples contain extended chain molecules and the change in the shape observed on strain relaxation must be due to a change in shape of the constituent molecules. Meyer proposed that in rubber this may arise from the mutual attraction of double bonds, in other cases similar changes in molecular shape may be caused by Coulombic interaction of chain substituents. A detailed model of rubber elasticity based on such ideas was formulated by Fikentscher and Mark,[510] who assigned to the relaxed rubber molecules a spiral form stabilized by an intramolecular attraction of the double bonds in the chain backbone. (It is interesting to note that in this model the transition from the relaxed to the extended rubber chain, involving a change from intramolecular to intermolecular interactions, was analogous to the transition from the α-helical to the extended β-structure of polypeptides, a phenomenon which was not understood at the time.)

The proposal by Fikentscher and Mark was rejected by Whitby[511] and by Staudinger,[512] who pointed out that polymers such as polystyrene, with no strongly interacting groups, are elastic at elevated temperatures (or at ambient temperature when swollen with a low molecular weight solvent). They intimated that this elasticity is somewhat related to the polydispersity of the chain lengths but did not attempt to propose a model to account for this property. Staudinger had, in addition, an important stake in opposing a model which allowed a chain molecule to change its shape because his "viscosity law" required polymer molecules to behave like rigid rods. As late as 1934 Bender et al.[513] claimed that rubber elasticity is compatible with rigid rodlike molecules.

The crucial break in the understanding of rubber elasticity must be credited to a remarkable paper by Meyer, v. Susich, and Valko.[514] After reviewing all the evidence concerning the elasticity of polymeric material they concluded that neither the double bonds nor the crystallizability of *Hevea* rubber are essential conditions for long-range elasticity. They identified as the characteristic phenomenon the absorption of heat when rubber retracts, even after an extension insufficient to produce crystallization. This implied that the retractive force arises from a tendency to pass to a state of higher entropy rather than a state of lower energy. In considering chain molecules oriented by an applied load, they suggested that "a substantial portion of the heat content must be owing to deformational and partial or complete rotational motions perpendicular to the direction in which the molecule is extended" and concluded that this requires that the retractive force be proportional to the absolute temperature. A quantitative interpretation of the stress-strain behavior of rubber was thought to be hampered by the disturbing effect of crystallization.*

Almost simultaneously with the paper by Meyer et al., a paper on "The Physical Structure of Elastic Colloids" was published by Busse.[516] He reasoned that "the fact that elastic colloids can be deformed several hundred percent and then recover to their original shape implies the existence of structural units which also have this property." Specifically, "in long molecules vibrations . . . cause rotation around single valence bonds to produce a very kinky shape." Although it lacked the precision of the thermodynamic arguments of Meyer et al., Busse's paper introduced another important concept by writing "there is an interlocking of the fibrous units at a few places which prevents their slipping completely past each other."†

The suggestion by Meyer et al. that the retraction of an extended rubber specimen is an entropy-driven process opened up the possibility of deriving a theory of rubber elasticity. This task was undertaken in 1934 by Guth and Mark.[517] By using independently a model similar to that of Kuhn[370] (see pp.109–110), that is, assuming chains with fixed bond angles θ and with free rotation around the bonds, they derived from the probability distribution of the end-to-end displacement h for the retractive force F of an isolated chain molecule containing n bonds of length b at small extensions the relation $F = (kTh/nb^2)(1 - \cos\theta)/(1 + \cos\theta)$. They also assumed that "macroscopic functions of state are to a good approximation proportional to those of the

---

*In an angry response to the paper by Meyer et al. Staudinger[515] stressed that he had shown for a long time that the double bonds of rubber are not an essential condition of elasticity. He seems to have missed completely the revolutionary significance of the new interpretation of rubber elasticity.

†A paper by E. Karrer [*Phys. Rev.*, **39** 857 (1932)] entitled "A kinetic study of the elasticity of highly elastic gels" is unfortunately available only as a rather poor abstract. It contains the sentence "stretching eliminates longitudinal degrees of thermal freedom, therefore, heating."

macromolecules, where the number of macromolecules per unit volume would serve as a proportionality factor."

A more elaborate attempt to derive the elastic modulus from the probability distribution of chain extensions in relaxed and stressed rubber was undertaken by Kuhn[518] in 1936. His procedure involved the consideration, in the relaxed specimen, of three chains with root-mean-square end-to-end displacements oriented in three mutually perpendicular directions. The average in the change of entropy of these three chains when the specimen was extended was then assumed to be a fair approximation of the entropy change of the average chain. This procedure led for a rubber of density ρ consisting of chains with a molecular weight M to an elastic modulus $7RT\rho/M$, which was considered reasonably close to experimental data obtained by Meyer and Ferri.[519] These workers had obtained data on the temperature-dependence of the stress in rubber subjected to different extensions and had concluded that it is proportional to absolute temperature, once the thermal expansion of rubber is corrected for, as required by the assumption that the stress is an entropy effect. The correction for volume expansion was later carried out by Anthony et al.[520], who found that for moderate extensions the stress is closely approximated by the product of its temperature coefficient at constant relative extension and the absolute temperature proving that the retractive force is almost completely an entropy effect. Both Kuhn and Ferri and Anthony et al. used rubber vulcanized with 8% of sulfur, and it is strange that neither commented on the effect that this vulcanization would have on the elastic modulus, although it had been well known for a long time that an increasing sulfur content leads to increasingly stiff rubbers. In fact, Bresler and Frenkel[520a] criticized the papers of Guth and Mark[517] and of Kuhn[518] by pointing out that "in vulcanized rubber separate chains are linked together by sulfur bridges . . . one can speak only of free sections of the chain."* They suggested that Kuhn arrived in his calculation at a rubber modulus close to the experimental value as a result of a compensation of errors—while the sulfur crosslinking would tend to increase the modulus, the barriers to free rotation would have the opposite effect.

Guth and Mark, as well as Kuhn, were careful to point out that the assumption of free rotation in their statistical theory was to be considered only as an approximation of the physical reality. We should recall, however, that data available at the time[367] greatly underestimated the magnitude of the hindering potentials, so that the assumption of free rotation appeared to be even better than it was. But the question whether any change of energy was involved in rotation around a chemical bond was important not only for the theory of the equilibrium but even more for the theory of the dynamics of rubber elasticity. The latter aspect was considered first by Bunn,[521] who pointed out that the

---

*In a paper that escaped the attention of Bresler and Frenkel, Kuhn[518a] stated in 1938 that in vulcanized rubber the modulus should be inversely proportional to the molecular weight of *sections of the chain between crosslinks.*

around a single bond attached to a double bond is much lower than for rotation around the bonds in a saturated hydrocarbon and suggested that this may be the cause of the rapidity with which extended samples of polyisoprene or butadiene copolymers retract. Aleksandrov and Lazurkin[521a] were responsible for an important advance in the understanding of the dynamics of rubber elasticity by pioneering studies of the response to periodic stresses. They reported data for loading frequencies of 1 to 2000 cycles per second over a temperature range of −180 to 200°C.

16. The Nonideality of Polymer Solutions

Another phenomenon that was eventually interpreted in terms of the statistical coil model for polymer molecules concerned the striking deviation of their solutions from ideal solution behavior. In 1911 Duclaux and Wollman[522] measured the osmotic pressure of cellulose nitrate solutions and noted that it rises much faster than the solution concentration. They wrote that this might have been considered analogous to the deviation from ideal gas behavior expressed by van der Waals' equation of state if the deviation had been observed in concentrated solution, but "unfortunately the anomaly of osmotic pressure is already considerable at very low concentrations." They concluded that the effect must be due to the colloid nature of the solute. A more extensive study was published in 1914 by Caspari[523] on osmotic pressures of rubber in benzene solution. He pointed out that Perrin had shown that "particles large enough to be microscopically visible obey the osmotic laws of Raoult and van't Hoff," so that there was no reason to suppose that colloids should be exempt from those laws, and "deduction of molecular weights from osmotic data is least likely to be illusory when colloid solutions . . . follow the Raoult-van't Hoff relation in regard to the concentration and temperature coefficient." He also found that the strong upward curvature of plots of osmotic pressure against solute concentration was similar to that of viscosity-concentration plots and that the osmotic pressure increased when the rubber solution was "deviscified" by boiling. This, he believed, indicated that osmotic pressure depends "on the physical condition of the solutions. This phenomenon is evidently connected with their colloidal nature." Since "there can hardly be any doubt" that rubber consists of two phases, its osmotic behavior must be a consequence of this feature.

In 1912 an impressive experimental paper by Posnjak[524] (with a theoretical appendix by Freundlich) described the measurements of swelling pressures exerted by partly swollen rubber in contact with various solvents. The data could be fitted by swelling pressures proportional to c^k, where c was the rubber concentration and the exponent k varied from 2.5 to 3.3.* This investigation encouraged Ostwald and Mündler[525] to formulate a theory in which the osmotic pressure of "disperse systems" was represented as the sum of an ideal solution term and a swelling pressure term : $\Pi = RTc/M + kc^n$, where k and n were adjustable constants.

In 1928 Adair[526] showed that osmotic pressures of aqueous hemoglobin solutions can be represented by $\Pi(V - nb) = nRT$ where the excluded volume b of the hemoglobin is 173.5 liters per mole. Data for synthetic polymers could not, however, be fitted by this simple relation, which appeared to be applicable only to globular protein particles. Haller[527] tried to reinterpret the two-term expression of Ostwald and Münder by suggesting that the term to be added to the ideal osmotic pressure resulted from the vibration and rotation of micellar aggregates. A more plausible theory was advanced by Schulz[528] who pointed out that freely moving particles cannot exert a swelling pressure on the osmotic membrane, as assumed by Ostwald. Schulz also suggested that the excluded volume per solute particle changes with the concentration of the solution due to a change in osmotic swelling and concluded that Π/c should be linear in c and extrapolate to RT/M. In a second paper Haller[369b] showed that the old osmotic data of Duclaux and Wollman[522] and Caspari[523] actually yielded linear plots of Π/c against c, a fact that these authors had surprisingly overlooked. Haller's paper, published in 1931, considers for the first time the separate contributions of enthalpy and entropy to the solution process by pointing out that the heat of mixing is frequently unfavorable for a polymer-solvent system so that the driving force for the solution must be entropic. Haller also formulated for the first time the second osmotic virial coefficient in terms of an excluded volume integral but added that "since we are still far from a detailed understanding, we have to forego an evaluation of the integral."

Between 1935 and 1940 a systematic effort was made by Meyer and his students to gain a better understanding of the osmotic behavior of solutions containing chain molecules. The first study in this series by Meyer and Lühdemann[529] used as solutes discrete molecular species with molecular weights that ranged from 500 to 800. They argued that rotation around bonds and chain deformations produce "an intramolecular Brownian motion which produces a pressure proportion to c^2." The decrease in the "apparent molecular weight" of the solute (derived from the ideal solution law) was then interpreted on "the assumption that portions of molecules can act as kinetic units." In fact, "when the molecules approach one another so that their mean separation is no larger than the separation of their chain ends . . . they behave like many particles." Clearly this was a regression from Haller's insight.

The question whether sections of flexible chain molecules may behave like

---

*This formulation obviously cannot be valid over the entire range of c; for instance, for completely unswollen rubber the swelling pressure must tend to infinity.

osmotically active kinetic units was brought by Mark to the attention of Hückel. In considering this problem, Hückel[530] wrote,

> I myself as well as other theoretical physicists to whom I posed the question at first expected such a result. A closer consideration shows, however, that this idea is wrong . . . since the portions of molecules are not mutually interchangeable.

Nevertheless, Hückel suggested that the intramolecular mobility of a chain molecule may make a decreasing contribution to the entropy as the solution concentration is increased because

> inner motions of atoms or groups of atoms within a molecule depend on whether the molecule is free or surrounded by other molecules . . . so that a change in concentration also involves a change in this inner motion. Particularly in the case of chain molecules one may suppose that such a change in inner motions produces a relatively small change of energy but a relatively large entropy change since such chains may assume a large number of configurations which differ little in energy.

He suggested the measurement of heat capacities to test this interpretation.

Thermodynamic data collected by Boissonas[531] on solutions of long-chain esters showed clearly that whereas the solution enthalpies yielded scattered values with no dependence on the chain length of the solute, solution entropies always exceeded the ideal solution values and became larger as the molecules increased in length. This result and earlier indications that the deviation of polymer solutions from ideality are largely due to an excess entropy of dilution was unknown to Guggenheim when he conjectured[532] that athermal solutions produced at constant volume may be nonideal because of a difference in the size of solvent and solute molecules. A quantitative theoretical study of this effect was carried out by Fowler and Rushbrooke,[533] who considered mixtures of spherical particles with dumbbell-shaped particles occupying a volume twice as large. Using a lattice model to evaluate the number of distinguishable arrangements of the mixture, they cautioned that such "calculations apply strictly to a crystalline solid and only by inference to a liquid mixture . . . the error in this approximation is unfortunately unknown."

The Fowler and Rushbrooke study apparently had come to Meyer's attention when he proposed a qualitative explanation of the excess entropy of dilution in solutions of chain molecules (an effect for which he had by now collected a great deal of data) by the use of a model in which these chains are placed on a lattice.* In his words,[534]

> In the condensed phase . . . *the shape and orientation of a deformable molecule depends also on its neighbors.* Athermal mixing with a substance of low molecular weight diminishes this restraint; shapes and orientations which were excluded in

---

*Meyer pointed out that the nonideality of solutions of synthetic polymers is not simply a function of molecular size because the effect is much smaller with globular proteins.

the pure state become possible. The mixture is thus characterized by an increase in the number of complexions* and of entropy which will be referred to as "the effect of internal mobility."

It should be emphasized that Meyer's approach was fundamentally different from Hückel's. Although he used the term "mobility," it was not the restraint on the *motions* of chain molecules but the static restraint on the arrangement of these chains in concentrated systems that reduced the entropy. Meyer considered, in particular, the two limiting cases of very high and very low chain concentration. At high concentration of chain molecules, the activity of the solvent should be independent of the molecular weight of the polymer, for instance, in the concentrated rubber-toluene system the toluene behaved as if the solute had a molecular weight of 400. In dilute solutions the connectivity of polymer segments produced a large decrease in distinguishable complexions.

A quantitative statistical-mechanical theory of solutions of chain molecules was developed independently by Huggins[535] and Flory.[536†] Counting the distinguishable arrangements of chain molecules placed on a lattice and assuming that the occupancy of a given site could be approximated by the overall fraction of occupied sites, they arrived at a solvent activity a_1 given by

$$\ln a_1 = \ln \phi_1 + \left(1 - \frac{1}{x}\right) \phi_2 + \chi \phi_2^2$$

where ϕ_1, ϕ_2 were volume fractions of solvent and solute, x was the number of lattice sites occupied by the polymer, and $RT\chi\phi_2^2$ was the molar enthalpy of dilution (with entropy contributions due to the limitations of the model). The result conformed to the observation that the solvent activity is insensitive to the chain length of the polymer in concentrated solutions and gave a good representation of equilibria between dilute polymer solutions and a highly swollen polymer phase. Flory also pointed out that in poor solvent media, at incipient precipitation, polymer solutions would exhibit ideal solution behavior. This concept was to play an important role in the development of polymer solution theory. Two conclusions of this original theory, one of the major landmarks in polymer science, proved to be erroneous. The theory predicted that plots of the reduced osmotic pressure Π/c against solution concentration should have slopes independent of the chain length of the polymer. It also suggested an estimate of chain flexibility which clearly led to improbable results. For instance, Huggins[535a] arrived at the conclusion that the flexibility of polyethylene or polyoxyethylene is equivalent to that of a chain that consists of rigid segments of 15 backbone atoms, whereas cellulose derivatives consist of chain molecules with a "high degree of randomness of orientation at the oxygen bridge between the rings."

*The number of distinguishable geometrical arrangements of the molecules.
†Flory has stated[410] that he learned of Huggins' work at the Colloid Symposium held at Cornell University in June 1941. He told Huggins that he was working on the same problem and was encouraged by the older man to publish his results independently.

17. Polyelectrolytes

During the 1920s most of the interest in the effect of ionic charges on the behavior of macromolecules centered on the properties of proteins. The difficulties encountered in the early development of this field are alluded to in the introduction of an influential monograph published by Loeb[537] in 1922.

> As long as chemists continue to believe in the existence of a special colloid chemistry differing from the chemistry of crystalloids, it will remain impossible to explain the physical behavior of colloids in general and proteins in particular.

He was referring here to a widespread belief that acids and bases were not to be thought of as reacting stoichiometrically with proteins but that all such interactions were due to ion adsorption on the surface of the colloid particle. A review that summarized the physical chemistry of protein solutions, published by Cohn[538] in 1925, shows, however, that by that time it was understood that titration curves of proteins could be rationalized in terms of the content of amino acids with acidic or basic side chains, provided account was taken of the effect of the net protein charge on the ionization equilibria of these groups.

It was also observed that the viscosity of solutions increases when acid or base is added to isoelectric protein but that this viscosity decreases again if the amount of acid or base is more than equivalent to the protein groups with which they interact. This effect was interpreted by Pauli[539] as the result of higher hydration of the ionized protein—an excess of acid or base supposedly repressing this ionization. An alternative theory proposed by Loeb,[540] was based on the suggestion made by Procter[541] as early as 1914 and 1916 that the swelling of gelatin gels is governed by a balance between osmotic forces that result from an excess in the ion concentration within the gel and the elastic cohesion of the gel. Thus Loeb believed that "the viscosity of protein solutions is due to submicroscopic protein particles (capable of occluding water) since the amount of water occluded is regulated by the Donnan equilibrium." This model bears a striking resemblance to modern polyelectrolyte theory which requires only a substitution of the rubberlike elasticity of the polymer chain molecule for the "cohesive forces of the colloid particle." A third model was proposed by Meyer,[509] who suggested that in the isoelectric protein the mutal attraction of cationic and anionic charges stabilizes a helical structure of the polypeptide chain which is more compact than the form assumed at lower or higher pH values. He intimated that such a change in the shape of the chain molecule may constitute the basis for muscle contraction.

Scattered studies were also carried out on acidic chain molecules of biological origin. Hammersten[542] published a long paper that dealt with DNA (referred to at the time as thymus nucleic acid). He was unaware of the fact that DNA consists of long chain molecules but thought that it was "definitely" a tetranucleotide. He believed that its titration could be explained by four ionization constants with $10^4 K_a$ values of 43, 2.2, 0.5, and 0.07. The osmotic pressure of the acid was more than 20% lower than expected for the hydrogen ions (assuming that the tetranucleotide aggregated, so that its contribution to the osmotic pressure was negligible) but Hammersten speculated that "the smaller ions can slip in and out of cavities of the larger ions" so that some of their osmotic activity is lost. Thomas and Murray[543] found that the pH dependence of the solution viscosity of gum arabic was similar to that of proteins, and Briggs[544] found that its apparent ionization constant decreased when the solution was diluted; this he ascribed to "a readily reversible size of the micelle accompanied by a shift in the tendency to ionize."

As we have already commented, Staudinger was convinced that studies of synthetic "model" polymers would help to clarify the behavior of natural macromolecules. As Staudinger and Trommsdorff[303] saw it, "there are few natural products of high molecular weight for which work on a more easily analyzable model substance appeared as necessary as for proteins." Poly(acrylic acid) was chosen as the "model" because it was felt that study of a polymeric acid should precede any investigation of a more complex amphoteric polymer. The study was concerned mainly with solution viscosity and led to the discovery that for unneutralized poly(acrylic acid) η_{sp}/c assumes extremely large values at high dilution, falls precipitously with increasing concentration, and rises again after passing through a minimum. On addition of base the viscosity rises sharply but it decreases again if more base is added than required to neutralize the polymer or when salts such as sodium chloride are added. It is remarkable that these dramatic changes of viscosity at constant polymer concentration did not shake Staudinger's conviction that the polymer chain molecules are unable to change their shape, that they behave as rigid rods. Staudinger and Trommsdorff argued that the viscosity increase produced by ionization (when the polymeric acid was diluted or when base was added) reflected the formation of a structured system as a result of electrostatic forces* and cited the dependence of the apparent viscosity on the rate of shear as support for this view.

In 1938 and 1939 W. Kern, one of Staudinger's students, published a series

---

*"An acid anion is surrounded by Na ions and a group of Na ions will again interact with the carboxylates of the acid anion . . . this results in a mutual positioning of the acid anions . . . to which we refer as swarm formation."

of papers on poly(acrylic acid) which defined many of the characteristic features of polyelectrolyte solutions.[545] He found that in dilute solution the ionization constant of the carboxyls increased with increasing concentration so that the hydrogen ion concentration was proportional to $c^{2/3}$ rather than $c^{1/2}$, as it would have been with a weak monocarboxylic acid. Potentiometric titration revealed that the K_a decreased by two orders of magnitude as the added base approached equivalence to the polymeric acid and Kern interpreted this effect as the result of strong energetic interactions between the ionizing groups. At the same time neither the titration curve nor the electrical conductance of the polymeric acid at varying degrees of neutralization changed with the chain length of the polymeric acid. This showed that "the high molecular acid anions form a wide mesh network which influences the macroviscosity but does not hinder the motion of hydrogen ions."* As for the viscosities of unneutralized or neutralized poly(acrylic acid) solutions with or without added salts Kern believed that he could express the data as a product of two factors, one depending only on the polymer chain length, the other only on the ionic environment.

Kern also carried out osmotic measurements that yielded important results. He found that the activity coefficients of sodium counterions γ_{Na} decreased with an increasing degree of neutralization of the polymeric acid from 0.73 to 0.14 and suggested the relation $\gamma_{Na} = \gamma_{Na}^\circ K/(K + c_{Na})$ to represent his data. Since this expression was analogous to that which described a buffer system, he referred to the partly neutralized polymeric acid as an "osmotic buffer." Many years later this concept would be reintroduced without any acknowledgement of Kern's priority.

---

*The distinction between macroscopic and microscopic viscosity was made first by Arrhenius in 1886 when he observed that the conductivity of electrolytes was similar in water and dilute gelatin gels. A similar observation concerning diffusion was made by Graham as early as 1862.[182] A beautiful study of electrolyte conductance in sucrose solutions, which showed that it decreases more slowly than the fluidity, was published by Green[546] in 1908.

18. Scientists in a Time of Crisis

Although this book is not intended to discuss the lives of the scientists who participated in the development of polymer science, the impact of the European political upheaval between the two world wars on the lives of some of them should be mentioned, at least in brief outline.

Dr. Magda Staudinger, Herman Staudinger's widow, donated her husband's papers to the Deutsches Museum in Munich, where a carefully catalogued "Staudinger-Archiv" was established. From this it emerges that two of Staudinger's actions during the first world war, while he was a professor in Zürich, earned him the enmity of the nationalist extremists in Germany. After the entry of the United States into the war in 1917 Staudinger delivered a lecture in which he tried to analyze the enormous industrial potential that had accrued to the Allies and concluded that in their resources they were now twice as strong as the Central Powers. He found it "incomprehensible that the German Government should not have done all in its power to keep that country out of the war." According to his widow, Staudinger sent a memorandum in that sense to the German High Command in which he counseled a peace on the best possible terms. Also in 1917, Staudinger wrote a memorandum for the International Red Cross* in which he argued that although wars may have been "justified in the past since they provided the only means to enlarge the scope of life, today this is no longer so" because of the vast potential of technology which, if properly employed, could ensure a bright future for mankind. He expressed his revulsion against the way in which technology was intensifying the horrors of war. Turning specifically against the use of poison gases, the manufacture of which had unfortunately become so cheap, he wrote, "In all previous European wars . . . it would have been impossible to use them as agents of destruction, even if such a criminal idea had occurred to an army commander."

Although Staudinger had not accused any of the belligerents but had protested generally against gas warfare, this passage elicited a vitriolic attack on

---

*Rev. Internat. Croix-Rouge, **1**, 508 (1919).

him by F. Haber,* who is credited with having persuaded the German army to use poison gas. In a letter of extraordinary bitterness he accused Staudinger of having provided Germany's enemies with arguments for a vindictive peace treaty. He claimed that poison gas was used first by the French, that it did not contravene international law, and that regarding its victims "the special suffering was disproved by reports of American and English medical services." In a passage that must strike today's reader as particularly callous Haber wrote to Staudinger that "where your chemical knowledge gave you opportunities to magnify the impression by exaggeration, you made diligent use of it."

The strong feelings excited by the Red Cross publication made themselves felt in 1925 when an attempt was made to block Staudinger's appointment to a professorship in Freiburg. Staudinger had to appeal to Wieland, whom he was to succeed, to use his influence against this resistance. He visited Freiburg, where he explained his attitude on gas warfare "to several gentlemen," apparently to their satisfaction.† Shortly before this visit Staudinger wrote a second paper for the *Revue Internationale de la Croix-Rouge* [**7**, 694 (1925)] which had a strikingly different flavor from his first discussion of gas warfare. He now claimed that the French *criminel* in the 1919 paper was a mistranslation of the German *unheilvoll* (disastrous). It is, however, hardly credible that had there been such a misprint, Staudinger would have never mentioned it in his correspondence with Haber, who was particularly offended by the implication that he had been an accessory to a criminal activity. In fact, he now wrote that "if war cannot be avoided in the future, poison gases . . . would not be more inhumane if it is possible to apply moral considerations in this context." He went on to cite statistics to prove that lasting damage from gas poisoning was less frequent than that caused by firearms. It appears that this paper was designed to pacify critics of Staudinger's earlier stand. Nevertheless, with the coming to power of Hitler's government in 1933 a ruling was issued that "because of his political past" Staudinger was not to be allowed to participate in international meetings. In fact, he was forbidden to attend a meeting in Rome, but was allowed to take part in the Faraday Society meeting in 1935 on condition that he travel without his wife and "explain her absence in a manner which will not be detrimental to the prestige of the Reich."‡

Staudinger saw himself as a victim of the Nazis, but his suspicion that K. Hess, an enthusiastic supporter of the regime, conspired against him in various ways (even trying to replace him in Freiburg) may have been exaggerated.

W. O. Herrmann, who had pioneered work on poly(vinyl acetate) and had discovered poly(vinyl alcohol), was fired by his employer in 1936 because of "political unreliability." He wrote in his memoirs[419c] that he was forbidden to take leave of colleagues since this might have led to undesired unrest in the laboratory.

*Staudinger Archiv. #DII 11.4b.
†The pertinent correspondence is in Mrs. Staudinger's possession. See also Staudinger Archiv. #D II 11.27.
‡Staudinger Archiv. #D II 17.18

With the rise of the Hitler movement in 1932 the positions of Mark and Meyer at the Ludwigshafen laboratory became precarious. As Mark recalled[546a], he was told by the director of the company that the political situation would preclude any chances of his advancement (both because he was an Austrian citizen and because he had a Jewish parent) and that he would be well advised to look for an academic position. A similar approach was made to Meyer, who left to become a professor at the University of Geneva. Although Mark seems to have enjoyed pleasant personal relations in Ludwigshafen, the animosity of powerful men in the hierarchy of the IG Farbenindustrie is exemplified by a letter to Staudinger from G. Kränzlein, leader of the polymer program in the Höchst laboratory,[546b] which contains the following passages:

> You make a mistake to wrangle with Jews. . . . You have no need for polemics with Jews, this gives them too much honor. Avoid and ignore this gang. . . . We keep our distance from Jews . . . it is your duty to stop mentioning them.

Mark accepted a chair at the University of Vienna where he rapidly initiated a program of polymer research. This activity came to an end when the Germans seized Austria in March 1938. Mark was arrested and interrogated about his friendship with the former chancellor, Dollfus, who was murdered by Nazis in 1934. He was eventually released and was able to arrange for the return of his passport (using a bribe equal to his yearly salary). Luckily he had been approached a few months earlier by a director of the International Paper Company, who suggested that he come to the company's research laboratory in Hawkesbury, Ontario. In view of the new situation Mark accepted this offer happily. Two years later he moved to the Polytechnic Institute of Brooklyn to build up the first academic teaching and research program of polymer chemistry in the United States.

The excesses of the German government caused a number of other scientists mentioned in this history to leave for America. Among those working on synthetic polymers were E. Guth, R. Simha, and E. Valkó. Even larger was the number of refugees prominent in protein research, such as M. Bergmann, L. Michaelis, and F. Haurowitz.

Some refugees who were working in Britain at the time war broke out in 1939 were interned as "enemy aliens" on the Isle of Man. From there M. Perutz was sent to an internment camp in Canada from which he was able to return to Cambridge in 1941. F. R. Eirich was shipped for internment in Australia, and although he was eventually released for work at the University of Melbourne, he could rejoin his family in England only in 1943.

M. Berthelot

A. Kekule

E. Fischer

H. Staudinger

K. H. Meyer

H. Mark

W. H. Carothers

W. Kuhn

G. V. Schulz

P. J. Flory

H. W. Melville

K. Ziegler

G. Natta

Part 3.   1942–1960

19. Free Radical Polymerization

In the period between the two world wars the understanding of the polymerization of vinyl compounds in the presence of sources of free radicals was gradually placed on a firm basis, beginning with Staudinger's investigations and culminating in the work of Schulz and Flory. However, many problems concerning the kinetic steps of initiation, propagation, chain transfer, and termination, remained to be clarified. Also, such attempts as had been made to explain copolymerization were misleading and the special features that distinguished emulsion polymerization, a process of vast technological importance, were yet to be analyzed.

In the early studies of the kinetics of the polymerization of vinyl compounds benzoyl peroxide was commonly used as the initiator. However, the formation of free radicals from this compound proved to be complex (with the radicals produced inducing further peroxide decomposition) and the process depended strongly on the solvent.[547] Research in this field was therefore greatly simplified by the introduction of initiators that yield free radicals at a rate independent of the medium. The use of triphenylmethyl azobenzene by Schulz[548] first drew attention to azo compounds as polymerization initiators and the demonstration that azo-bis-isobutyronitrile yields initiating radicals at essentially the same rate in a variety of solvents[549] soon made it the initiator of choice in kinetic studies, in particular because the rate of decomposition could be followed conveniently by the volume of nitrogen evolved.

Although it was generally assumed that initiating radicals are incorporated into the polymer chains, this point was definitely proved only when tribromobenzoyl peroxide was used in the initiation of styrene polymerization. Price and Tate[550] reported that one initiator fragment was attached to each polymer chain and concluded that termination must involve disproportionation rather than a combination of chain radicals.* However, a careful study by Bevington

---

*In 1948 Kern and Kämmerer[550a] published in Germany an erroneous report that the use of p-bromobenzoyl peroxide initiator led to polystyrene in which each polymer chain contained four bromine atoms.

et al.[551] who used a ^{14}C-labeled initiator to determine the number of catalyst fragments on polymer chains, showed that styrene polymerization is terminated by chain radical combination; by contrast, in the polymerization of methyl methacrylate disproprotionation was found to predominate and to become increasingly important with rising temperature.

In a number of German patents granted from 1939 to 1943 it was reported that the addition of reducing agents to peroxide-intiated polymerization led to a pronounced acceleration of the reaction rate.[552] In a review of the German development in this field Kern[553] recounted that the action of the reducing agents was originally believed to be due to an elimination of oxygen which inhibited the polymerization. However, when great pains were taken to remove the last traces of oxygen, the reducing agents still produced their accelerating effect, and this led Kern to propose that the interaction of peroxide with the reducing agent leads to the generation of free radicals, in analogy to a suggested mechanism for a redox-initiated autoxidation chain reaction.*

In September 1945 the Faraday Society held a "General Discussion on Oxidation" that contained a section on electron transfer reactions. At this meeting Bacon[554] disclosed extensive studies of redox-initiated polymerizations that had been carried out since October 1940 in the laboratories of Imperial Chemical Industries. This report was followed by an account from Evans' laboratory of a detailed study of vinyl polymerizations in aqueous solutions that contained hydrogen peroxide and ferrous ion.[555] Haber and Weiss[555a] had proposed that the Fe^{2+}-catalyzed decomposition of H_2O_2 involves a chain reaction in which the hydroxyl radical is an intermediate and it was now shown that the reaction of OH· with a vinyl monomer competes with the chain reaction producing oxygen from H_2O_2. Similarly, polymerization was induced by the addition of a reducing agent to a persulfate solution,[556] and here also a free radical intermediate was postulated. It should be noted that the meeting took place a short time after the end of the war, so that none of the speakers was aware of the work on redox-initiated polymerizations carried out in Germany.

The rate of free radical initiated polymerizations when a steady-state radical concentration is attained was shown by Schulz and Husemann[431] and Flory[439] to be proportional to the ratio of the rate constant for chain propagation to the square root of the rate constant for chain termination, $k_p/k_t^{1/2}$. This ratio could be evaluated if the rate at which radicals were formed and the efficiency with which they add to the monomer were known, but it was realized that steady-state measurements could not yield k_p and k_t separately. The solution

---

*Kern gave a detailed account of the history of these studies in a company report dated May 23, 1944 (Archives of Farbwerke Hoechst, File 167). Laboratory experiments in Leverkusen during the second world war demonstrated that styrene-butadiene rubbers produced at low temperatures with the use of redox initiators had superior properties (F. Holscher, "Kautschuke, Kunststoffe, Fasern," BASF publication No. 10, 1972, p. 33). However, as a result of the Yalta agreement between the Allied Powers, Germany was not permitted to continue with the production of synthetic rubber (considered a strategic war commodity) after the war and the "cold rubber" technology was developed in the United States

of this problem is due largely to Melville, but the manner by which he arrived at this important result is rather curious. Melville's early work had dealt with the kinetics of gas reactions and it was therefore natural for him to write in 1937:[557]

> It is a rather remarkable fact that although the study of polymerization reactions has increased enormously within the past few years, gas-phase polymerization has received scant attention. By performing such experiments in the gas phase, all the exact technique of gas kinetics is at once available to unravel the complex sequence of reaction types which make up a polymeric reaction.

Melville proceeded to study the photochemically initiated polymerization of methyl methacrylate vapor. His strategy was to introduce hydrogen atoms into the system by the mercury-sensitized decomposition of hydrogen

> in the hope that free radicals would be produced in sufficient concentration to induce polymerization. By comparing the kinetics of this reaction with those of the direct polymerization (induced by irradiation in the absence of hydrogen) it was hoped that some light might be thrown on the mechanism of the latter process.

Melville found, as he expected, that the hydrogen atom initiated polymerization had a rate proportional to the square root of the light intensity, consistent with a biomolecular termination of chain radicals. There was no measurable after-effect when the light was switched off. By contrast, photopolymerization in the absence of hydrogen led to a deposition on the walls of the vessel of polymer which continued to react in the dark for many hours with monomer vapor. From this Melville concluded, incorrectly, that in this case polymerization did not involve free radicals but a successive addition of monomer to a chain with an "excited" double bond at the chain end.*

The dependence of the rate of the hydrogen atom initiated polymerization on the square root of the light intensity allowed Melville to use a technique that had been introduced in 1924 for the study of the photochemical oxidation of oxalate by iodine,[558] a chain reaction with the same dependence on light intensity. If the system was irradiated intermittently, with light and dark periods of equal duration, the ratio of rates observed under intermittent and steady illumination depended on the ratio of the dark period t to the lifetime of the kinetic chain τ, being ½ to $t/\tau \gg 1$ and $\sqrt{1/2}$ for $t/\tau \ll 1$. By varying the frequency of illumination through a rotating sector Melville found $\tau = 4.2$ milliseconds.

Before Melville embarked on his extended studies of polymerizations in the liquid phase Jones and Melville[559] experimented with another exotic technique for the estimation of lifetimes of kinetic chains in gas-phase polymerization. When two separate volumes of methyl acrylate vapor were irradiated it was expected that the polymerization rate would increase with increased

---

*He did not realize that the polymer deposited on the wall contained trapped radicals which could not diffuse to each other to terminate the chain reaction.

spacing of the two volumes because this would reduce the probability that chain radicals originating in these two volumes would interact to terminate the kinetic chain. This effect was actually observed and interpreted in terms of the radical lifetime τ using an estimate for the diffusion coefficient of the propagating chain radical. (This work was probably stimulated by the classical demonstration by Paneth and Hofeditz[560] that the lifetime of methyl radicals could be determined from the rate of removal of a metal mirror as a function of the time between the radical generation and their encounter with the mirror. However, while this time was well defined for radicals transported in a gas stream, estimates of the diffusion coefficients of polymer chains propagating in the gas phase were necessarily highly uncertain.)

The study of gas-phase polymerization hardly justified Melville's hopes that it would advance an understanding of the reaction mechanism, but the introduction of the rotating sector technique led to an important advance in the investigation of liquid-phase polymerizations. Burnett and Melville[561] were first to apply the method to the polymerization of vinyl acetate and showed that a knowledge of the rate of chain initiation, the polymerization rate, and the lifetime of the propagating radical yielded the propagation and termination rate constants k_p and k_t. Later, other techniques were developed in which τ could be obtained from the polymerization velocity in the initial phase after the start of illumination, before the steady state was attained, or from the aftereffect following the termination of illumination. A variety of methods was used to follow the polymerization by an increase in the solution viscosity[562] and changes in volume, [563] refractive index,[564] or dielectric constant.[565]

The availability of the rotating-sector technique for the separate determination of the rate constants for chain propagation and termination allowed Matheson et al.[566] to provide unambiguous proof that the "gel effect" in free-radical polymerization is due to a sharp decrease in the rate constant for termination with an increasing polymer concentration. The availability of electron spin resonance spectroscopy made it possible for Bresler et al.[567] to prove the same point by demonstrating that when polymerization is carried to a high polymer concentration the free-radical concentration may increase ten thousand-fold.

In spite of the large technological importance of copolymerizations, these processes were poorly understood before the second world war. In considering copolymerizations by the free-radical mechanism, Wall[446h] had ignored earlier suggestions [442,446g] that the rate at which a monomer molecule adds to a propagating chain might depend on the nature of the radical at the chain end. Yet this dependence was clearly implied by the observation made as early as 1930 by Wagner-Jauregg[446] that maleic anhydride which polymerizes by itself with great difficulty is highly reactive in copolymerization. In 1944 Alfrey and Goldfinger[568] and Mayo and Lewis[569] considered independently the copolymerization of two monomers characterized by four propagation rate constants. Using the ratio of the two types of chain end radicals in the steady state, they arrived at an expression for the relation between the composition of the monomer mixture and the composition of the copolymer in terms of the two relative rates with which a chain terminating in a given monomer residue adds another monomer of the same kind rather than the comonomer.* The large number of "monomer reactivity ratios" determined in the next few years[570] led to the early conclusion[571] that "there is no close relation between the rate of polymerization of two monomers by themselves and their relative tendencies to react with free radicals and enter a copolymer" and that "a general order of activity of monomers towards radicals is complicated by the tendency of some pairs to alternate in copolymerization. This seems sometimes due to steric effects, at other times to dipole effects."

These qualitative observations led Alfrey and Price[572] to formulate a quantitative theory according to which the specification of a "general monomer reactivity" parameter Q and a polarity parameter e for each monomer would allow the estimation of the reactivity ratio of any monomer pair. Specifically, in the copolymerization of monomers A and B the reactivity ratios were predicted as $r_A = (Q_A/Q_B) \exp[-e_A(e_A - e_B)]$ $r_B = (Q_B/Q_A \exp[-e_B(e_B - e_A)]$. This implied that $r_A r_B \leq$ as had, in fact, been observed.[571]

A theory of the rate of free-radical copolymerizations was developed by Walling[573] in 1949. This rate was expressed as a function of parameters obtainable from the homopolymerization of the two monomers, the two monomer reactivity ratios, and the ratio ϕ of the rate constant for chain termination by the interaction of two dissimilar chain radicals and the geometric mean of the rate constants for termination by each chain radical with its own kind. Walling's data led to surprisingly large estimates of ϕ (e.g., $\phi = 40$ for the copolymerization of styrene with methyl acrylate) which suggested that chain termination is also sensitive to polarity effects. Copolymerization rates often passed through a deep minimum as the composition of a monomer mixture was varied; for instance, the addition of 2% styrene reduced the polymerization rate of vinyl acetate a hundredfold, since styrene radicals react at a relatively very slow rate with the vinyl acetate monomer.

Flory had already pointed out in 1937[439] that the length of molecular chains produced by free radical polymerization may be limited by chain transfer to the monomer. This effect is particularly large with vinyl acetate, and its polymerization was found, in fact, to lead to products with molecular weights that depended on the initiator concentration only when this concentration was very large.[574] The dependence of the styrene polymerization rate and the chain length of the polymer obtained on the dilution of the monomer with a variety of

---

*In Germany the principles of copolymerization were poorly understood. In 1943 Wolz of the IG Farbenindustrie reported on the drift in the copolymer composition with conversion in the butadiene—acrylonitrile system. He found that the monomer, which is present in small amounts, is enriched in the copolymer and concluded "that this cannot be explained by separate monomer activation." He wrote that only the acrylonitrile is "activated" by the initiator and that "what reaction takes place between the activated acrylonitrile and the butadiene, whether there is transfer of the activation energy or complex formation, cannot be decided by our experiments." (Hoechst Archives, File 165.)

solvents was studied by Suess et al.,[575] who found that the reaction rate was independent of the nature of the diluent but that different diluents led to strikingly different molecular weights of polystyrene. To explain this result they used a tortuous argument that involved a reversible interconversion of "active" and "inactive" polymer chain ends. Schulz et al.[576] found that the product of the polymerization degree and the polymerization rate fell off sharply with an increasing dilution of styrene and concluded that at a low monomer concentration "another process" must be added to the interaction of two chain radicals to limit the chain length. It is striking that they could not bring themselves to accept chain transfer, proposed by Flory,[439] as the explanation of their data. Neither did Breitenbach and Maschin,[577] when they found that carbon tetrachloride reduces sharply the chain length of polystyrene and leads to a polymer containing chlorine.

Only in 1943 were old literature data interpreted in a rational manner when Mayo[578] showed that, in the presence of chain transfer agents, plots of the reciprocal degree of polymerization against the ratio of the chain transfer agent and monomer concentrations should be linear, with a slope equal to the ratio of the rate constants for the reaction of the chain radical with the chain transfer agent and the monomer, respectively. He reported such plots for a wide variety of chain transfer agents. Ironically, he did not include in his survey the only class of chain transfer agents which was of great technical importance at the time: The recipe adopted by the U.S. Office of Rubber Reserve during the second world war for the manufacture of styrene–butadiene rubber by emulsion polymerization[579] contained dodecyl mercaptan but the function of this ingredient was not understood. According to Marvel,[580] one of the leaders of this enterprise, it was commonly referred to as OEI, "one essential ingredient";* in the literature of the time one finds generally the term "regulator."

During the early development of synthetic rubber the presence of a crosslinked fraction presented a serious technical problem. This fraction was originally broken down by thermal treatment,[581] but in 1937 a German patent[582] disclosed that the difficulty could be eliminated by the addition of diisopropyl xanthogen disulfide("diproxide"). It was observed that mercaptans are formed from this reagent and that the pH-dependence of its "regulating" action parallels that of mercaptan formation. This led to the proposal[583] that mercaptans be used as regulators.†

In the United States there was a widespread notion that the formation of crosslinks in butadiene copolymerization depended on the 1,2-addition of the diene and that mercaptans somehow inhibit this reaction.[584] Surprisingly, the true nature of the role of the mercaptans was not recognized for some years

---

*Harkins still used the term OEI in his important 1947 paper.[603b]
†Graulich wrote in 1943 that "the action of diproxide must be due to the decomposition of mercaptan and the addition of sulfur to the molecule" (Farbwerke Hoechst Archives, File 163). Mercaptans were not used in Germany as "regulators" in the production of rubbers during World War II[581] because of a shortage of fatty acids needed for their synthesis.

after it was demonstrated[585] that their addition to olefins in the presence of oxygen (or peroxides) is a chain reaction:

$$RSH + O_2 \longrightarrow RS\cdot + HO_2^-$$
$$RS\cdot + R'CH{=}CH_2 \longrightarrow R'\dot{C}HCH_2SR$$
$$R'\dot{C}HCH_2SR + RSH \longrightarrow R'CH_2CH_2SR + RS\cdot$$

Only in 1946 did Snyder et al.[584] show that mercaptans are powerful chain transfer agents in the polymerization of styrene and that this is the basis of their utility in the preparation of diene rubbers.

In 1933 an experiment was carried out by Houtz and Adkins[586] the result of which could not be properly understood at the time. They polymerized styrene in a solution of a polystyrene which had been previously isolated and dried, and compared the solution viscosity of the resulting polymer with that of a mixture of the original polystyrene with the polymerization product of styrene obtained in the absence of polymer. They found that the product from the styrene polymerization in the polymer solution yielded more viscous solutions and concluded that "polystyrene after precipitation and drying retains the capacity to add styrene with the formation of chains of greater length." This result was later reinterpreted by Flory[439] as evidence that polymers may act as chain transfer agents; the formation of free radicals by the withdrawal of an atom from the polymer chain could lead to monomer addition and the formation of branches extending from the macromolecular backbone. The polymerization of vinyl acetate leads to a particularly pronounced branching of this type, and if the acetyl group acts as the chain transfer agent the branch can be detached from the polymer by solvolysis. This explanation was offered by Wheeler et al.[587] for the fact that solvolysis of poly(vinyl acetate) and reacetylation of the poly(vinyl alcohol) derived from it leads to a product of lower molecular weight than the original polymer.* Moreover, branches propagating from a polymer backbone can recombine to form a gel network, as observed in a number of cases.[589]

If radicals produced on the chain backbone can initiate branches composed of the same monomer residues they should also be able to serve as initiators of branches chemically different from the polymer backbone. Structures of this type had been made on occasion without being recognized. As early as 1913 Matthews[590] reported that "improved products are obtained if in place of adding polymerized styrene to rubber, the rubber is dissolved in styrene and the solution is subjected to conditions under which styrene polymerizes," but the time was not ripe for speculation on the significance of this observation.

---

*The solvolysis and reacetylation of poly(vinyl acetate) was investigated by Staudinger and Warth,[588] who claimed that "the mean degree of polymerization remains essentially unchanged." This conclusion was important to Staudinger since the retention of the degree of polymerization after the derivatization of polymers was used as one of the arguments against the concept of polymers as molecular aggregates.

Fourteen years later Ostromislensky[590a] was granted a patent, based on the same idea claiming among the advantages of the polymerization of styrene in the presence of dissolved rubber a "decreased sensibility to shock and a greater flexibility." Ostromislensky's patent stimulated research at the Dow Chemical Company, where Amos et al.[590b] obtained in 1954 a patent in which they claimed that polymerization of rubber solutions in styrene leads to "linear interpolymers."* Bacon et al.[590c] reported in 1938 that maleic anhydride adds to *Hevea* rubber and Campagnon and Le Bras[590d] found that milling rubber with various monomers changes profoundly the rubber characteristics. However, Carlin and Shakespeare,[591] who polymerized *p*-chlorostyrene in the presence of poly(methyl acrylate) and found that after hydrolysis the benzene-insoluble fraction contained chlorine, were first to interpret their results as a consequence of chain transfer, leading to side branches of poly(*p*-chlorostyrene) on the poly(methyl acrylate) backbone. Alfrey and Bandel[592] introduced the term "graft polymer"; to optimize grafting efficiency they used chlorine-containing polymers (styrene–vinylidene chloride copolymers) as chain transfer agents. They also showed that graft polymers with a nonpolar backbone and polar branches (or vice versa) are efficient emulsifiers. At the time the possibility of synthesizing graft polymers excited much enthusiasm because it was viewed as demonstrating the wide scope for "molecular architecture" to design materials with unusual properties. In a careful study of graft polymerization Smets and Claeson[593] found that principles similar to those which govern copolymerization also apply to these processes. For instance, vinyl acetate could not be grafted to polystyrene since addition of vinyl acetate to a radical on a polystyrene backbone would reduce the resonance stabilization of the radical.

After the development of electron spin resonance spectroscopy, it was demonstrated that free radicals produced in polymers exposed to ionizing radiation could build up to high concentrations and persist for a long time.[594] These "trapped radicals" were able to initiate the polymerization of a variety of monomers leading to graft polymer synthesis, a process studied with particular intensity by Chapiro.† Grafting could be achieved by exposure of the pre-irradiated polymer to a monomer[595] or by allowing the irradiated polymer to react with oxygen to form peroxides, the thermal decomposition of which initiated the grafting reaction.[596] In the case of Teflon, which is particularly resistant to penetration by monomers, a variation of experimental conditions could lead to a restriction of grafting to the Teflon surface or to grafting to the

---

*Later it was recognized that this process leads to the grafting of polystyrene side chains on the rubber molecules resulting in a product which aids good cohesion between the polystyrene and rubber phases essential for improved mechanical properties.

†In 1953 I attended a meeting at the Brookhaven National Laboratory at which possible uses of radioactive waste products for polymer applications were to be discussed. At that meeting R. B. Mesrobian stated that one should concentrate on processes that can be conveniently accomplished only by ionizing radiation.   He suggested two applications: grafting to irradiated polymers

entire Teflon sample.[597] The modification of the surface properties of polymers by grafting excited particular interest. It is significant that the 1958 IUPAC symposium on macromolecules (*J. Polym. Sci.*, **34**) contained nine contributions on the synthesis of graft polymers and their properties.

Attempts to carry out free-radical-initiated polymerizations in a disperse phase were made first in an effort to produce synthetic rubber and the incentive for this approach was clearly a desire to imitate the production of a rubber latex in the *Hevea* plant (see p. 84).* Early patents did not appreciate the distinction between "suspension polymerization," in which relatively large droplets of monomer suspended in water were converted to polymer beads, and true emulsion polymerization which turned out to have characteristic features different from those of polymerization in bulk.

Suspension polymerization is of great industrial utility because it leads to polymer in the form of small beads that are easily handled. It has also the great advantage that the continuous phase—water—constitutes a heat sink that makes it easy to maintain temperature control and that the viscosity of the system does not increase as the polymerization is carried to essentially complete conversion. Initially, attempts were made to keep the polymerizing droplets separate by vigorous stirring of the suspension,† but it was soon realized that the partly polymerized drops will aggregate unless a "suspension agent" is added to the system. At first a variety of water-soluble polymeric materials which acted as "protective colloids" was used,[599] but later it was found that a number of finely divided insoluble solids may serve the same purpose.[600]

It was soon found that emulsion polymerization offered the important advantage of allowing, in contrast to polymerization in a homogeneous system, polymers of high molecular weight to be obtained at high reaction rates. The cause of this behavior was long obscure. I have noted that Staudinger and Frost[420] speculated in 1935 that some special property of "the surface of the colloid particle" was responsible for the increased reactivity. As late as 1946 two publications[601,602] suggested that the high molecular weight was due to the high viscosity at the polymerization site, which would lead to the reduced rates of chain termination demonstrated by Norrish and Smith.[444]

A detailed interpretation of the mechanism of emulsion polymerization emerged from a series of studies carried out from 1942 on in the laboratories of the U.S. Rubber Company under contract with the U.S. Office of Rubber

---

and the possible polymerization of crystalline monomers into which free radicals had been introduced by γ-irradiation. During the next year he presented a paper at the IUPAC symposium on macromolecules held in Torino which may have stimulated interest in radiation-induced graft polymerization.

*It was recognized only much later that the polyisoprene in the *Hevea* plant is not derived from isoprene but by a complex enzyme-catalyzed metabolic sequence.[598]

†This process was patented (W. Bauer and H. Lauth, Ger. Pat. 656,134, January 29, 1938) for the suspension polymerization of acrylates and methacrylates.

Reserve. W. D. Harkins of the University of Chicago acted as a consultant in these investigations and formulated[603a] the following conclusions in 1945 and 1946:

1. Most of the polymer is formed in soap micelles swollen with a mixture of polymer and monomer.
2. Monomer droplets stabilized by absorbed soap act merely as a "storehouse" of monomer which diffuses through the water into the swollen micelles.*
3. The number of polymer particles formed increases with the initial soap concentration.

A beautiful and most revealing account of the entire course of these studies (in which X-ray diffraction was used to study the swelling of micelles, surface tension measurements to follow the disappearance of free soap anions during polymerization, and electron microscopy to characterize the size of the polymer particles) was published by Harkins in 1947.[603b]

Stimulated by Harkins' conclusions, Smith and Ewart[604] in the following year formulated a theory of the kinetics of emulsion polymerization. Although they considered various special cases, they concluded that under the conditions used in the preparation of synthetic rubber or in the emulsion polymerization of styrene the monomer-polymer particles within the soap micelle are likely to contain no more than one propagating chain because the existence of two free radicals within the small volume of the particle would lead to rapid chain termination. Thus entry of a radical produced by the initiator in the aqueous phase will alternately start and stop a chain polymerization so that half the particles will be polymerizing at any time. The number of particles will increase until their surface is sufficient to adsorb all the soap; it will become larger when the initiator or the soap concentration is raised. The predictions of the theory were in good agreement with the data on emulsion polymerization of styrene; also, with a knowledge of the number of particles in which polymerization takes place, the reaction rate could be interpreted in terms of the rate constant for chain propagation.[605] Although later studies showed that the Smith-Ewart scheme is not generally applicable to all emulsion polymerizations, their paper represents a major advance in the understanding of this process.

According to the Smith-Ewart mechanism, it was expected that it would be possible to obtain a latex with a near uniform particle size if polymerization were carried out in a "seed latex" with the soap concentration adjusted to prevent new polymer particles from forming. Vanderhoff et al.[606] showed that

---

*Similar conclusions were reached in Germany as early as 1938 by H. Fikentscher [H. Fikentscher, *Angew. Chem.*, **51**, 433 (1938), H. Fikentscher, H. Gerrens, and H. Schuller, ibid., **72**, 856 (1960)]. He realized that polymerization did not occur in the monomer droplets but in "the aqueous phase," and did not specify the soap micelles as the polymerization site, possibly because he used monomers with a higher water solubility than the styrene used by Harkins.

this could, in fact, be accomplished and that the relative rate of growth of latex particles decreases with their increasing size leading to a sharpening of their size distribution. A latex with a particle size of such uniformity could be obtained that the spherical particles assembled spontaneously on drying into hexagonally close-packed crystal-like structures exhibiting light diffraction.[607]

The availability of convenient sources of γ-radiation from Co^{60} led in the 1950s to attempts to induce polymerization in crystals of vinyl derivatives (see footnote on pp. 176-177). It was found that many monomer crystals could be converted to polymers, but the data were at first difficult to interpret because initiation, propagation, and possible radiation damage of the polymer were taking place at the same time. This difficulty was overcome by carrying out the irradiation at low temperatures at which radicals were introduced into the crystal but no polymerization could take place before the sample was allowed to warm outside the radiation source. The first kinetic study of such a "postpolymerization" was carried out by Morawetz and Fadner[608] on polyacrylamide. It showed that polymer chains grow at a decreasing rate for many months after preirradiation, with little decay of the radical concentration observed by ESR spectroscopy. However, because of the lack of molecular mobility in crystals, the concept of chain propagation in the crystalline phase encountered widespread skepticism, and it was suggested that the reaction might proceed because of partial melting of the crystal during the exothermic process. Eventually the important role of the crystal lattice was proved by the finding that potassium acrylate polymerizes after irradiation many orders of magnitude more rapidly than the sodium salt[609] and that butadiene can be polymerized in urea clathrate crystals to yield exclusively the linear 1,4-*trans* polymer.[610] More dramatic was the discovery by Okamura and his collaborators[611] that various cyclic monomers such as trioxane polymerize under ionizing radiation only in the solid state and yield polymer crystallites highly oriented relative to the crystallographic axes of the parent monomer.

In one of a series of publications devoted to the chemistry of diacetylenes Bohlmann and Inhoffen[612] reported in a footnote that these compounds polymerize *spontaneously* at temperatures well below their melting points. The intense coloration of the product led them to suspect, erroneously, that they were dealing with a highly condensed aromatic structure, and it was only in 1971 that Wegner[255a] explained the nature of the reaction. Some diacetylene derivatives were later shown to be convertible from single monomer crystals to macroscopic single crystals of the polymer.

In a remarkable paper published as early as 1934, Chalmers[436b] noted that "the decomposition of high polymers has many features which suggest a chain mechanism of inverse character to that involved in their generation." Taylor and Tobolsky[613] arrived, on the basis of data for the heat of polymerization and the low activation energy for chain propagation, at the conclusion that polymerization and depolymerization should be in equilibrium at moderately elevated temperatures. Experimental proof was then provided by Mesrobian and Tobolsky,[614] who showed that in the presence of free radicals the polymerization of styrene and the depolymerization of polystyrene lead to systems of the same solution viscosity. Ironically, the polymerization–depolymerization equilibrium had been observed some years earlier by investigators who failed completely to understand the significance of the phenomenon. Snow and Frey[615] found that the rate at which olefins copolymerize with sulfur dioxide vanishes at what they called "a ceiling temperature." Puzzled by this phenomenon, they noted that "the closest analogy is found in photosynthesis and fermentation, both of which involve living cells" and speculated that their observation might perhaps be due to a side reaction in which a polymerization inhibitor is formed. The true significance of the "ceiling temperature T_c" as the temperature at which the polymer and monomer are in equilibrium was pointed out by Dainton and Ivin[616] in 1948. They pointed out that T_c must therefore be equal to the ratio of the heat and the entropy of polymerization.*

In the same year two studies were published by the U. S. National Bureau of Standards in which the products of the pyrolysis of high polymers under a high vacuum were analyzed by mass spectroscopy. Both reported a striking variation in polymer behavior: While polyisobutene and polystyrene yielded high proportions of the parent monomer, polyethylene chains seemed to be broken at random. From this, Madorsky et al.[617] concluded that "fragmentation will be determined by the frequency of low energy carbon-carbon bonds and by the steric hindrance to the escape of fragments caused by the side chains," whereas Wall[618] understood that pyrolysis products are determined by the relative rates of depolymerization and chain transfer. In England Grassie and Melville[619] carried out a detailed study of the pyrolysis of poly(methyl methacrylate) by following concurrently the monomer yield and any change in the chain length of the residual polymer. Their data indicated that depolymerization is initiated at the chain ends and that up to a molecular weight of 5×10^5 any chain that begins to depolymerize has a high probability of being completely converted to the monomer. The chain character of the degradation was proved by demonstrating that polymerization inhibitors also inhibit the depolymerization.

On the basis of all this experimental evidence Simha et al.[620] formulated a detailed theory of depolymerization. They also pinpointed the factors that determine how many monomer molecules will be liberated after a break in the chain backbone before the "dezipping" comes to an end because of chain transfer, radical recombination, or disproportionation.

In 1953 it was reported by Charlesby[621] and by a group from the General Electric Company[622] that some polymers (notably polyethylene) are crosslinked when exposed to ionizing radiation, while other polymers [such as poly(methyl methacrylate)] are degraded by a similar treatment. It was soon realized that the course of the transformation under ionizing radiation corre-

---

*Because ΔH_p and ΔS_p are generally negative, the polymer is unstable above T_c. In the transformation of the cyclic S_8 to the linear sulfur polymer ΔH_p and ΔS_p are positive and the polymer is stable only *above* a "floor temperature" of 159°C.[616]

lates with the behavior of polymers in pyrolysis; polymers with quaternary carbons in the chain backbone, which yield large amounts of monomer on heating tend to depolymerize, whereas polymers with low monomer yields on thermal degradation, tend to crosslink under irradiation.[623,624] It was conjectured that in vinyl polymers a C—H bond is broken and that the hydrogen atom then abstracts a tertiary hydrogen so that two radicals are formed close enough to each other to facilitate crosslink formation.[623] In polyethylene, radicals were believed to migrate by repeated hydrogen abstraction until two of them could combine to a crosslink.[625] The incorporation of a few styrene residues into a polyisobutene chain was shown to provide considerable protection against degradation by ionizing radiation,[626] because of the tendency of aromatic structures to dissipate the absorbed energy.

20. Ionic and Coordination Polymerization

The catalytic action of acids and bases for polymerizations was known for a long time before an effort was made to study these processes in detail. Thus Berthelot in 1866 described the sulfuric-acid-catalyzed polymerization of styrene,[43] Wislicenus found in 1878 that iodine induced the polymerization of vinyl methyl ether,[68d] and in 1877 Butlerov[72] used BF_3 to polymerize propylene. Base catalysis, which seems to have been introduced by Wagner[44] in 1868 for the polymerization of styrene, played a crucial part in early attempts (1910–1912) to produce a synthetic rubber.[63,64,68a] This process was later subjected to analysis by Ziegler and his students.[447–451,627]

In some of his experiments Staudinger used tin tetrachloride as a catalyst for the polymerization of styrene,[302] α-methylstyrene,[628] and cyclopentadiene.[629] In 1926 he wrote that

> there is a complete lack of a systematic investigation of substances producing the polymerization of unsaturated compounds and this is why we tested the most varied halogen derivatives to see which induce the formation of high molecular polycyclopentadiene.

He noted that halogen compounds which tend to form complexes with organic compounds are often polymerization catalysts but was puzzled by their inability to induce the polymerization of acrylates and vinyl bromide. Staudinger felt sufficiently uncertain of the principles that govern catalysis that he had his collaborator Bruson include even alkali and alkaline earth halides in the survey of compounds to be tested for their ability to catalyze polymerization. It is puzzling that Staudinger never mentioned in his writings the formation of carbonium ions which had long been demonstrated. Walden[630] discovered in 1902 that triphenylmethyl chloride solutions in liquid SO_2 containing $AlCl_3$ or $SnCl_4$ exhibit a high conductivity and concluded that "the color produced from two colorless substances points to a chemical process." Baeyer[631] described the adduct of $(C_6H_5)_3CCl$ and $SnCl_4$ as a "carbonium compound" and represented it by $(C_6H_5)_3C\sim ClSnCl_4$ and Hantsch[632] first used the term "carbonium salts" in an extensive study in which the role of the solvent on their ionization equilibria was characterized by spectroscopy. The notion that carbonium ions might be reaction intermediates had also been advanced[633] to explain why certain compounds exhibited the same dependence on the dielectric constant of the solvent in their rates of isomerization and racemization.

The first suggestion that carbonium ions are involved in acid-catalyzed polymerizations is due to Whitmore,[634] who wrote in 1934:

> Those olefins which add a molecule HX most readily also polymerize most readily. In each case the first step is the addition of a hydrogen ion to the extra electron pair of the double bond. . . . one carbon is left with only six electrons. It is thus positively charged and can undergo changes characteristic of an atom with a deficiency of electrons.

This idea was specifically applied to the acid-catalyzed polymerization of isobutene in which the hydrogen ion addition to the monomer produced the trimethyl carbonium ion and initiated a chain reaction because "in the same way that hydrogen ion adds, the positive tertiary butyl group can add to isobutene."

The explanation of the action of Friedel-Crafts catalysts presented a more difficult problem. Hunter and Yohe[635] postulated that a pair of electrons is donated by the double bond of a vinyl monomer to $AlCl_3$ leading to a species

$$\begin{array}{c} \overset{..}{H}\ \overset{..}{R} \\ H\!:\!C\!:\!C\!:\!H \\ \overset{..}{Cl}\!:\!\overset{..}{Al}\!:\!\overset{..}{Cl} \\ Cl \end{array}$$

in which the carbon with only six valence electrons tends to add further monomer molecules. A similar scheme was later suggested by Price[636] for the $SnCl_4$-catalyzed polymerization of styrene, and although he realized that this would lead to an increasing charge separation as the chain grew in length he did not seem to regard this feature as an insurmountable difficulty.

The problem of the mechanism by which Friedel-Crafts catalysts initiate polymerization was eventually solved by Polanyi's discovery[637] that water is a cocatalyst in the polymerization of isobutene by BF_3. Later Evans and Meadows demonstrated[638] that rigorously dried BF_3 does not polymerize isobutene. They concluded that traces of water lead to the reaction $BF_3 + H_2O \rightarrow BF_3OH^- + H^+$ and that the hydrogen ion then adds to isobutene to produce the initiating trimethyl carbonium ion. Similarly, Fontana and Kidder[639] found that highly purified $AlBr_3$ does not polymerize propylene unless HBr is added. The cocatalyst, however, need not be a source of hydrogen ions. Wertyporoch[639a] had shown that $AlBr_3$ solutions became conductive on the addition of alkyl halides, which implied the formation of carbonium ions, and Plesch[640] found that when styrene is polymerized by $SnCl_4$ in the presence of an alkyl bromide, the alkyl residue is incorporated into the polymer. A later study of this type of cocatalysis by Kennedy and Thomas,[641] who used radioactive methyl chloride in the $AlCl_3$-induced polymerization of isobutene, showed that only 27% of the polymer chains carried the cocatalyst fragment suggesting that the remaining polyisobutenes were initiated by a chain transfer process.*

For a number of reasons cationic polymerization proved to be much more difficult to study than polymerizations initiated by free radicals. This emerges clearly from reports of two meetings devoted to the current status of this field, held in 1949 at Trinity College, Dublin,[643] and in 1952 at the University College of North Staffordshire.[644] In summing up this second meeting, Dainton acknowledged the important advances that had been made but concluded, nevertheless, that the many unanswered questions and contradictions presented "a crude and confused mass."

Some of the polymerizations, notably that of isobutene, were heterogeneous so that data were difficult to reproduce and interpret. A considerable experimental difficulty was caused by the need to control minute traces of water in the polymerizing systems. The characteristics of polymerizations at very low temperature were difficult to study kinetically, although Fontana and Kidder[639] had developed an ingenious technique in which the reaction was followed by the amount of solvent boiled off. Chain termination could involve a number of processes; expulsion of a hydrogen ion, combination with the counterion, chain transfer with the solvent,[645,646] or abstraction of a hydride ion from a "dead" polymer leading to branching.[647] In addition, the presence of carbonium ions tended to lead to isomerizations. Most workers reported that the molecular weights of polymers were independent of the catalyst concentration, as would be expected for a process in which termination cannot involve the interaction of two propagating chains, but even on this point agreement was not unanimous.[648] The most striking feature, an increase in the polymerization rate with a decreasing reaction temperature, was emphasized by the discoverers of polyisobutene;[649] it reflects (in contrast to free radical polymerization) the fact that termination of the kinetic chain has a higher activation energy than chain propagation.

Although the existence of carbanions was established in 1921 by Hantsch,[632] who studied the sodium salt of triphenylmethane, a clear proposal of an "anionic chain polymerization" was formulated only in 1948 by Beaman.[650] He selected for study methacrylonitrile, expected to be particularly susceptible to the attack of anions, and showed that sodium in liquid ammonia, Grignard reagents, and triphenyl carbanion were efficient polymerization initiators. He also noted that the susceptibility to anionic initiators decreased

---

*The action of cocatalysts in polymerization by Friedel-Crafts catalysts was foreshadowed by a German patent, issued as early as 1933,[642] which claimed that the BF_3-catalyzed polymerization of olefins to oils is improved by the use of hydrogen halides or alkyl halide "accelerators."

with decreasing electronegativity of the double-bond substituent. This point was defined more quantitatively by Higginson and Wooding,[651] who were the first to study the kinetics of anionic polymerization. They found that acrylonitrile could be polymerized by relatively weak bases such as methoxide (pK 17), but that very much stronger bases (e.g., triphenyl carbanium, pK 33) were needed for the polymerization of butadiene or styrene. The pronounced sensitivity of ionic polymerization to the polarity of the double-bond substituent also provided a convenient method of characterizing the polymerization mechanism: Landler[652] and Walling et al.[653] reported in 1950 that equimolar mixtures of styrene and methyl methacrylate yield a copolymer that contains mostly styrene with cationic initiators and mostly methyl methacrylate with anionic initiators, whereas free radical polymerization leads to copolymers that contain the two monomer residues in similar concentration.

In 1926 Staudinger speculated[629] whether the nature of a polymerization catalyst influences the structure of the polymer produced, but at that time methods to resolve this question were not available. The first clear demonstration that the polymerization conditions determine the nature of the polymer was provided in 1950 by Ziegler et al.,[654] who found that polymerization of butadiene by butyl lithium in dimethyl ether and in benzene solution leads to 58% and 85% 1,4-addition, respectively, while catalysis by metallic sodium yields only 20% 1,4-addition. Then in 1956 Staveley[655] made the truly sensational announcement that a lithium dispersion in isoprene leads to a polymer in which the overwhelming majority of the monomer residues is incorporated, as in the *Hevea* plants, 1,4-*cis* into the chain molecule, so that the product may be described as a "synthetic natural rubber."* This striking control of the polymer microstructure is destroyed by solvents that contain oxygen,[656,657] and it is restricted to Li and its alkyl compounds, whereas analogous compounds of other alkali metals are ineffective.[658] Anionic polymerization also produced stereoregular methyl methacrylate polymers, and here the stereospecificity again depended on the solvent medium.[659] In dimethoxyethane the isotactic polymer was obtained, whereas polymerization in hydrocarbons yielded the syndiotactic product.

In 1936 workers in the Du Pont laboratories reported a green compound of naphthalene and sodium[660] for which "the formula $C_{10}H_8Na_2 \cdot C_{10}H_8$ correctly represents the proportions of sodium and naphthalene while avoiding the question of structure." This compound was found to catalyze polymerization[661] and it was eventually recognized by Weissman and his associates to be the sodium salt of a naphthalene radical anion.[662] As M. Szwarc recalled,[663] a conversation with Weissman about the properties of this radical anion led him to the idea that it should donate an electron to styrene to form the species ·CH_2—$\bar{C}H\phi$ which, he thought, might initiate a simultaneous anionic and free-radical polymerization. The experiment[664] showed, however, that sty-

---
*A similar result was obtained almost at the same time, as we shall see, with a Ziegler-Natta catalyst. (See p. 191).

rene radical anions produced in this manner combined rapidly to a di-anion, which polymerized by the anionic mechanism. Szwarc realized that under carefully controlled conditions chain termination could be avoided so that the carbanion ends of the propagating chains would retain their activity after disappearance of the monomer.* This was proved by adding fresh styrene after exhaustion of the original monomer, leading to renewed polymerization. More important, when isoprene was added after the polymerization of all the styrene the polymer obtained was insoluble in both polystyrene and polyisoprene solvents, indicating that a block copolymer had been obtained. This discovery led to a great deal of industrial activity and, in particular, to Milkovich's development of thermoplastic elastomers that consist of block copolymers in which a rubbery block is flanked by two glassy blocks.[666]

It seems that none of the early workers who used Szwarc's technique for the synthesis of block polymers knew that anionic polymerization of ethylene oxide and its homologs, with sequential addition of the two monomers, had previously been shown to lead to block polymers useful as detergents.[667] By using an initiator with one or two alkoxy groups, diblock or triblock polymers were obtained.

The anionic polymerization of vinyl monomers initiated by aromatic radical anions led to two other results of great importance: The two carbanion chain ends could be reacted with reagents to introduce specific end groups.[668] Also, as pointed out some years before by Flory,[459] a polymerization in which a fixed number of chains is initiated simultaneously and in which the chains cannot be terminated, should produce a very narrow Poisson distribution of chain lengths. Although unusual care is required to meet these conditions, Siriani et al.[669] succeeded in obtaining by Szwarc's technique a poly(α-methylstyrene) with a ratio as low as 1.03 for the weight- and number-average molecular weights. This approach has remained the unique method for the preparation of polymers with such narrow chain-length distributions.

The discovery of coordination catalysts for the low-pressure polymerization of ethylene by Ziegler and his collaborators is one of the most fascinating stories in the history of polymer science.† Although Ziegler had carried out pioneering work on anionic polymerization of butadiene between 1928 and 1934 (see pp. 135–136, 185), he had no special interest in the preparation of polymeric materials. The investigations that were to make him, by a series of unforeseen accidents, a towering figure in polymer science started with a study of alkyl-lithium compounds.[672] He found that ethylene could add to these compounds to extend the alkyl chain and that the lithium alkyls tended to decompose into α-olefins and LiH. Ziegler wondered whether this decomposition

---

*The potentially long-lived nature of the active chain ends in anionic polymerization had been previously demonstrated by Abkin and Medvedev[665] in an elegant experiment. They interrupted the sodium-catalyzed polymerization of butadiene by pumping off the unreacted monomer and showed that polymerization was resumed at the same rate when the butadiene was reintroduced.

†Beautiful accounts of the manner in which Ziegler arrived at his epochal discovery may be found in his Nobel Prize address[670] and in the account written with his chief collaborators.[671]

ible but a study of the problem was hampered by the poor solubility of LiH. At this point Ziegler was prepared to abandon the study but his collaborator Gellert insisted on "one last attempt" with the recently discovered LiAlH$_4$, which had the desired solubility.[670,672] It was found that this compound added ethylene readily and that triethylaluminum reacted with ethylene even faster than tetraethyllithium aluminum. When, however, the ratio of ethylene to Al(C$_2$H$_5$)$_3$ was increased the length of the alkyl chains could not be extended beyond a relatively low value and this was ascribed to a spontaneous scission:

$$\text{>Al—(CH}_2\text{CH}_2)_n\text{—H} \rightarrow \text{>Al—H} + \text{CH}_2\text{=CH—(CH}_2\text{CH}_2)_{n-1}\text{—H}$$

Three of Ziegler's doctoral students were then assigned the task of studying the factors that limit chain propagation in the interaction of trialkylaluminum with olefins. Martin[673] found that the reaction of tripropylaluminum with propylene "forms under all conditions exclusively the dimer and it is impossible to obtain, as with ethylene, high polymers." This indicated to Martin a thermal instability of organic aluminum compounds with branched alkyl residues. Next, Holzkamp[674] embarked on a study of the reaction of tripropylaluminum with ethylene. Here the insertion of ethylene into the alkylaluminum bond should lead exclusively to chains with an odd number of carbon atoms, whereas the displacment of an alkyl chain by ethylene should produce chains that contain an even number of carbons:

$$\text{>Al—CH}_2\text{—(CH}_2\text{CH}_2)_n\text{—H} + \text{CH}_2\text{=CH}_2 \rightarrow \text{>AlH} + \text{CH}_2\text{=CHCH}_2\text{—(CH}_2\text{CH}_2)_n\text{—H}$$

$$\text{>Al-H} + n\text{CH}_2\text{=CH}_2 \rightarrow \text{>Al(CH}_2\text{CH}_2)_n\text{—H}$$

Thus the ratio of chains with an even and odd number of carbon atoms obtained after decomposition of the alkylaluminum compounds should yield the ratio of the "displacement" and "propagation" reaction rates. Holzkamp found, as expected, that the displacement reaction became relatively less important as he reduced the reaction temperature from 110 to 90°C but a further reduction to 70°C led to a completely unforeseen result: Only propylene and butene were obtained, indicating that the displacement reaction was now the dominant process. This surprising result was eventually traced to the use of one specific autoclave as the container for the reaction, and a painstaking study led to the implication of traces of nickel salt which had been dissolved from the stainless steel when it was cleaned with acid and were reduced by the alkyl aluminum compounds to colloidal nickel. Holzkamp concluded that this nickel catalysis of the displacement process was responsible for the earlier failure to obtain very long alkyl chains in the reaction of triethylaluminum with ethylene and wrote that "it seemed attractive to investigate as many other metals as possible." Ziegler was particularly interested in catalysts which would lead to the conversion of ethylene into butene in the presence of triethylaluminum.

This was to be the thesis problem of Breil[675] who was asked "to test systematically the entire periodic system of elements."[671] It is ironic that this effort, which was aimed at the discovery of catalysts that would favor displacement over chain propagation should have led, instead, to the much more important opposite finding that triethylaluminum in the presence of certain transition metal compounds catalyzes the polymerization of ethylene to high polymers. This was observed first with zirconium acetylacetonate and later with the particularly effective titanium compounds which allowed the polymerization to proceed even at atmospheric pressure and room temperature.* Breil suggested that catalysts for the displacement reaction are colloidal metals, whereas polymerization catalysts are formed by metal compounds only partially reduced by the organo-aluminum to lower oxidation states. He noted that according to its IR spectrum the polyethylene produced by his catalysts contained a very small number of methyl groups so that it "probably has mostly a linear structure.†" He also found that this material softened at 130–150°C, a temperature substantially higher then polyethylene prepared by the high-pressure process, but did not relate this property to the structure of the chain.

In searching the literature, Breil discovered that M. Fischer assigned to IG Farbenindustrie a patent that described a strikingly similar process for the polymerization of ethylene as early as 1943.[676] At that time it was desired to discover methods by which ethylene could be polymerized to lubricating oils and Fischer's discovery that "the formation of liquid polymers is reduced in favor of solid substances when the catalytic aluminum chloride is used in conjunction with titanium tetrachloride" seems to have been more of a disappointment than an intimation of a new route to the synthesis of polyethylene. Whether or not Fischer anticipated the discovery in Ziegler's laboratory, Breil stated in his thesis: "It seems to me obvious that experiments of this kind led to the formation of a catalyst similar to mine." This was based on the assumption that triethylaluminum was likely to have been formed from aluminum chloride and ethylene under Fischer's reaction conditions.‡ The Badische Anilin und Soda Fabrik (BASF), successor to IG, also adopted this stand in threatening to contest Ziegler's patent.[677] The controversy was eventually settled and BASF obtained a license.

G. Natta, the second central figure in the discovery of the potentialities of coordination polymerization, had in many ways a background different from Ziegler's. Whereas Ziegler had been rather reluctant to move from a university

---

*Actually, Holzkamp had already noted the production of polyethylene in the presence of chromium acetylacetonate. However, because his duplicate experiments led to contradictory results, Ziegler attached no importance to this observation.[671]
†In a patent application filed in 1951 (U.S. Pat. 2,816,883, Dec. 12, 1957) DuPont claimed linear polyethlene produced by free radical polymerization at 40–100°C under 5000–20,000 atmospheres.
‡Dr. H. Martin informed me in July 1983 that recent experiments in Mülheim suggest that no triethylaluminum would be formed under the conditions of Fischer's patent.

to The Kaiser Wilhelm Institute for Coal Research* because he was afraid that he might be forced to veer from pure research to product-oriented work,[670] Natta derived great scientific satisfaction from his close association with the Montecatini Company, which also supplied him with most of his research staff. In his work before 1950 Natta carried out numerous crystallographic studies (with 56 publications in this field) and acquired extensive experience in heterogeneous catalysis.† In 1932 he visited Freiburg, and although his objective was to learn from Seemann about electron diffraction he became acquainted with Staudinger on that occasion. In 1952 he heard Ziegler lecture on the polymerization of ethylene in the presence of trialkylaluminum compounds and was, even at this stage, sufficiently intrigued by this work to start a program along that line in his laboratory.

Natta also sent some of his collaborators to Ziegler's laboratory to observe the progress of these researches. When Ziegler discovered the low-pressure polymerization of ethylene Natta had these men instructed in the process and also arranged an agreement between Ziegler and Montecatini. Early in 1954 Natta asked his research group to attempt polymerization of propylene by the Ziegler catalyst, hoping that it would lead to a linear polymer of high molecular weight and expecting the polymer to be a rubber. The result of this experiment was a material that seemed rubbery at first but on fractionation was found to contain, to Natta's intense surprise, a highly crystalline portion. On December 10, 1954, he submitted a short paper to the *Journal of the American Chemical Society*[679] in which he reported that "using various heterogeneous catalysts to be described elsewhere we have synthesized linear crystalline polypropylene, poly-α-butylene and polystyrene." He specified the repeat period of polypropylene as 6.50 Å and added that "each portion of the principal chain included in the elementary cell corresponds to three monomer units." The evidence led him to "attribute to the new crystalline polymers a structure in which at least for long portions of the principal chain all the asymmetric carbons have the same configuration," and because of the steric requirements of the chain substituents "a spiralization of the principal chain must take place."‡ He suggested the term "isotactic" for the stereoregularity observed in these polymers.

The notion that a high degree of stereoregularity would be required to allow a vinyl polymer to crystallize was implicit ever since Staudinger[311,312] attributed the amorphous nature of polystyrene to the randomness of the configurations of the tertiary carbon atoms along the chain. This principle was later

---

*Later renamed "The Max Planck Institute for Coal Research."
†Commemorative articles by P. Pino[678] are a valuable source of information on G Natta.
‡Natta's manuscript was originally rejected by the referee on the grounds that the nature of the catalyst had not been disclosed. This judgment was overruled by Flory, an editor of the *Journal*, on the grounds that the paper was of most unusual significance.[680] Natta's close association with Montecatini was responsible for the lack of experimental detail characteristic of most of his publications. When Natta lectured at a symposium on "stereo-specific syntheses" at the Polytechnic Institute of Brooklyn in April 1958 many of his lecture slides were illegible and some of the audience suspected that this was intentional.

modified only for poly(vinyl alcohol), in which the relatively short chain substituents allow crystallinity in spite of a stereoirregular structure.[681] The first suggestion that stereoregularity might be influenced by the conditions under which a polymer was prepared is due to Huggins,[682] who thought that low-temperature polymerization should favor either sequences with similar configuration or a regular alternation of configurations. In 1948 Schildknecht et al.[683] reported that although vinyl isobutyl ether polymerizes to a rubber when BF_3 is used as a catalyst, heterogeneous catalysis by the ether adduct of BF_3 leads to the formation of a crystalline product with a 6.20 Å repeat period. The authors wrote that "as fully extended vinyl chains correspond to multiples of 2.54 Å, the chains must not be fully extended" and proposed an alternating configuration of the asymmetric centers.* The use of the heterogeneous "alfin" polymerization catalysts introduced by Morton[686] (e.g., consisting of sodium isopropoxide and allyl sodium) was also later shown to lead to isotactic polystyrene[687] but Morton had not been aware of having obtained a crystalline product. Isotactic polypropylene was probably also obtained with the use of supported metal oxide catalysts[688] but in the original patent applications†, filed before that of Natta's group, the first claimed merely a rubbery polymer of high molecular weight, the second a "solid" fraction insoluble in pentane. Neither patent mentioned crystallinity.

One may wonder why Ziegler has not arrived at polypropylene right after his discovery of low-pressure polyethylene. H. Wesslau,‡ one of Ziegler's students who had worked on the $TiCl_4$-trialkylaluminum catalyst, obtained a solid in an attempt to polymerize propylene, but when the material seemed to have a higher melting point than polyethylene he felt sure that he was mistaken because he could not believe that branching would raise the melting point of a paraffin. H. Martin‡ believed that the reason for the failure to obtain polypropylene in early experiments at Mülheim may have been due to the low stereoregularity of polymers produced at that time; since a low solubility was used as a criterion of polymer formation these polypropylenes may have been missed. It does not seem improbable, however, that the failure of propylene to polymerize beyond the dimer in the presence of trialkylaluminum (in such striking contrast to ethylene)[672,673] prejudiced Ziegler and his collaborators against the possibility that their new catalyst could lead to high molecular weight polypropylene.· Natta did not tell Ziegler about his discovery when he visited him shortly after the crucial experiments, an omission that was resented as rather disloyal after all the assistance Ziegler's laboratory had provided to

---

*Although he described this proposal as "plausible," Flory[684] concluded even before Natta's work that an isotactic structure cannot be excluded for Schildknecht's polymer. The polymer was, in fact, isotactic.[685]

†A. Zletz (to Standard Oil of Indiana), U.S. Pat. 2,692,257 (Oct. 19 1954); Phillips Petroleum Co., Belg. Pat. 530,617 (Jan. 24 1955).

‡Interviews, July 1983.

·A bizarre story is told by McMillan:[688a] In the plant of Petrochemicals Ltd. polyethylene was produced by the Ziegler process. One day the plant ran out of ethylene and the technologist in charge asked for permission to use propylene instead. When a polymer was obtained, no one was surprised and nobody thought of filing a patent.

Natta's research team.[689] Undoubtedly Natta's relations with Montecatini limited his freedom of frank scientific exchange.

The brilliant research team assembled by Natta made rapid progress in the study of stereoregular polymers so that only seven months after submission of the initial note to the *Journal of the American Chemical Society* Natta could send to *Makromolekulare Chemie* a paper[685] in which he defined the chain conformation of isotactic polypropylene, poly(α-butene), and polystyrene, announced the preparation of the first syndiotactic polymer (1,2-polybutadiene), the 1,4-polybutadienes with a prevalent cis or trans structure, and an amorphous high molecular weight rubbery polypropylene. He also showed that IR spectroscopy can be used to characterize the crystallinity of stereoregular polymers. However, although this impressive paper refers to "improved catalysts" that lead to high stereoregularity and implies that different catalysts are needed for different products, no data are included about their nature.

The possibilities inherent in coordination polymerization were not, of course, lost on the U.S. rubber industry. Whereas Natta's group worked with the more easily available butadiene, the polymerization in the Goodrich laboratories of isoprene with a coordination catalyst allowed Horne[689a] to announce the synthesis of *cis*-1,4-polyisoprene which apparently resembled natural rubber even more closely than the product obtained by Staveley at Firestone.[655]

In 1956 Price et al.[690] discovered that a racemic mixture of propylene oxide could be polymerized by a heterogeneous catalyst to a crystalline product with the same melting point as the polymer derived from the pure l-monomer. The polymer obtained from the racemic monomer had no optical activity suggesting that "it consists of a mixture of all-d and all-l polymer molecules." Thus the catalyst had to have two kinds of sites, each highly selective, for one of the enantiometic monomer species. A similar observation was made in Natta's laboratory[691] with 4-methyl-hexene-1. Natta now also discussed the mechanism of polymerization by what has come to be known as Ziegler-Natta catalysts and suggested that it involved "insertion of monomeric units between the catalyst and the growing polymer chain in such a way that the CH_2 group is connected with the catalyst." Because groups that originally formed part of the alkylaluminum moiety were found in the polymer, Natta concluded that the polymer is attached to the aluminum atom. This surmise was reinforced when a soluble bimetallic complex of a bicyclopentadienyl titanium compound with a trialkyl- or triarylaluminum was used in studies[692] which exemplify the wide range of techniques employed in Natta's laboratory. Nevertheless, the conclusions were challenged by Breslow and Newburg[693] on the grounds that groups attached to the titanium and aluminum atoms exchange readily and that some catalysts that contain only a titanium compound are known. Their belief that the polymer is attached to the titanium appears to be shared by most workers in this field.

The astonishing activity in Natta's laboratory in the five years following the discovery of stereospecific polymerization is documented by 170 papers, published in that period.* They contain crystal structures of stereoregular polymers and some catalysts, the discovery that *cis*- and *trans*-CHD—CHCH$_3$ yield spectroscopically distinct stereoregular polymers,[695†] descriptions of methods for varying the degree of stereoregulation, specification of catalysts for the stereospecific polymerization of vinyl ethers by a cationic mechanism,[697] copolymerization by coordination catalysts, studies of the kinetics of coordination polymerization, with the discovery that hydrogen can control the polymer chain length,[698] and studies of the solution properties of stereoregular and stereoirregular polymers. The 1963 Nobel Prize in Chemistry, which Natta shared with Ziegler, was a most richly deserved recognition of their contributions.

---

*Natta and Danusso[694] have edited a collection in which the more important papers are given in full, the others in abstract form. Papers published in Italian, French, or German were translated to English.
†Later it was found[696] that monomers of the type RCH=CHOR' yield "threo diisotactic" polymers in which the R and R' are on the same side of the plane that contains the chain backbone if represented in the all-trans form.

21. Advances in Polycondensation and Ring Opening Polymerization

Two important developments in polycondensation were achieved during the second world war, one in Germany, the other in England, but wartime secrecy kept them in wartime from public view. In 1947 Bayer[699] published an impressive account of the development, under his direction, of polyurethanes and polyureas from diisocyanates and dihydroxy or diamino compounds, respectively. His paper makes it clear that the original motivation for this research was a desire to circumvent Du Pont's Nylon patents, but the scope of the application of these new polymers was much broader than anticipated. They were found to be useful for the production of fibers, bristles, lacquers, corrosion-resistant coatings, foams, and adhesives. The most unexpected result was the rubberlike elasticity exhibited by some of these materials. Because it was assumed that the elastic behavior requires the production of a crosslinked network, Bayer et al.[700] produced a polyurethane in which 2-nitrodiphenyl-4,4'-diisocyanate was used to join polyester prepolymers with OH end groups. The plan was to reduce the nitro groups and use the amino functions to crosslink the linear chains. To the surprise of the investigators the elastic behavior was observed even without this modification. They concluded that "the alternation of linear and bulky (*sperrig*) units plays an important role in the elastic properties." Nevertheless, Bayer and Müller,[701] when reviewing in 1960 the work on polyurethanes carried out at Farbenfabriken Bayer, concluded that crosslinking is essential if an elastic product is to be obtained. Yet in the same year Katz[702] was granted a patent for Du Pont in which it was shown that elastic polyurethanes are obtained without crosslinking by coupling noncrystallizable "soft" and crystallizable "hard" prepolymers. These linear polymers owe their elasticity to the crystallization of the "hard" segments that act as quasi-crosslinkages below a characteristic melting point.

In February 1943 Whinfield[703] gave a lecture to the Faraday Society on natural and synthetic fibers in which he extolled "the outstanding genius of this great American chemist" Carothers. Nothing in Winfield's presentation could have led his audience to suspect that three years earlier he had discovered an important gap in Carothers' work: whereas Carothers had concluded, after studying a large number of polyesters, that this class of polymers has melting points too low to be useful for textile fiber applications, Whinfield and Dickson[704] found that poly(ethylene terephthalate) melts at the surprisingly high temperature of 256°C. It has been claimed[705] that this discovery was not accidental but the result of acute scientific reasoning,* a contention that is highly questionable. Whinfield's[706] account stresses only the effect of chain symmetry which made him expect that the polyterephthalate would be crystalline but the polyisophthalate amorphous. Actually, poly(ethylene isophthalate), although difficult to crystallize, was later shown[707] to melt at 240°. The cause for the high melting point of poly(ethylene terephthalate) was established[708] only six years after the Whinfield–Dickson patent as being due to the low entropy of melting resulting from a high chain rigidity.†

In 1955 Magat and Strachan[709] obtained a patent on a technique of polymerization which was to have far-reaching consequences. They showed that by the use of highly reactive pairs of bifunctional reagents, one soluble in water, the other in an organic phase, polycondensation could take place at the interface under mild conditions, even at ambient temperatures. The idea was not entirely new since German patents[710] had already claimed the synthesis of polyurethanes from bischloroformates in organic solvents and diamines in water solution, but the inventors of these processes seem to have missed the scope of the method. The principles that govern such polycondensations were outlined by Witbecker and Morgan[711] and Morgan and Kwolek,[712] who pointed out, in particular, that while one-phase polycondensations are "limited to intermediates and polymers that are stable under the conditions usually required" such a limitation is avoided in the interfacial reactions. This is particularly significant if the polymer aimed at has a melting point so high that extensive decomposition would result from an attempt to have the polycondensation proceed in the melt. Shashoua and Eareckson[713] gave a striking demonstration of the opportunities provided by interfacial reactions. With terephthaloyl chloride and ethylene diamine they synthesized a high molecular weight polyamide that melted at 455°C.

In 1960 Meerwein[714] published a summary of the remarkable studies which had been carried out during the second world war under his guidance on the ring-opening polymerization of tetrahydrofuran. Meerwein used as catalysts salts of trialkyl oxonium, R_3O^+, with anions such as BF_4^-, $SbCl_6^-$, and $FeCl_4^-$, which he and his students had discovered in 1937 and 1939,[715] or

---

*"It was an invention deliberately sought and found."
†The melting point T_m is the ratio of the enthalpy and the entropy of melting. The enthalpy of melting of poly(ethylene terephthalate) is similar to that of aliphatic polyesters.

"complex acids" such as HBF_4 or $HSbCl_6$. This work came as a surprise because it had been generally believed, ever since Carothers' classical studies, that the relatively strain-free, five-membered ring compounds could not be converted to linear chain molecules. The polymerization of caprolactam, involving an acid or base catalyzed opening of the seven-membered ring, had first been observed by Schlack[408a] but his procedures led to a very slow process. A great improvement resulted from the discovery that the sodium salt of caprolactam acts as a powerful polymerization catalyst.[715a] The base-catalyzed polymerization of caprolactam was then studied in the 1950s in great detail by Wichterle and his collaborators.[715b]

The early postwar period also saw a renewed interest (see p. 137) in the conversion of α-amino acid-N-carboxyanhydrides (NCA's) into polypeptides. Early in 1946 the first of a long series of papers that dealt with polypeptide syntheses was submitted for publication from Katchalski's laboratory. This paper described the preparation of poly(L-lysine), the average chain length of which was estimated by end-group analysis as 32. The paper was rejected, and before the work was published as a note late in 1947 (the full paper appeared a year later[715c]) Woodward and Schramm[716] published a note with the catchy title "Synthesis of Protein Analogs" in which they claimed to have produced a polypeptide with a molecular weight exceeding a million.* This erroneous estimate was based on the high viscosity of their polypeptide in benzene solution, later shown[906] to be a result of extensive association. Polypeptides with molecular weights in excess of 10^5 became accessible only when Blout et al.[717] introduced polymerization initiated by a methanol solution of alkali hydroxide. Katchalski and Sela[718] suggested that "their primary importance . . . will probably remain as simple synthetic polymer models." This concerned not only the conformation of polypeptide chains but also studies in which polypeptides were used as antigens or as enzyme substrates.[718a]

But polypeptide studies had also another motivation. The Courtalds Company had assembled in Maidenhead, England, a group of outstanding scientists who studied, under the direction of Bamford, the synthesis of poly-α-aminoacids and their structures by spectroscopic and crystallographic techniques.[719] It is obvious that in going to the expense of this broad program Courtaulds hoped that the polymers would form valuable fibers, particularly because dyeing and other characteristics were expected to be similar to those of silk. Economic factors prevented the commercial success of this venture, but Maidenhead became the source of a wealth of scientific information.

---

*Woodward and Schramm apparently were unaware of the excellent study of α-amino acid-N-carboxyanhydride polymerization by Meyer and Go.[458] In response to my inquiry concerning the original rejection of his paper, E. Katchalski-Katzir wrote to me as follows: "Several years thereafter I found out from Professor Woodward himself that he was the referee at the time and that he did not believe that we had succeeded in synthesizing a high molecular weight peptide. The 1947 and 1948 publications on poly-lysine were readily accepted and even praised by the Editor, as a result of Woodward's recommendations. . . . Woodward and I became good friends."

As shown in a later chapter, certain conformational characteristics of proteins could be duplicated with synthetic polypeptides. At this point it should be noted only that the kinetics of the propagation of the polypeptide chain exhibit characteristics suggesting that chain conformation exerts a pronounced effect on the reaction rate. Lundberg and Doty[720] found that the use of oligopeptides as initiators of NCA polymerization was much more effective if the conformation of the amino acid residues in the oligopeptide was the same as in the NCA. Also, a racemic NCA polymerized much more slowly than the pure enantiomer. It was concluded that the helical conformation characteristic of poly-α-amino acids made up of residues of the same configuration propagates more rapidly. This observation was extended by Ballard et al.,[721] who found that the addition of lithium perchlorate, which disrupts the helical conformation, eliminates the stereospecificity of polypeptide propagation.

We are meeting here a phenomenon entirely new in the study of synthetic polymers; that is, the reactivity of a chain molecule controlled by its shape in solution.

22. The Rise of Molecular Biology

The Annual Report for 1938 of the Rockefeller Foundation contains the following passage by Warren Weaver, director of the Natural Sciences Program:

> With the aid of the precise tools furnished by the modern laboratory, explorations are reaching deeper and deeper into the living organism and are revealing many facts about the structure and behavior of its minute intercellular substances. And gradually there is coming into being a new branch of science—molecular biology—which is beginning to uncover many secrets concerning the ultimate units of the living cell.

Although the expression "molecular biology" was new, Weaver had for some years persuaded the Rockefeller Foundation to earmark a large proportion of funds expended on the physical sciences to support this research area. Even in 1933 he wrote that "the time is ripe to help stimulate significant advances by bringing to bear on the basic problems in biology the powerful quantitative techniques of mathematics, physics and chemistry." Since then he was responsible for the support, among other biology-related researches, of the construction of ultracentrifuges in Svedberg's and other laboratories and of crystallographic studies of the structure of proteins and their building blocks by Astbury and Pauling. His far-sighted policy was undoubtedly an important contributing factor in the acceleration of the development in this field.

Two concerns are central to molecular biology: The first is the structure and function of proteins; the second, the molecular mechanism by which the hereditary message is transmitted. The spectacular advance in this field may well come to be viewed in historical perspective as the most significant event of the quarter century following the second world war. Although this development does not fall within the scope of "polymer science," it exercised a continuing fascination for a number of the leading scientists who worked with synthetic polymers and stimulated some of their most striking discoveries. It should therefore be presented, at least in rough outline, in the context of this book. Because our story covers the period only up to 1960, the dramatic development of molecular biology unfortunately cannot be carried to the time of its fruition. It is also beyond the scope of this book to describe much of the human drama associated with the emergence of these new approaches to understanding the essential constituents of living cells. All too often those who considered themselves practitioners of what they viewed as a superior "molecular" approach ignored or offended older pioneers of biochemistry and biology. I recall a meeting at which Heidelberger, a towering figure in the history of immunochemistry, was patronized by a "molecular biologist" whose arguments were eventually shown to have been wrong. When Delbrück[754] wrote that "Listening to the story of modern biochemistry . . . the cell is a sack full of enzymes" his condescension was bound to be painfully resented by those who had laid, by great ingenuity, the foundation of enzymology. It became almost axiomatic that "basic" biological research could be conducted only by working with the simplest organisms such as viruses and bacteria, and I remember contemptuous remarks about those "studying pretty birds."

Proteins

In spite of intensive efforts to arrive at a fundamental understanding of proteins, all work along that line was long hampered by the inadequacy of methods for amino acid analysis, which were both laborious and inaccurate. This situation began to change only in 1945, with the introduction of partition chromatography, on columns and on paper, by Martin and Synge.[722] Chromatography of protein hydrolyzates on columns of crosslinked polystyrene beads was then perfected by Moore and Stein[723] into a rapid, reliable method of amino acid analysis which soon became a standard tool of the protein chemist.

The highly resolved X-ray diffraction pattern furnished by protein crystals (demonstrated first by Bernal and Crowfoot[275]) made it most likely that the amino acid sequence in a protein was uniquely defined, but for a long time no method could be conceived by which this sequence would be determined. Bergmann and Niemann[724] interpreted the compositions of various proteins as indicating that each amino acid occurs at a periodic spacing along the polypeptide chain and related this regularity to Svedberg's suggestion that all proteins are multiples of a basic unit with a molecular weight of 34,000 (see p. 140). We see here how scientists of that period were straining to discover "order" in a molecule of a seemingly improbable complexity. The presumed "regularity," however, was soon revealed as an illusion based on an inaccurate amino acid analysis.[725]

When Sanger embarked in 1945[726] on the study of insulin, which was to earn him the Nobel Prize for 1958,* his objectives appeared at first to have been limited to the determination of amino end groups with the use of a reagent

---

*He received a second Nobel Prize in 1980 for developing a method of sequencing DNA.

that produced colored products with primary amines. At that time the molecular weight of insulin was in doubt, and the most commonly accepted value was one that turned out to correspond to the dimer of the protein. On that basis Sanger concluded that insulin contains four polypeptide chains, two with a glycine (A-chains) and two with a phenylalanine (B-chains) amino terminal residue. Later he split the disulfide bonds that linked the two types of chain and separated them, noting that the glycine and phenylalanine terminal fractions appeared to be homogeneous. By 1951 Sanger and Tuppy[727] were ready to attempt a determination of the sequence of the 30 amino acid residues in the B-chain by splitting it by acid or enzyme catalysis, separating the fragments by two-dimensional paper chromatography, and working out a sequence consistent with all the fragments. Later a similar analysis established the sequence of the 18 amino acids in the A-chain,[728] and finally the position of the three disulfide bonds was determined.[729] Thus Sanger accomplished within ten years a feat that had previously been considered beyond reach.

One important result of this work was the demonstration that there was no apparent order in the amino acid sequence. Once the feasibility of sequencing was demonstrated similar analyses were attempted for other proteins, and by 1960 Moore, Stein, and their collaborators[730] could announce the sequence of 124 residues in the chain of the enzyme ribonuclease with the position of the four intramolecular crosslinks.* Another spectacular application of Sanger's technique followed on the discovery of Pauling et al.[731] that the hemoglobin of patients suffering from sickle-cell anemia is slightly less acidic than normal hemoglobin. Overall amino acid analyses seemed to reveal no significant difference between the two species. However, when Ingram[732] subjected the two hemoglobins to enzymatic degradation and separated the fragments by paper chromatography and electrophoresis he could easily locate the peptide that was different. He could then show that the pathological condition was produced by the change of a single glutamic acid to valine in each of the identical halves of hemoglobin composed of 287 amino acid residues.

Attempts to clarify the principles that result in the folding of polypeptide chains into the compact particles of globular proteins were pursued vigorously, even before it became apparent that it is feasible to determine the amino acid sequence. In a 1943 review in which Huggins[733] proposed structures for the α- and β-forms of fibrous proteins he expressed the view that "there is every reason to believe that the same principles apply" to the structure of globular proteins. Later Pauling and his collaborators collected meticulously crystallographic data on amino acids and peptides, leading to the celebrated paper by Pauling et al.[734] in which it was shown that only two helical structures of polypeptide chains allow intramolecular hydrogen bonding in a way that satisfies all the geometric constraints. Only one of these helices, the α-helix with 3.6 amino acid residues per turn and a pitch of 1.5Å, was later observed. Pauling et

---

*In 1950 Haurowitz[730a] had still written "it would be hopeless to endeavor to fully characterize peptide chains containing 100 or more amino acids."

al. predicted that "this helix also contributes an important structural feature to hemoglobin, myoglobin and other globular proteins."* In proposing the α-helix, Pauling stressed† that to deduce this result from X-ray diffraction data alone would have been a formidable task, so that an imaginative building of molecular models was needed to provide an essential supplement to the work of the crystallographer. This suggestion exercised a profound influence on Crick and Watson in their work on the structure of DNA.

The principle that the folding of polypeptide chains into the unique conformation of the globular protein is a spontaneous process determined by the amino acid sequence was first enunciated in 1950 by Caldwell and Hinshelwood,[735] who believed that this conformation represents a free energy minimum.‡ The original suggestion of Mirsky and Pauling[482] that globular proteins owed the stability of their "native state" to hydrogen bonding was later supplemented by a consideration of "salt bridges" between anionic and cationic side chains.[736] Finally, in 1959, Kauzmann[737] proposed that the cohesion of nonpolar sidechains in an aqueous medium, "the hydrophobic bond," makes the dominant contribution to the stability of the native structure. This prediction was eventually strikingly confirmed by crystallography. Stabilization by intramolecular crosslinkages was demonstrated by White and Anfinsen[738] to be a secondary process that followed the specific folding of the polypeptide chain: When the four disulfides of ribonuclease were reduced, a large proportion of the eight thiols could be oxidized to the original disulfides during careful renaturation, although there were 105 ways in which the thiol groups could have reacted with one another.

Only X-ray crystallographic analysis could, in principle, lead to a detailed description of how the polypeptide chain is folded in the native structure of globular proteins. Yet the difficulties of such an undertaking seemed insuperable. In 1946, eight years after Perutz had started his work on hemoglobin and some years after he had been joined by Kendrew's study of the myoglobin structure, Astbury,[739] the pioneer in the crystallographic investigations of fibrous proteins, wrote pessimistically:

> Though it would be rash to set an upper limit to the complexity of structures that might be handled by these methods, it must be confessed that the present outlook with regard to the detailed solutions of protein crystals is not very hopeful.

Later, when unforeseen developments enabled him to bring his work to a brilliant conclusion, Kendrew was to remark in his Nobel Prize address

---

*This prediction was to prove correct for hemoglobin and myoglobin, whereas α-helices were eventually shown to contribute little to the structure of most proteins.
†Lecture at the IUPAC meeting in Stockholm, June 1953.
‡The current view is that it is determined kinetically and is not necessarily the structure with maximum stability.

My almost total ignorance of this method was fortunate in that it concealed from me the extent to which contemporary x-ray crystallographic techniques fell short of what was needed to solve the structure of molecules containing thousands of atoms; it was a case of ignorance being bliss.*

According to Bragg's testimony,[740] the relative optimism in regard to the solution of the structure of hemoglobin, a molecule with about 5000 scattering atoms,† was sustained in the early years by the unjustified assumption that the polypeptide chains were arranged in some regular order. In the case of myoglobin, in particular, Bragg et al.[741] felt justified to conclude that "there appears to be a real simplicity in the chain structure." A summary of Perutz's early work on hemoglobin[742] describes the shape of the molecule merely as being approximated by a cylinder with a diameter of 57 Å and a height of 34 Å, but it points out also that the symmetry of the crystal requires the molecule to be composed of two identical halves. In addition, a comparison of the scattering patterns of hemoglobin crystallized from water and salt solution could be interpreted as showing that the water contained in the crystal was present between the hemoglobin molecules which were compact bodies free of solvent. In regard to its internal structure it was assumed[743] "that the protein molecule consists for the greater part of parallel polypeptide chains in a hexagonal packing," although the low intensity of the X-ray reflections suggested "corner turning, nonuniformity in the distribution of side chains and, perhaps, meandering of the chains from strict parallelism."

It was well known that the solution of a crystal structure could be greatly facilitated if one could observe the change in the relative intensity of diffraction spots when heavy atoms were attached to the molecule without perturbing the structure. This approach was suggested as early as 1939 by Bernal,[744] yet even in 1948 Kendrew and Perutz[745] were convinced that "even the heaviest of heavy atoms would make a negligible contribution to the reflections from so large a unit cell." In this they may have been influenced by Crowfoot Hodgkin's half-hearted attempt[267] to compare the diffraction from crystals of zinc and cadmium complexes of insulin, which led to no useful result.

This conviction changed in 1952 when Perutz determined the absolute intensity of the X-ray reflections from a hemoglobin crystal and found that they were much smaller than he had assumed. He was led to conclude that the attachment of a heavy atom might after all produce the desired change in the diffraction pattern, but he did not know how to bring about such an "isomorphous replacement." A happy accident provided the solution. He received the reprint of a paper that showed that when the two thiol groups of hemoglobin react with sodium *p*-chloromercuribenzoate, the oxygen-binding heme groups

---

*A similar sentiment was expressed by Perutz when I asked him in 1967 what he had in mind when he started to study the hemoglobin structure in 1938. He answered simply, "I did not understand the problem."
†Scattering from hydrogen atoms is negligible.

interact with one another in a manner similar to that in the unmodified hemoglobin. This suggested to Perutz that the molecular structure remained unchanged. An experiment confirmed this and yielded the hoped for changes in the intensities of the diffraction spots. "Madly excited," Perutz fetched Bragg to show him the pictures that meant that the problem of protein structures was almost sure to be solved in time.*

Perutz and his collaborators[746] soon demonstrated that attachment of p-chloromercuribenzoate or silver atoms to the two thiol groups of hemoglobin leads to consistent changes in the relative diffraction intensities. The application of the same method to myoglobin seemed more difficult at first because this protein contained no thiol groups, but an empirical survey of a large number of possible heavy atom ligands in Kendrew's laboratory led to the discovery of several that were bound to discrete points on the myoglobin molecule.[747] In fact, since these heavy atoms were linked to different sites, they led to a greater simplification of the structure analysis than was obtained for hemoglobin, so that the myoglobin structure became the first protein crystal structure to be solved. Because it was assumed that α-helices would be a prominent feature, a solution at 6Å resolution was first carried out by Kendrew et al.[748] in 1958. This yielded a structure containing thick rods with sharp bends which was interpreted as suggesting that 75% of the polypeptide chains assumed the α-helical form. The lack of order in the orientation of the helical segments was one of the chief surprises. Two years later, when the structure was refined to a 2 Å resolution,[749] it could be confirmed that the rodlike sections had all the characteristics of Pauling's α-helix. Moreover, while Pauling and Corey[750] had concluded that one sense of the helix should be preferred but could not decide between the alternatives, an important result of the crystal structure analysis was the demonstration that the α-helical segments in myoglobin are all right-handed.

The hemoglobin structure, obtained in Perutz's laboratory[751] at a somewhat lower resolution, was published side-by-side with the paper on myoglobin. This protein consists of two pairs of different subunits, both of which turned out to have structures strikingly similar to that of myoglobin. It should be noted that the final stages of the structure determinations of myoglobin and hemoglobin were aided to a crucial degree by the existence of electronic computers to process the enormous volume of data, a development that surely could not have been foreseen when Perutz embarked on his venture in 1938. Even so it was a fortunate accident that these two proteins have a large fraction of their polypeptide chains wound in a α-helix, a large feature that was easily recognized.† Crowfoot Hodgkin, an equally brilliant investigator, was to spend many more years before she solved the structure of insulin which presented, in spite of its small size, much more severe technical problems.

---

*This account is based on a letter[267] from M. F. Perutz to Dr. Crowfoot Hodgkin and Bragg's[740] account of the early history of protein crystallography.
†Later studies showed that a large content of α-helices is quite untypical of proteins.

Nucleic Acids

The study of the transmission of hereditary characteristics was dominated for a long time by geneticists, who in their attempts to establish rules for inheritance had little concern for the physical or chemical identity of the gene. In his 1933 Nobel Prize address Morgan,[752] the father of modern genetics, put it quite bluntly:

> What are genes? . . . are we justified in regarding them as material units; as chemical bodies of a higher order than molecules? Frankly, these are questions with which the working geneticist has not much concern . . . at the level at which genetic experiments lie it does not make the slightest difference whether the gene is a hypothetical unit or a material particle.

Yet there were exceptions to this attitude. In 1937 Muller,[753] the second major figure in genetics research of this period, published a provocative essay "Physics in the attack on the fundamental problems of genetics." He argued that

> an understanding of the properties of genes would bridge the main gap between inanimate and animate. Such a study would be of intense interest from the point of view of physics, physical chemistry and organic chemistry. . . . So peculiar are these properties that physicists often deny the possibility of their existence. . . . I am therefore making this plea to physicists in the hope that they will interest themselves in these problems.

At that time Muller believed that the gene was a protein,

> an active arranger of material . . . the analogy to crystallization hardly carries us far enough . . . there are thousands of genes in every nucleus and each gene has to reproduce its own specific pattern.

In a prophetic passage Muller suggested that "one possible line of approach might be through the study of x-ray diffraction patterns." He thought that tobacco mosaic virus crystals suggested another approach because "this material has the properties of a gene."

Unknown to Muller, a brilliant young physicist had already embarked on a lifetime involvement in the problems of genetics. Max Delbrück's[754] interest in biology was first aroused by the suggestion that

> Just as we find features of the atom, for instance, its stability, which are not reducible to mechanics, we may find features of the living cell which are not reducible to atomic physics. . . . Bohr's suggestion of a complementary situation in biology,* analogous to that of physics, has been the prime motive of my interest in biology . . . theories should be formulated without fear of contradicting molecular physics.

*See N. Bohr's famous lecture "Light and life" [*Nature*, **131**, 421, 457; *Naturwiss.*, **21**, 245 (1933)].

According to the autobiography published with Delbrück's Nobel Prize lecture thirty-five years later, his intention to explore ways in which his training could be applied to biological problems "was helped by the rise of Nazism which made official seminars less interesting. A small group of physicists and biologists began to meet privately beginning about 1934." These unconventional contacts led to the publication in 1935 of a "green pamphlet" by Timoféeff-Ressovsky, Zimmer, and Delbrück,[755] in which a geneticist and two physicists combined their expertise to explore the possible physical basis of mutation.* Timoféev-Ressovsky† surveyed the experimental background of mutagenesis. Zimmer formulated a theory of X-ray-induced mutation in which he concluded that it involved a "hit" of an ion-pair on a "target," the size of which he estimated in 1943 to be 1.8×10^{-17} cm^3. Delbrück wrote a section on an "atomistic-physical model of gene mutation" in which he proposed that since the "stability of configuration must be high, the genes participate in metabolism only as catalysts." In a passage reminiscent of Mark and Meyer's scruples concerning the use of the term "molecule" for the species in polydisperse polymers (see p. 80) Delbrück insists that this term, which applies to a collection of particles with identical properties, should not be used for a gene because "the gene molecule has in each living creature a single representative."‡ He therefore used the term "*Atomverband*" (association of atoms) and proposed that mutation involves its isomerization induced thermally or by an ion pair produced by ionizing radiation. It is interesting that he specifically excluded gene reproduction from his speculation. He made two noteworthy suggestions:

1. If gene mutation is thermally induced, a high frequency should be accompanied by a low temperature-dependence.
2. The use of monochromatic ultraviolet irradiation "should be capable to isolate certain types of mutation."

The views of Zimmer and Delbrück, published in the *Nachrichten der Gesellschaft der Wissenschaften zu Göttingen*, seem to have made little impact at the time. Eight years later, however, these ideas were brought to the attention of a much wider audience. Schrödinger, who was then Director of the Institute of Advanced Research in Dublin, was required as part of his duties to present a series of semipopular lectures, and feeling that quantum mechanics was hardly a suitable subject, he chose on this occasion as his title "What Is

---

*For those interested in the detailed history of the part of molecular biology concerned with the mechanism of heredity the truly magnificent monographs by Fruton[137] and Olby[756] are highly recommended.

†Timoféev-Ressovsky, a Russian pioneer of radiation biology, worked in Berlin 1924–1945. In 1937 he ignored an order to return to Moscow. When Soviet troops captured Berlin he was arrested and sentenced to ten years of hard labor. His gruesome experiences are recounted by Solzhenitsyn whom he befriended in prison. (A. Solzhenitsyn, The Gulag Archipelago, Vol. 1, Harper and Row, New York, 1984, pp. 598–600, 603, 604.)

‡This is a rather amazing statement: Delbrück apparently thought only of unicellular organisms, although he did not make this clear.

Life?" Scientific contacts in wartime were quite restricted and Schrödinger based his presentation largely on the proposals made by Delbrück, whom he had known well when they both worked in Germany. The lectures, published in the following year,[757] made a profound impression on a number of those who became prime actors in the emerging understanding of the physical basis of heredity (Ref. 756, pp. 246–247). The impact produced by this little book is hard to understand today. Schrödinger exhibited no knowledge of chemistry (beyond the existence of molecules, referred to as "associations of atoms"), never mentioned nucleic acids or enzymes, and referred to proteins (citing Haldane) only as the probable basis of genes. His calculation of the large number of isomers available to "a well organized association of atoms" was hardly novel: Fischer[758] had calculated in 1916 that in a "typical protein," consisting of 30 amino acid residues of 18 kinds, these residues could exist in more than 10^{27} sequences and Wrinch[759] had suggested in 1934 that genes are proteins in which the amino acid sequence accounted for their specificity. It is true that whereas Delbrück's emphasis had been on the mechanism of mutation Schrödinger concentrated on the general problem of the transmission of the genetic message and introduced the provocative term "aperiodic crystal." But what exactly did he mean by that term? It certainly was not a prophetic allusion to the principle of DNA replication because the problem of replication is mentioned nowhere in the book. And how did genes perform their function? All we are told (p. 21) is that "incredibly small groups of atoms, much too small to display statistical laws, play a dominating role in the very orderly and lawful events within a living organism. They have control of the large-scale features." To confuse the issue the number of atoms involved is estimated on page 32 as a million (based on Zimmer's estimate of the size of the "target" in X-ray-induced mutation), whereas on page 49 the gene is said to contain "1000 and possibly much less" atoms. How are we then to understand the fame and influence of this book? Two explanations come to mind:

1. If a man with Schrödinger's fame as a physicist concluded that the time was ripe for physicists to concern themselves with the basic problems of biology, this carried weight. By contrast, the appeal to physicists by Muller,[753] a geneticist, was discounted in particular because Muller's belief in long-range forces operating during cell division seemed physically unacceptable.
2. It is also likely that Schrödinger's expression "aperiodic crystal" conjured in the minds of Luria, Crick, Watson, and Benzer possibilities that went well beyond Schrödinger's meaning.

In attempts to represent genes as physical particles three problems had to be confronted:

1. What is their chemical nature?
2. What is the mechanism by which they duplicate?
3. What is the mechanism by which they exert their function?

In answer to the first problem, it was long the prevalent view that genes were proteins. Nucleic acid was generally discounted as a possible carrier of the genetic message since it was believed to be "too simple a material and too uniform," whereas "the diversity of proteins presents no such difficulty."[760] This judgment was based on the belief that in thymonucleic acid (later referred to as desoxyribose nucleic acid, DNA)* the four bases, adenine, thymine, guanine, and cytosine, are attached in a regularly repeating sequence to the polymer backbone. Yet in spite of some claims to the contrary,[761] Gulland et al.[762] concluded in 1945 that this idea was a historical accident based on the old belief that DNA was a tetranucleotide. "Had the size of nucleic acid molecules been realized at an early date, it is doubtful whether this hypothesis would have gained such a firm hold."

Two examples may be offered to illustrate the length to which investigators would go to discount evidence that might suggest a genetic role for DNA. In 1938 Caspersson and Schultz[763] pointed out that the wavelength dependence of the efficiency with which ultraviolet light inactivates bacteriophages (which were generally thought of as models of genes) matches the absorption spectrum of DNA. But they concluded that "the property of a protein which allows it to reproduce itself is its ability to synthesize nucleic acid." Three years later Hollaender and Emmons found a similar wavelength dependence for the mutagenesis of fungal spores but cautioned that "it is dangerous to over-emphasize the importance of nucleic acid in the study of radiation effects on living cells. It is possible that the nucleic acid transfers the absorbed energy to protein."[764]

The crucial development that should have settled once and for all the identity of DNA as *the* carrier of genetic information was the result of work extended over more than twenty years in Avery's laboratory at the Rockefeller Institute. In 1932 and 1933 Alloway[765] demonstrated that when a filtered extract from one genetic type of pneumococci is allowed to act on another type of this microorganism, a hereditary transformation is produced yielding cells with membranes that contain polysaccharide of the same kind as the cells from which the extract was obtained. The chemical nature of the "transforming principle" was obviously a matter of greatest interest and became the subject of a most meticulous study published in 1944 by Avery, MacLeod, and McCarty.[766] They showed that the active substance was stable in the presence of enzymes which degrade proteins, the polysaccharide of the cell membrane, and ribonucleic acid (RNA). On the other hand, it was rapidly deactivated by an enzyme which degraded DNA. Although these results might have convinced most workers that the "transforming principle" *was* DNA, Avery et al. expressed themselves with greatest caution: "If the results of the present study are confirmed, the nucleic acids must be regarded as possessing biological specificity the chemical basis of which is as yet undetermined." In spite of the great care they had used to eliminate the role of substances other than DNA,

---

*For simplicity we use the designation DNA in what follows, whatever designation was used in the original text.

they still considered the possibility that "the biological activity . . . is not an inherent property of the nucleic acid but is due to minute amounts of some other substance."

According to Wilkins,[767] "the work of Avery, MacLeod and McCarty . . . even in 1946 seemed almost unknown, or, if known, its significance was often belittled." This negative attitude was undoubtedly the result of the prevalent notion that DNA contains a monotonously repeating sequence of tetranucleotide residues that leaves no possibility for specificity. Yet not everyone shared this opinion. Boivin and Vendrely[768] concluded in 1946 that Avery's group had *proved* that DNA is the sole genetic substance and they were able to demonstrate transformation in another microorganism. In the following year Boivin[769] bravely proclaimed that "each gene can be traced back to a macromolecule of a special DNA." In regard to the chemical nature of this specificity "one has to envisage the possibility of a primary — or, more likely, a secondary— structure by which DNA's differ from each other." Apparently Boivin still found it difficult to assign the DNA specificity to a varying nucleotide sequence and preferred to think of it as the result of a variation in the DNA conformation.

When reading the text of the famous lectures presented by Chargaff[770] in June 1949 and published in 1950 one can have little doubt that the laborious studies of DNA carried out under his direction were motivated by a suspicion that the specificity discovered by Avery et al. implied a variation in the *composition* of DNA derived from different sources. In particular, he was wary of the hypothesis that the four bases occur in equal numbers in DNA because "there is nothing more dangerous in the natural sciences than to look for harmony, order, regularity, before the proper level is reached."* In fact, Chargaff soon found that DNA samples from different organs of any given species had the same composition but that this composition did not correspond to stoichiometric equivalence of the four bases and it also differed widely between different organisms. In a characteristically cautious comment he wrote, "it is noteworthy—whether this is more than accidental cannot be said—that in all DNA examined so far the molar ratios . . . of adenine to thymine and of guanine to cytosine was not far from 1." Chargaff's purist approach is emphasized by the comment (with an implied criticism of others working in this field) that "terms such as 'template' or 'matrix' or 'reduplication' will not be found in this lecture" because "an ounce of proof weighs more than a pound of prediction."

Once Chargaff had established to his satisfaction that the tetranucleotide theory was dead he clearly searched for a way to characterize the distribution of the various nucleotide residues in the DNA chain. Any attempt to determine nucleotide sequences was out of the question—only a quarter century later would such a task become technically feasible. The next best approach was to subject DNA to enzymatic degradation and compare the base composition of

---

*This was the fallacy in Svedberg's speculation about protein structures (p. 140) and in Bergmann's and Niemann's assumption of a periodicity of amino acid residues (p. 198).

the residual chain molecules at various stages of the process. Using this procedure, Zamenhof and Chargaff[771] showed that the nucleotides were distributed unevenly along the DNA chain, with regions rich in adenine and thymine more resistant to breakdown.

The final dramatic proof that nucleic acids are the sole carriers of the genetic message was provided by studies of viruses. Studies of the events that follow the infection of bacteria by a virus suggested to Northrop[772] that "the nucleic acid may be the essential autocatalytic part of the molecule as in the transforming principle . . . the protein may be necessary only to allow entrance into the host cell." This concept was soon proved by Hershey and Chase,[773] who used bacteriophages whose protein and DNA were labeled by ^{35}S and ^{32}P, respectively. They found that after the infection of the bacteria only the phosphorus label entered the cells and that only the phosphorus label was found in the viral progeny. Four years later similarly striking proof was provided for the ribonucleic acid (RNA) constituent of the tobacco mosaic virus. After demonstrating that simple mixing of the viral RNA and protein results in a reconstitution of the virus particle, Fraenkel-Conrat[774] made a hybrid by using RNA and protein from different strains of the virus; its progeny contained protein that was immunologically identical with the protein of the strain from which the RNA was obtained, although the infected cells had never been in contact with this protein.

The first crystallographic study of fibers prepared from DNA was published by Astbury and Bell[775] in 1938. They interpreted their data as suggesting that the purine and pyrimidine bases are stacked on top of one another at a spacing of 3.4 Å, and because they thought that the fiber repeat was eight times as long they concluded that the crystal structure supported the hypothesis that the DNA chain contains a regular repetition of the four nucleotides. In a lecture presented by Astbury[776] eight years later (after the work of Avery et al. on the "transforming principle") he concluded that these regular sequences might be interrupted by "some chemical peculiarity that might confer specificity." The most influential part of this work was the emphasis placed on the close correspondence between the 3.4 Å spacing of the DNA bases and the similar extension of amino acid residues in the β-form (see p. 83) of fibrous proteins. Astbury asserted "at the risk of dabbling in numerology" that this correspondence must have "a deep significance."

This statement produced a strong impression on Hinshelwood when this great explorer of chemical kinetics turned his attention to kinetic aspects of life processes. In an extended study of bacterial cells grown under a variety of conditions he found that the DNA content of the cells was independent of their rate of proliferation, whereas the RNA content was proportional to their rate of growth. This observation, coupled with Astbury's suggestion, led Caldwell and Hinshelwood[735] to propose that RNA acts as a template for protein synthesis, "a process analogous to crystallization guides the order in which the various amino acids are laid down," while in its turn "the protein molecule governs the order of nucleotide units" in acting as a template for RNA synthesis.* This was the first explicit suggestion in which protein synthesis and nucleic acid duplication invoked spatial complementarity as the guiding principle; the opposing view, that protein synthesis involves a process akin to crystallization in which amino acids are attracted to a protein spread out as a two-dimensional sheet, still had influential partisans as detailed in Haurowitz's[777] monograph published in the same year as the paper by Caldwell and Hinshelwood.† In speculating how the specific folding of the polypeptide chain is accomplished, Haurowitz concluded "that the cell must contain some unchangeable pattern which acts as a template . . . we have to accept its existence because we have hardly any alternative explanation for the production of specific and identical protein molecules."

The discovery of the double-helical structure of DNA by Crick and Watson has been described by Olby[756] in great detail. It is a fascinating story, quite untypical of the manner in which major scientific discoveries are made. Watson came to Cambridge late in 1951, a year after he had received his doctorate with a thesis on X-ray inactivated bacteriophages. By his own testimony he had absolutely no knowledge of crystallography. Yet, as he said eleven years later in his Nobel Prize address, he felt sure at the time that "the main challenge in biology was to understand gene replication. . . . This meant solving the structure of DNA."[779] In Cambridge Watson met Crick, twelve years older, whose education had been interrupted by the war and who was at the time working on a Ph.D. thesis dealing with protein and polypeptide crystallography. Watson succeeded in infecting Crick with his enthusiasm for an attempt to solve the DNA structure. Their optimism was based on the hope that a model-building approach, analogous to that used by Pauling in his discovery of the α-helix of polypeptides, might also be successful in this case. The fact that Pauling had for many years gathered structural information that served as the basis for his prediction, whereas neither Crick or Watson had had any experience with DNA, did not seem to trouble the two enthusiasts.

Experimental work on DNA fibers had been initiated at King's College somewhat before Watson's arrival in Cambridge. This work was directed by Wilkins, a physicist, but since he had little expertise in X-ray diffraction Rosalind E. Franklin, an experienced young crystallographer, was added to the staff in January 1951. From the very beginning her relations with Wilkins, who tried to treat her as an assistant rather than as an independent investigator, were extremely difficult. At the same time Franklin, who had been trained along established lines to aim at the solution of crystal structures deductively from the X-ray diffraction data, was clearly contemptuous of the speculations that guided Crick and Watson's approach and resentful of Watson's attempt to use her data.

---

*Ironically, Caldwell and Hinshelwood did not assign any significant function to DNA.
†Ten years earlier Pauling and Delbrück[778] argued on theoretical lines that directed synthesis is more likely to involve the production of a spatially complementary molecule than the duplication of a molecular pattern.

For a considerable time none of the investigators realized that available diffraction patterns of the sodium salt of DNA represented a superposition of two structures. Franklin and Gosling[780] were first to discover that an A-structure obtained at a relative humidity of 75% changed reversibly to a B-structure at higher aqueous vapor pressures. The diffraction patterns of the two forms showed clearly that the phosphate groups were placed on the outside of the DNA, in contact with water filling the interstitial spaces.* Pauling asked Wilkins for the X-ray photographs obtained at King's College but was turned down; Pauling and Corey[782] then proposed, on the basis of inferior data, a DNA structure that consisted of three interwoven helices with a core of hydrogen-bonded phosphate groups and with the purine and pyrimidine bases extending to the periphery. This positioning of the bases was considered plausible because "this would permit the nucleic acid to interact vigorously with other molecules" and presumably provide for the specificity of DNA. Surprisingly, Pauling and Corey were unconcerned with the fact that their structure was incompatible with the ionization of the phosphate groups at physiological pH, a feature that quickly persuaded Crick and Watson that the structure could not be correct.

Franklin concentrated her work on the A-structure, which yielded a much richer diffraction pattern, while Watson and Crick's work was helped immeasurably when they learned about the discovery of the B-structure.† Although this yielded few diffraction spots, their pattern revealed unmistakably the helical nature of the structure and the crucial helical parameters; a 34-Å repeat with the nucleotide bases stacked 3.4 Å apart in the fiber direction. The helix diameter of about 20 Å was within the range estimated from published electron micrographs.[784] For some time Watson resisted, in his model-building efforts, the possibility that the DNA backbone might lie on the outside of the helix since he could not see how an irregular sequence of purine and pyrimidine bases could be accommodated in the helix interior. Eventually two earlier observations provided a clue: Gulland and Jordan[785] had obtained DNA titration data showing that the nucleotide bases were protonated only in much more acidic solutions than expected and interpreted this as indicating that these bases participate in intramolecular hydrogen bonding. The second crucial information was Chargaff's[770] finding that the adenine/thymine and guanine/cytosine ratios were invariably close to unity. In fact Watson found that the structures produced by hydrogen-bonding adenine to thymine and guanine to cytosine had the same overall length and could provide the rungs of a "spiral staircase" structure between two DNA backbones without requiring any regularity in the sequence of the bases attached to either chain. The final informa-

---

*The paper that contained these conclusions was submitted for publication before Watson and Crick obtained their double-helical structure but was published only some months after the Watson-Crick[781] initial papers in *Nature*.

†According to Watson's account,[783] he was first shown the diffraction photograph of the B-structure by Wilkins. Detailed information concerning this structure was obtained later by Crick through an indiscretion on the part of Perutz.

tion came from Crick's realization that the space group of the DNA crystals required the two chains of the double helix to run antiparallel to one another.

The double-helical model of DNA published by Watson and Crick[781a] in April 1953 contained the memorable sentence: "It has not escaped our notice that the specific pairing we have postulated immediately suggests a possible copying mechanism for the genetic material." That the specific pairing of adenine with thymine and guanine with cytosine implies that each strand of the double helix may function as a template for the synthesis of the complementary chain was explicity stated in a second note in *Nature*[781b] and in a detailed summary of the work published in the following year.[786] Franklin and Gosling[787a] immediately confirmed that the structure postulated by Watson and Crick was consistent with their diffraction data for the B-form and showed later[787b] that the diffraction from the A-form also indicated a double-helical structure, although with a somewhat different geometry.

The unorthodox manner in which a problem with the importance of the DNA structure was solved raises questions that deserve a brief comment. First it should be emphasized that the subsequent distinguished careers of Crick and Watson demonstrated amply that their spectaculor success in 1953 was no "flash in the pan." In fact, when they and Wilkins were awarded the Nobel Prize in chemistry in 1962 Crick[788] spoke "On the genetic code" and Watson[789] on "The involvement of RNA in the synthesis of proteins," thus demonstrating that both had moved on to conquer new fields. Franklin had moved from King's College at the time the DNA double helix appeared in print. Her important contributions to the solution of the DNA structure were summarized by Klug,[790] who inherited her research notes when she died of leukemia in 1958. Whether her methods would have led her alone to a solution of the problem is doubtful and she was well aware of this uncertainty.[780] On the other hand, her direction of the work on the structure of the tobacco mosaic virus, published shortly after her death,[781] testifies to her scientific brilliance.

Watson and Crick's bold conjecture concerning the mode of DNA replication (in such striking contrast to the somewhat excessive caution which stopped both Avery and Chargaff from spelling out all the implications of their observations) received in 1958 two striking experimental confirmations. Meselson and Stahl grew bacteria in a medium enriched with ^{15}N and added at a given time an excess of nutrient with the natural nitrogen isotope composition. They argued that the "semiconservative replication" postulated by Watson and Crick should lead to DNA species with only three discrete densities, corresponding to double helices with 2,1 and no heavy chains. The result of ultracentrifuge analyses in a density gradient by Meselson et al.[792] fully substantiated this expectation.*

---

*In reminiscing about this famous experiment, Meselson and Stahl[793] recollect that the original idea was to centrifuge particles of two different densities in a medium with an intermediate density, with the expectation that one would "sediment," the other "float." According to the account I heard

The second dramatic demonstration of DNA replication was provided in Kornberg's laboratory[794] by the enzyme-catalyzed synthesis of DNA in the presence of a DNA "primer." Particularly impressive was the demonstration that the adenine/thymine content of the newly formed DNA corresponded in each case to that of the primer.

The notion that genes control the synthesis of proteins (and, in particular, enzymes) was suggested at various times long before the chemical nature of the gene became known. It was clearly expressed in 1941 by Beadle and Tatum[795] as a result of a study in which they identified enzymes, involved in a specific step of a metabolic sequence, which were eliminated in the mutation of an organism. Yet the nature of this control was not understood; Beadle[795b] thought that "the gene's primary and possibly sole function is in directing the final configuration of protein molecules," implying that it was the folding of the polypeptide chain that was being directed. The DNA double helix of Watson and Crick, which allowed any sequence of nucleotides along the chain molecule, naturally led to the notion that this sequence somehow "codes" for the sequence of amino acids in a protein. Thus it was now only this sequence which would be directed; there was no conceivable way in which DNA could determine the three-dimensional structure of the globular protein and it had to be assumed that the specific folding of the polypeptide chain was a spontaneous process.

When Crick[796] formally proclaimed the "sequence hypothesis" in a lecture "On protein synthesis" presented in 1958* he conceded that there was little experimental evidence to back it up but argued that "once the central role of proteins is admitted there seems little point in genes doing anything else" than directing their synthesis. This, however, raised two questions:

1. How can the sequence of four nucleotides code for the sequence of twenty amino acids?
2. By what mechanism can an amino acid "recognize" this sequence?

Although the solution of these problems had to await a period beyond that treated in this book, it should be noted that Crick formulated two important principles in 1958:

---

from Vinograd, this was still the expectation when the DNA of Meselson and Stahl was centrifuged in a CsCl solution. When it was found that the three DNA species collected unexpectedly in bands, it was Vinograd, a man who had carried out research with the ultracentrifuge for many years, who realized that the CsCl had formed a density gradient in the ultracentrifuge cell and that each DNA moved to a location at which its density was matched by that of the medium. This account differs substantially from that of Meselson and Stahl,[793] who wrote that they "made use of an analytical ultracentrifuge in Vinograd's laboratory" but could not recall how they "first came to realize that after only a few hours of centrifugation the necessary density gradient would be established."

*Crick thought at the time that the folding of antibodies might possibly not be controlled by the amino acid sequence; apparently he was following Pauling's ideas on this subject (see pp. 146–147).

1. Information can be transmitted only from nucleic acids to proteins, never from one protein to another or from a protein to a nucleic acid.*
2. An "adapter" interacting specifically with a given nucleotide sequence must be required to carry an amino acid to the appropriate nucleic acid site.

In 1955 Grunberg-Manago and Ochoa[797] discovered an enzyme that catalyzed the conversion of nucleoside diphosphates to polyribonucleotides. Unlike Kornberg's DNA-synthesizing enzyme, this enzyme could catalyze not only the synthesis of chains that contained all four nucleotide residues of natural RNA but could also produce chains with any desired nucleotide composition. It is ironic that when Ochoa[798] presented in 1959 the Nobel Prize lecture that described his work in this area he had no idea that within a few years these polyribonucleoties would serve to solve the problem of the manner in which DNA codes for an amino acid sequence.

Rarely has a branch of science advanced so explosively as molecular biology in the time discussed in this chapter. Carlson[799] quotes Muller, the discoverer of X-ray-induced mutation, as confiding to him in 1959, "I had always known that some day there would be a chemical basis for the structure and function of the gene, but I never believed that I would live to see it." Similarly, the rate at which this field would expand in the future was grossly underestimated. Pontecorvo[800] expressed himself with great caution when he said in 1958 that "the possibility of translating a polynucleotide code into a polypeptide . . . is no longer inconceivable." On the other hand, the successes achieved may have led some to entertain exaggerated expectations. Lederberg[801] paid an extravagant compliment to synthetic polymer chemists when he said:

> If the ingenuity and craftsmanship so successfully directed at the fabrication of organic polymers . . . were to be concentrated on the problem of constructing a self-replicating assembly . . . I predict that the construction of an artificial molecule having the essential function of primitive life would fall within the grasp of our current knowledge.

---

*As long as T. Lysenko, a former protegé of Stalin's, had a dominant position in the USSR, the role of DNA in heredity was not allowed to be admitted in countries under Soviet influence. When I visited Prague in 1957 a journalist asked me what I considered the most striking development in macromolecules. When I mentioned DNA, he blushed and asked me to speak about "a less political subject." By 1963 Lysenko's power had ended and I was taken in Czechoslovakia to see a play in which the hero was a biochemist working on DNA!

23. The Impact of Spectroscopy

It was only around 1940 that spectroscopic methods gradually became an essential tool of the chemist, making in time an important contribution to the understanding of macromolecules. Ultraviolet spectrometers, which were to become widely available, provided a convenient method for the nondestructive determination of the aromatic component in minute copolymer samples. Yet only in 1946 was a spectroscopic determination of the composition of styrene–butadiene rubbers described in the literature.[802] It represented a striking improvement in convenience and reliability over older methods based on elemental analysis or on chemical reactions like those by which carbon-carbon double-bond concentrations were determined.

Ultraviolet spectroscopy made a variety of important contributions to the study of biological macromolecules and their analogs. The high precision determination of the base composition of DNA carried out in Chargaff's laboratory,[803] which was to play such an important role in the discovery of the double-helical DNA structure, would not have been possible without a combination of chromatography with UV spectroscopy. In 1954 Thomas[804] described the intensification in the absorption band of DNA when its solutions were heated or subjected to other conditions leading to "denaturation"; that is, a disruption of the double-helical structure. This spectral change then became the standard method of observing the "melting" of DNA. Three years later Felsenfeld and Rich[805] showed that the decreased UV absorption that accompanied the association of polyadenylic and polyuridylic acid (poly-A and poly-U) can be used to determine from a "spectroscopic titration" the stoichiometry of the helical complex. By the end of the decade Ross and Sturtevant[806] had used the decrease in absorption to follow the kinetics of the double-helix formation from poly-A and poly-U. At the same time Tinoco[807] formulated a theory to show that the interaction of light-induced dipoles in bases stacked in a double-helical structure was expected to lead to a decrease in the extinction coefficient in good agreement with data obtained with DNA and polyribonucleotide complexes.

Infrared spectroscopy proved to be a particularly powerful tool for the study of polymers. In fact it may be argued that its importance to polymer science was even greater than to the organic chemistry of compounds of low molecular weight, since a number of problems of polymer characterization were inaccessible to any other experimental approach. As early as 1940 Fox and Martin[808] concluded that the absorption in a region characteristic of methyl groups indicated a substantial branching in polyethylene prepared by the high-pressure process. Thompson[809] interpreted the IR spectrum of polyethylene as suggesting a methyl/methylene ratio of 1:50 and drew attention to the possibility of using the spectrum for the estimation of other "imperfections" in the polyethylene chain, such as the presence of carbonyl groups (whose concentration correlated with the dielectric loss factor) and various types of unsaturation. We have already noted (p. 188) that IR spectroscopy was crucial a few years later in establishing the difference between the unbranched polyethylene prepared by Ziegler's process and the branched chains characteristic of high-pressure synthesis.

A particularly important analytical application of IR spectroscopy concerned the characterization of diene polymers. Before the advent of this technique unsaturation in rubbers was determined by halogenation with iodine monochloride,[810] whereas pendant vinyl groups were usually differentiated from double bonds in the chain backbone by their faster epoxidation by perbenzoic acid. It is interesting to note that Kolthoff and Lee[811] described this method as late as 1947, well after Thompson[809] had demonstrated that external and internal double bonds can be conveniently distinguished in IR spectra. The important advance, however, came in 1949 when two laboratories[812] reported quantitative methods for the spectroscopic determination of not only pendant vinyl groups but also of *cis* and *trans* double bonds in backbones of butadiene polymers and copolymers. This characterization was not only important as an indication of the dependence of the structure of rubbers on polymerization conditions but was, of course, essential to a comparison of synthetic *cis*-1,4-polyisoprene with natural *Hevea* rubber (pp. 185, 191).

The ambivalence with which spectroscopists viewed studies of polymers was expressed in 1950 by Sutherland[813] when he contrasted the rigor with which the vibrational spectrum of ammonia lent itself to the most detailed interpretation with the limitations of information that could be derived from polymer spectra. He felt that "the most that can be stated at present about a large organic molecule or polymer is that it does or does not contain certain chemical groups and even this statement has frequently to be hedged with qualifications." Yet he conceded that "such information can be of vital importance," a conclusion which was certainly in order in view of the results that had already been obtained on polyethylene and diene polymers. At the time Sutherland expressed these views, however, information of quite a different kind was becoming available from studies of the IR dichroism of crystalline polymers. Ambrose et al.[814] found that this dichroism suggested that CO and NH groups lie parallel to the fiber axis in porcupine quills but perpendicular to it in feathers, suggesting to them a folded α-form of the polypeptide chain with intramolecular hydrogen bonds (see p. 83) for the quills and extended chains with intermolecular hydrogen bonds (the β-form) for the feathers. An important result was obtained with synthetic polypeptides that could assume the α- or the β-structure, depending on the solvent from which films were cast.[815] Elliott and Ambrose[816] found that the vibrational frequency of the CO stretching mode was significantly higher in the α-form and this allowed them to determine that this form in which "all the hydrogen bonds of the peptide links are intramolecular" retains its stability in dilute solutions of inert solvents. They also recorded the qualitative observation that higher concentrations of these polypeptides in such media exhibit pronounced flow birefringence. The full significance of this phenomenon had to await the discovery of the α-helix,* but the observation of Elliott and Ambrose was the first demonstration of the persistence of the rigid α-structure in suitable solvents.

In the introduction to a series of papers in which a full analysis of the IR spectra of polymers was to be attempted, Liang et al.[817] listed in 1956 three impediments to such a task.

1. No rotational spectra from which precise information on molecular geometry could be deduced could be obtained.
2. No Raman spectra existed to supplement information from IR spectroscopy.†
3. It was conceded that a calculation of normal vibrations of a large polymer molecule with an asymmetric repeating unit is a task from which even the most industrious of us shrinks."

Nevertheless, in view of the importance of the problem, spectral analyses were to be carried out for the crystalline portion of semicrystalline polymers. By extrapolation of data from samples with a varying degree of crystallinity, spectra of the crystalline polymer with a unique conformation were to be deduced. With additional information from dichroism and from partially deuterated samples a group theoretical analysis was then to yield a more complete interpretation of the IR spectra. This approach was used first by Krimm et al.[818] on polyethylene and later carried out on a number of other crystallizable polymers.[819] The results could in some cases be used to confirm conclusions arrived at on the basis of chemical or crystallographic evidence. For instance, with poly(vinyl chloride) the spectrum was consistent with a head-to-tail addition of monomer residues and a syndiotactic structure of the chain in the crystalline regions of the polymer.[820‡] In the cases of poly(vinylidene chloride)[820]

---

*The IR dichroism of polypeptides interpreted by Ambrose et al. as their "folded" conformation was equally consistent with the α-helical structure.
†This changed much later when laser sources became available.
‡Although the crystallinity of poly(vinyl chloride) as conventionally prepared is very low, when vinyl chloride is polymerized in a urea clathrate complex the steric constraints lead to the pure syndiotactic product.[610] It was therefore important for Krimm[820a] to check some of his conclusions by the use of this material.

and poly(vinyl alcohol)[821] the dichroism allowed a choice to be made between alternative structures proposed on the basis of crystallographic analysis.

The IR spectroscopy of cellulose was greatly aided by the finding[822] that hydroxyl groups in the amorphous portion could be selectively deuterated. The dichroism of the O—H stretching vibration could then be used to amend hydrogen-bonding proposals made on the basis of crystallographic studies.[823] In particular, such spectroscopic data were valuable for the interpretation of the change in the cellulose crystal structure produced by mercerization.[823,824]

In some cases IR spectroscopy was able to furnish evidence for the polymerization mechanism. The production of distinct diisotactic polymers obtained by the stereospecific polymerization of *cis*- and *trans*- CHD = CHCH$_3$[695] could not have been discovered without IR spectroscopy. On the other hand, spectroscopic evidence proved the identity of polysulfones obtained from *cis*- and *trans*-2-butene.[825] A particularly useful application of IR spectroscopy concerned the analysis of the complex changes produced by the thermal or photoinitiated degradation of polymers. In a pioneering study of this type, Burlant and Parsons[826] showed that the pyrolysis of polyacrylonitrile involves the transformation

$$\begin{array}{ccccccc}
& CH_2 & & CH_2 & & & CH_2 & & CH_2 \\
\diagdown \diagup & & \diagdown \diagup & & \diagdown \diagup \longrightarrow \diagdown \diagup & & \diagdown \diagup & & \diagdown \diagup \\
CH & & CH & & CH & CH & & CH & & CH \\
| & & | & & | & | & & | & & | \\
CN & & CN & & CN & C & & C & & C \\
& & & & & \diagup \diagdown\!\!\diagdown & & \diagup \diagdown\!\!\diagdown & & \diagup \diagdown\!\!\diagdown \\
& & & & & N & & N &
\end{array}$$

The clarification of this process assumed great importance later when it was found that the initial product of pyrolysis could be dehydrogenated to "Black Orlon," a graphitic fiber of outstanding thermal stability.

Nuclear magnetic resonance spectroscopy was discovered in 1946, and it was immediately recognized that narrow resonance peaks could be expected only for molecules reorienting rapidly in the imposed magnetic fields, whereas immobilized molecules would exhibit a broad absorption band. It was therefore surprising for Alpert[827] to find that rubber yielded the narrow resonance characteristic of mobile fluids and that the width of this resonance did not broaden appreciably when the rubber was slightly vulcanized. Alpert's conclusion that this behavior is "probably due to rotation of parts of the chain" may constitute the first clear evidence of the intensity of microbrownian motions in rubbery polymers.* Holroyd et al.[829] reported in 1951 that the temperature-dependence of the NMR resonance of various polymers revealed a narrowing

---

*During the same year that Alpert's observations were published (1947) F. Grün[828] showed that gases diffuse through rubber only slightly less rapidly than through water, another indication of microbrownian motion in the rubber.

of the absorption band close to the temperature at which a change in the thermal coefficient of expansion and a change in the heat capacity were observed; that is, in the range in which the polymer changed from the glassy to the rubbery state. In semicrystalline polymers the molecular mobility is, in general, much higher in the amorphous than in the crystalline fraction and this was revealed by a superposition of a narrow and a broad NMR absorption. Wilson and Pake[830] suggested that the degree of crystallinity of polymers could be estimated from the relative areas of these two contributions to the NMR spectrum, but this approach has not proved reliable. Gutowsky and Meyer[831] detected for *Hevea* rubber no compound structure of the NMR spectrum at any temperature, presumably because even the crystallites are highly mobile, whereas Fuschillo et al.[832] found that in polyamides even the motions in the amorphous fraction are highly restricted. Slichter and McCall[833] pointed out that the separation of the NMR spectrum into narrow and broad components is generally rather arbitrary, but they demonstrated a strikingly higher mobility in the crystalline portion of branched polyethylene, compared with the crystallites in the linear species.

From our vantage point it may seem surprising that no high-resolution NMR spectra of polymers were reported before 1959, eight years after such spectra were obtained for small molecules.[834] In fact, Slichter's 1958 review[835] on "The study of high polymers by nuclear magnetic resonance" never even mentions the possibility of obtaining high-resolution spectra. Why should this have been so? The difficulty, as outlined by Naylor and Lasoski,[836] was twofold: First, it was believed that "the inherent line widths of polymeric protons are fairly large," limiting the attainable resolution. Second, at a time when deuterated organic solvents were not available "the difficult problem of finding a good solvent whose NMR spectrum does not obscure that of the polymer" imposed an additional restriction. The first objection was shown to be invalid by Bovey et al.,[837] who obtained reasonably well resolved polystyrene spectra in carbon tetrachloride solution. In their pioneering work they arrived at a number of important conclusions:

1. The width of the NMR peaks was independent of the polymer chain length, showing that line broadening is governed by segmental motions, not the tumbling of the molecule as a whole.

2. Only for degrees of polymerization of more than about twenty, did the NMR spectrum attain the characteristics of the high polymer, suggesting that the conformational restrictions of the polymer are fully effective only at this point.

3. Although the α-hydrogen peak in polystyrene was not seen because of spin-spin splitting, it could be observed if the β-hydrogens were replaced by deuterium.

4. Denatured proteins yielded well resolved spectra but the rigid native structure broadened the spectrum so much that no individual resonance peaks could be observed.

In introducing this paper, Bovey et al. expressed the belief "that such investigations may prove of value in obtaining new kinetic and configurational information concerning polymer chains, a rather modest assessment of the manifold uses of high-resolution NMR spectroscopy made eventually by the polymer chemist.

The second paper by Bovey and Tiers[838] introduced the application of NMR spectroscopy to the characterization of the stereoisomerism of vinyl polymers. When Natta and his collaborators discovered techniques for the synthesis of stereoregular polymers they characterized them by crystallography. This method was applicable only to polymers with a stereoregularity sufficiently high to lead to partial crystallization. Yet it was obvious that even polymers that cannot crystallize can vary in their tacticity, which might be described in the simplest case, as suggested by Coleman,[839] by the probability that the addition of a monomer unit to the progagating chain leads to the formation of an "isotactic placement" of a pair of monomer residues. Yet no satisfactory method existed to derive this probability from experimental data, although Morawetz and Gaetjens[840] had tried to deduce the "microtacticity" of a vinyl polymer from the kinetics of its reaction, and Fox et al.[659] had used an empirical correlation of IR spectra of poly(methyl methacrylate) samples to describe their stereoisomerism. The NMR method, demonstrated by Bovey and Tiers on poly(methyl methacrylate), allowed an analysis of the relative concentrations of the stereoisomeric triads and dyads of monomer residues. Moreover, the difference in the symmetry of isotactic and syndiotactic dyads made it possible to assign unambiguously a spectrum to one of the two structures, so that the characterization of the tacticity became independent of crystallography. Finally, a comparison of the dyad and triad distributions made it possible to decide whether the configuration of a new asymmetric center during the propagation of a vinyl polymer chain is wholly governed by the configuration of the preceding asymmetric center or whether a more complex control of the steric configuration of the chain is involved.

Naylor and Lasoski[836] introduced another application of NMR spectroscopy to the description of polymer chains. Although a great deal of information had been collected to show that vinyl polymerization proceeds predominantly by head-to-tail addition, the chemical behavior of poly(vinyl alcohol) suggested a small concentration of head-to-head units[841] but the method used in that case was not applicable to other polymers. Naylor and Lasoski used now ^{19}F NMR spectra to show the presence of a small fraction of monomer residues in poly(vinylidene fluoride) added head-to-head.

Yet another spectroscopic method, electron spin resonance spectroscopy, was used initially by polymer chemists for two purposes; that is, the study of radical concentrations in radical-initiated vinyl polymerizations and the formation and behavior of radicals produced in bulk polymers by exposure to ionizing radiation. Normally the radical concentration in free-radical polymerization is too low to be detected by ESR spectroscopy and methods such as the use of intermittent irradiation had to be used for its estimation (see pp. 171–

172). However, as first shown by Fraenkel et al.[842], in the polymerization or copolymerization of a divinyl monomer a sufficient number of radicals may be trapped in the crosslinked gel so that their concentration may be obtained from ESR spectra. A detailed study of such a process by Atherton et al.[843] described the build-up of the radical concentration during the irradiation of a methyl methacrylate-glycol dimethacrylate mixture and the decay of the radicals when the irradiation was discontinued. It was noted that the spectrum of these free radicals was the same as in poly(methyl methacrylate) exposed to X-rays. Another type of radical trapping was observed by Bamford et al.[844] during the polymerization of acrylonitrile, which leads to the occlusion of radicals in the precipitated polymer. Bresler et al.[845] constructed a highly sensitive ESR spectrometer which allowed them to follow the radical concentration during the homogeneous photopolymerization of vinyl monomers without crosslink formation. They found that at monomer conversions of about 60% radicals at the end of poly(methyl methacrylate) chains were remarkably stable in the dark, although the polymerization rate decayed rapidly. When the system was again illuminated polymerization resumed immediately. These observations were interpreted as indicating that short-chain radicals could propagate while the radicals at the end of long chain molecules were trapped in the interior of the macromolecular coils. It is well established that occlusion of the active chain ends in polymer coils reduces the rate of bimolecular chain termination,[846] but it is hard to see why it should interfere with monomer addition to a free radical long before the system reaches the glassy state and it is unfortunate that the provocative conclusions of Bresler et al. were never checked by other investigators.

ESR spectroscopy proved valuable in the interpretation of results obtained when radicals were introduced into acrylamide crystals by γ-irradiation at low temperature, followed by warming to a temperature at which polymerization could proceed in the solid state. Morawetz and Fadner[608] found that the rapid decay in the polymerization rate at low monomer conversions was due only to a minor extent to a decrease in the concentration of free radicals but had to be largely a result of the disruption of the crystals or the trapping of the active chain end in the polymer phase.

ESR spectra of radicals produced by the irradiation of solid polymers with ionizing radiation were reported first by Schneider et al.[847] The conversion of radicals trapped in polyethylene by the addition of a molecule of oxygen was followed by Abraham and Whiffen[848] by the change in the ESR spectrum. As expected, the decay of radical concentrations *in vacuo* and on exposure to oxygen was faster in branched than in linear polyethylene (Lawton et al.[849]), reflecting an easier gas diffusion in the less perfect crystallites.

24. Polymer Solutions

In the development of polymer science between the two world wars a number of essential points concerning polymer solutions were clarified. Most important, the concept that "colloidal solutions" could not be treated by physicochemical principles, was decisively defeated. Thus osmometry was accepted as a valid approach to the determination of the weight of the dissolved particle, whether it was a covalently bonded macromolecule or an association complex of such molecules. The validity of thermodynamic theory also allowed the determination of particle weights by equilibrium ultracentrifugation. Osmometry, however, was often unreliable and limited to a relatively narrow range of molecular weights, whereas ultracentrifuges were extremely expensive so that few laboratories could afford them.

Of great importance also was the suggestion that macroscopic hydrodynamics can be applied to the frictional behavior of polymer solutions. Even though Staudinger was wrong in his interpretation of the viscosity of these solutions, his introduction of the simple viscosimetric method greatly accelerated the development of polymer science and pointed to the need of placing the experimental data on a sounder theoretical basis. Svedberg's demonstration that the equilibrium distribution of macromolecules in the ultracentrifuge and a combination of the sedimentation rate with the diffusion coefficient led to the same molecular weights proved that hydrodynamic concepts could be applied to the motion of macromolecules through a solvent.

During the next twenty years impressive advances were registered in the understanding of both the thermodynamics and the hydrodynamics of polymer solutions. In addition, the development of the light-scattering method provided the polymer chemist with a powerful tool for polymer characterization. Thus the specification of average molecular weights and molecular dimensions became routine. An effort was made to devise methods of characterizing molecular weight distribution by avoiding time-consuming fractionations. A foundation was laid to the difficult field of polyelectrolytes. Last of all, two unexpected spectacular phenomena were discovered. Synthetic polypetides were found to undergo, with a change of solvent, a transition from a randomly coiled chain to a regular helical conformation and rigid rodlike polymers were shown to exhibit an equilibrium between a disoriented and an anisotropic solution phase.

The Shape of Chain Molecules

Once Kuhn's proposal[370] to represent chain molecules by the "random coil" model had gained general acceptance, two concepts with which he had dealt merely in a qualitative way had to be analyzed in detail:

1. Kuhn understood that the fixed bond angles would expand a chain beyond dimensions that characterized a "freely jointed" chain and had shown that with this feature the proportionality between the mean-square end-to-end displacement and the chain length would be retained. Any other restraints to chain flexibility, however, were treated merely by the introduction of the concept of a "statistical chain element" without any indication of the relation of this parameter to molecular properties.
2. Kuhn also appreciated the fact that the finite thickness of a chain would restrict the number of shapes available to the molecular coil and that this would lead to a chain expansion. He assumed that this "excluded volume effect" would cause the mean-square end-to-end displacement of chains to increase with a higher than the first power of the chain length but he estimated this exponent in an entirely arbitrary way. When Mark[370a] proposed exponents for polymers in good and poor solvents he seems to have selected them so as to fit experimental data rather than basing them on theoretical considerations.

The dependence of the end-to-end displacement of a paraffinic chain on the probability distribution of the internal angles of rotation was derived first by Oka[850] and later independently by Benoit,[851] Kuhn,[852] and Taylor,[853] who found that the mean-square end-to-end displacement of long flexible chains should be proportional to $(1 + <\cos \phi>)/(1 - <\cos \phi>)$, where ϕ is the internal angle of rotation taking the trans conformation as $\phi = 0$. Taylor[853b] was careful to stress that this result presupposes the absence of correlations between internal rotations and he pointed out that this condition does not apply to a paraffin hydrocarbon chain in which two successive gauche bonds with an opposite sense of rotation place two methylenes, as first pointed out by Pitzer,[854] prohibitively close to each other.

In 1951 Volkenstein suggested that the theoretical treatment of the flexibility of polymer chains is greatly simplified by approximating the continuous probability distribution of internal angles of rotation by a set of rotational isomers corresponding to the potential energy minima.[855] This approach would also allow a quantitative interpretation of the temperature coefficient of chain flexibility as the result of the increasing population of rotational isomers characterized by high energy as the temperature is raised. The work of Volkenstein and his collaborators remained, however, largely unknown to western scientists before an international symposium held in Prague in 1957. Although

Volkenstein presented there a brief summary of results obtained by the Leningrad school[856] to explain among other subjects how correlation between rotational isomers could be handled, the full impact of his work had to await the English translation of his 1959 monograph *Configurational Statistics of Polymeric Chains.*[857] In Leningrad, Volkenstein's method was used by Ptitsyn and Sharonov[858] to estimate the flexibility of various vinyl polymers using values derived from crystal structure data for bond angles and the geometry of the lowest energy rotational isomers. These were frequently unreliable; for instance, for polyisobutene a $(CH_3)_2C-CH_2-C(CH_3)_2$ bond angle of 114° was assumed, whereas later studies showed that this angle is expanded to a much higher value.*

A quantitative theory of the expansion of flexible chains due to the spatial interference of chain segments far from each other along the contour of the chain was developed by Flory[859] in 1949. Considering the swelling of an isolated molecular coil as leading to an equilibrium between the osmotic forces favoring the dilution of polymer segments and the rubberlike elasticity opposing coil expansion to less probable conformations he concluded that linear dimensions of the coil would be increased by an "expansion coefficient" α related to the molecular weight M of the polymer and the Flory-Huggins interaction parameter χ (see p. 156) by $\alpha^5 - \alpha^3 = C(\frac{1}{2} - \chi)M^{\frac{1}{2}}$, where the constant C was inversely proportional to the molar volume of the solvent. It was shown in later years that this result represented an excellent approximation to the behavior of polymer chains, although it was derived on the basis of a model in which the osmotic forces were treated as operating on continuous, spherically symmetrical, chain-segment densities. The functional dependence of α on the variables of the system had several important consequences. It predicted that in good solvents α would increase without limit with increasing M, becoming proportional to $M^{0.1}$ in the asymptotic limit of polymers with high molecular weight. It showed also that in bulk polymers in which the "solvent" consisted also of polymer molecules the excluded volume effect should disappear. Flory wrote that this deduction could be understood considering that "while an expansion of the molecule decreases the interference with itself, interference with surrounding polymer molecules is correspondingly increased." While it was difficult to see how this conclusion of Flory's could be confirmed experimentally at the time it was formulated, it was fully substantiated in recent years with the use of neutron scattering from melts containing polymers and their deuterated analogs.†

---

*On the basis of a suitable model compound and potential energy calculations Allegra Benedetti and Pedone [*Macromolecules,* **3**, 727 (1970)] reinterpreted the X-ray diffraction data on polyisobutene with the $(CH_3)_2C-CH_2-C(CH_3)_2$ bond angle 124°. The proposed potential energy minima of the internal angles of rotation were also quite different from those assumed by Ptitsyn and Sharonov.

†Although it is easy to accept Flory's argument for the disappearance of the excluded volume effect in polymer melts, the assumption that the flexibility of a polymer chain is the same when it is surrounded by small molecules and by other chain molecules is intuitively not easy to visualize. As

In a companion paper published during the same year Fox and Flory[860] considered the effect of temperature on the expansion coefficient α. Because the definition of the interaction parameter χ implied that it should be linear in the reciprocal temperature, they concluded that a temperature θ should exist at which $\alpha^5 - \alpha^3 = 0$; that is, the chains would assume "unperturbed dimensions" with their mean-square end-to-end displacements proportional to the chain lengths. This conclusion had two most important consequences. It led to the abandonment of athermal solutions as a state of reference in theories of polymer solutions; from now on the reference state was generally accepted to be represented by polymer solutions in "θ-solvents" (i.e., media in which polymer segments interacted more weakly with solvent molecules than with one another so that the effective mutual attraction of the segments compensated for their physically excluded volume). It also made it possible for the first time to interpret experimental data in terms of the intrinsic flexibility of polymer chains characterized by h_o^2/nb^2, where $\langle h_o^2 \rangle$ is the mean-square end-to-end displacement in a θ-solvent and n is the number of bonds of length b in the chain. The expansion coefficient could then be obtained from $\langle h^2 \rangle / \langle h_o^2 \rangle = \alpha^2$, where $\langle h^2 \rangle$ is the mean-square end-to-end displacement in a medium which interacts with the polymers more favorably than the θ-solvent.

Attempts to treat the effect of excluded volume on the expansion of a flexible-coil molecule in a manner that would take explicit account of the connectivity of the chain segments took two routes. By theoretical calculations the expansion coefficient could be derived by this method only for solvents close to the θ-temperature, where α^2 was found to be[861,862] linear in $(\frac{1}{2} - \chi) M^{\frac{1}{2}}$ in agreement with Flory's result. The development of electronic computers at that time also made it attractive to explore the problem by a computer simulation. This approach was pioneered by Wall et al.,[863] who generated chains on a tetrahedral lattice and recorded the distribution of end-to-end distances in nonintersecting chains. Unfortunately the fraction of these chains declined by 4% on each segment addition and for chains of 400 segments only one in 10^7 could be retained. Later this problem was partly overcome[864] by joining "successful" chains and this led to a mean-square end-to-end distance of nonintersecting chains proportional to the 1.18 power of the number of chain segments in remarkable agreement with the asymptotic behavior of Flory's formula.

For relatively stiff rodlike molecules the random flight model of molecular coils is obviously not applicable unless the rod is long in comparison with its mean radius of curvature. The dimensions to be expected with these "wormlike chains" were deduced by Kratky and Porod[865] in 1949. Their model consisted of chain segments joined at an angle that deviated only a little from 180°, so that for a small number of segments the directions of the terminal

---

we have seen (p. 107), one of Staudinger's arguments against the random-coil concept was the difficulty of reconciling it with the efficiency with which chain molecules in polymer melts fill the available space.

segments were correlated. The mean cosine of the angle between terminal segments decays then with the contour length L of the chain as $<\cos \gamma> = \exp(-L/a)$, where a is a "persistence length." If $L >> a$ the model behaves as a random coil in which a is half the length of Kuhn's "statistical chain element." Strangely enough, Kratky and Porod derived their expression in connection with a study of poly(vinyl bromide), which certainly does not fit the concept of a "wormlike chain." It may also be noted that the behavior of slightly flexible rodlike chains was treated previously by Bresler and Frenkel[865a] in terms of the bending modulus of the rod. The concept of "wormlike chains" later became important in the description of helical polypeptides and polynucleotides.

Colligative Properties

The use of osmometry was greatly advanced by the development of improved osmometer designs by Fuoss and Mead[866] in 1942 and by Zimm and Myerson[867] in 1946. Fuoss and Mead reported the measurement of a molecular weight as high as 5×10^5 and wrote that their value was in good agreement with that obtained in an ultracentrifuge. Surveys conducted in 1953 and 1955[868] showed, however, that in spite of the experimental advances agreement between different laboratories, to which the same standard polymer samples had been issued, was disappointingly poor. The difficulty was believed to rest with the variation in the amounts of low molecular weight polymer fractions that passed through the osmotic membranes used. Although Staverman[869] showed that molecular weights can be obtained from osmotic measurements with "leaky" membranes if the permeability of the membrane to the solute is known it is questionable whether his result can be applied to a polydisperse sample.

At the same time experimental advances extended the range of molecular weights which could be characterized by cryoscopy and ebulliometry to cover the region for which no satisfactory osmotic membranes were available. In 1951 I measured a number-average molecular weight of 9200 for polyethylene by boiling-point elevation,[870] and the use of thermistors soon extended the range accessible to ebulliometry[870,871] and cryoscopy[872,873] to 30,000–40,000.

At this time the nonideality of polymer solutions in the range of osmotic pressure measurements was also subjected to closer scrutiny. According to the theory of Huggins and Flory (p. 156), plots of the ratio of osmotic pressure and polymer concentration Π/c against c should have an initial slope (a second virial coefficient A_2) independent of the chain length of the polymer. As Flory[874] pointed out in 1945, however, this theory was based on the assumption that chain-segment densities were uniform throughout the solution, a concept clearly untenable at dilutions at which there was little interpenetration between the coiled macromolecular chains. He therefore proposed a model that contained spherical regions with a chain-segment density that corresponded to the average density within the swollen polymer coil and evaluated the excluded volume that would correspond to such a cloud of disconnected segments. Five years later Flory and Krigbaum[875] improved the model by using the results of the Flory-Huggins theory for the evaluation of the free energy required for the overlap of two of the segment clouds that represent polymer coils. The result showed that the function characterizing the ability of polymer coils to penetrate one another in dilute solution depends on the same combination of variables that controls the expansion of polymer coils beyond their unperturbed dimensions. As Flory pointed out, this is hardly surprising because the effect of excluded volume should be the same, whether it concerns the interaction of segments belonging to the same chain or to different chains. In particular, in a θ-solvent in which the excluded volume disappears, not only will the chains assume their unperturbed dimensions but also the second virial coefficient will disappear.

The Flory-Krigbaum theory predicted a much smaller value of A_2 than the Flory-Huggins treatment and this was in agreement with numerous experimental results. It predicted also that A_2 should decay slowly with an increasing molecular weight of the polymer. According to theory, strongly solvated long polymer chains were expected to have A_2 proportional to $M^{-0.2}$, but because osmotic measurements could not be performed with precision over much more than a tenfold range of molecular weights it was difficult to demonstrate this effect. Krigbaum and Flory[876] reported data in reasonable agreement with theory, but in numerous other osmotic studies no significant change in A_2 with changing molecular weight was observed.

Light Scattering

Raman[375c] showed as early as 1927 that "the Tyndall effect is related to the osmotic pressure," but his derivation of this relationship remained unknown to scientists concerned with the study of polymers. Only after the second world war, during which Debye acted as a consultant to the U.S. Government in the synthetic rubber program,* did his development of a theory of light scattering from polymer solutions become one of the most powerful tools of the polymer chemist. No one could have been better prepared than Debye to realize the full potential of this technique. As early as 1915 he had considered the scattering of X-rays by the electrons of randomly distributed atoms.[877] He showed how the atomic scattering intensity, proportional to the square of the number of the atomic electrons at small scattering angles, decays with increas-

---

*P. Debye, who had been since 1934 director of the Kaiser Wilhelm Institut in Berlin, was asked soon after the outbreak of the war in 1939 to relinquish his Dutch citizenship. This he refused to do and left instead for the United States, where in 1940 Cornell University appointed him head of the Chemistry Department.

ing angle because of destructive interference so that the angular dependence of the scattering intensity characterizes the spatial distribution of the electrons. Later[878] he used the same principle to interpret X-ray scattering from vapors in terms of the structure of small molecules, such as carbon tetrachloride and the two 1,2-dichloroethylene isomers.

The scattering from solutions involves an additional complication in that destructive interference is now dominated by intermolecular interactions. Debye[879] showed that this problem can be solved by applying Einstein's theory[880] of light scattering due to fluctuations of density in a one-component system and fluctuations of both density and concentration in systems of two components. Debye's papers demonstrated that the light scattering method may be used not only for molecular weight determination but also for the characterization of solute-solvent interaction by the second virial coefficient. He also pointed out that the angular dependence of the light-scattering intensity may be used to obtain the radius of gyration of the solute molecule and indicated the relation between this quantity and the length of randomly coiled chain molecules.

The first attempt at an experimental verification of Debye's theory was published in 1946 by Doty et al.,[881] who compared the molecular weight of polystyrene fractions derived from Debye's relation with values obtained by osmometry. When the two did not match they wrote that "because of present experimental uncertainties we attach no significance to the difference between the two methods of measurement." They were soon apprised by Debye that the difference was indeed significant because light scattering would yield the weight-average molecular weight. Zimm and Doty[882] confirmed this experimentally by measurements of two polystyrene fractions and their mixture.

The first use of the increase in the destructive interference of the scattered light with an increasing scattering angle to derive the dimensions of a dissolved polymer was reported in 1946 by Stein and Doty.[883] They estimated the mean-square radius of gyration of cellulose acetate fractions from the ratio of the light intensities scattered at angles θ and $\pi-\theta$ and concluded that the chains "are rather extended up to a molecular weight of about 80,000." Two years later an important paper published by Zimm[884] described a greatly improved apparatus for the measurement of light scattering and introduced the double extrapolation method to zero concentration and zero-scattering angle (the "Zimm plot") to eliminate both effects due to solution nonideality and to destructive interference of the scattered light. This provided a more reliable general method for deducing the mean-square radius of gyration of a macromolecule from the angular dependence of the light-scattering intensity and justified the statement that

> although the average spatial extention of the molecules of a polymeric material affects most of its properties, light scattering from dilute solutions seems to be the only way this dimension can be unambiguously determined.*

*The restriction to dilute solutions has been removed in recent years by the introduction of neutron-scattering techniques.

Zimm demonstrated that the expansion of polystyrene chains increases with the solvent power of the medium and found that their dimensions in all solvents exceed by a considerable factor the dimensions which would be expected for a random distribution of internal angles of rotation.

In the years that followed, the great versatility of the light-scattering method was demonstrated, theoretically and experimentally, in a variety of ways. As early as 1946 Ewart et al.[885] noted that the interpretation of light scattering from polystyrene solutions in solvent-precipitant mixtures (benzene-methanol) in terms of the molecular weight of the polymer have to take into account the increased concentration of benzene in the polymer domain. Kirkwood and Goldberg[886] and Stockmayer[887] developed rigorous expressions to show how data from polymer solutions in mixed solvents can be interpreted in terms of the preferential interaction of the polymer with one of the cosolvents. Similarly, light scattering from a ternary system that contained two polymers and a solvent was predicted and shown to yield information about the polymer-polymer interaction.[887–889] Light scattering was also demonstrated as a unique tool for the characterization of the *chemical heterogeneity* of a polymer sample. Whereas a chemically homogeneous polymer leads to no additional light scattering when dissolved in a medium that matches its refractive index, this is obviously not the case for a polymer that contains fractions of varying refractivity. Bushuk and Benoit[890] showed how the dependence of the light-scattering intensity on the refractive index of the solvent can be interpreted in terms of distribution of refractivities and therefore the composition of a polymer sample.

Light scattering also provided a striking method for the study of block copolymers (Benoit and Wippler[891]). In a triblock ABA chain molecule a solvent may be chosen to match the central B-block or the terminal A-blocks. These blocks then become "invisible" to light scattering. If the B-block is matched the light scattering reflects the behavior of two A-blocks connected by a "phantom" chain; if the A-blocks are matched light scattering yields information that reflects their effect on the expansion of the central block.

Apart from these specialized applications, the development of the light-scattering technique had a profound influence on polymer science in that it extended molecular weight measurements well beyond the range accessible to osmometry; molecular weight and second virial coefficient data also became much more reproducible between different laboratories.[868] The ratio of the osmotic number-average and the light-scattering weight-average molecular weight came to be used as a rough measure of polymer polydispersity. The possibility of comparing chain extension and molecular weight reliably became important in the formulation of a theory of intrinsic viscosity and in the discovery of helix-coil transitions. On the other hand, because the technique available at the time did not allow light scattering to be measured at angles smaller than 25°, extrapolation to zero scattering angle was unreliable for solutions of extremely large molecules. This limitation in the available light-scattering technique led Rice et al.[892] to conclude, incorrectly, that the molecular weight of DNA does not change during thermal denaturation; that is, that the strands of the double helix do not separate.

Intrinsic Viscosity

Ever since Staudinger suggested the use of solution viscosity for the characterization of the molecular weight of polymers this method, experimentally so simple, enjoyed great popularity, even among those who did not accept the reasoning on which Staudinger's "viscosity law" was based. As we have seen (p. 109), W. Kuhn[370] tried in 1934 to formulate a theory of intrinsic viscosity $[\eta]$, based on the random-coil model of chain molecules assuming that the solvent is immobilized within the coil, which therefore behaves like a sphere with a radius proportional to the end-to-end displacement of the chain. Because this would have led to $[\eta]$ proportional to $M^{1/2}$ if random flight statistics determined chain extension, Kuhn introduced an excluded volume effect due to the thickness of the chain, which he believed would raise the exponent a in $[\eta] = KM^a$ to values observed experimentally. A few years later Huggins[893] considered the problem, but assuming no hydrodynamic interaction between the atoms of a randomly coiled chain molecule concluded that $[\eta]$ should be proportional to M and commented that "long randomly kinked molecules should obey Staudinger's 'law.' " Huggins conceded that the neglect of the hydrodynamic interactions was unrealistic but thought, incorrectly, that "this may be corrected by using for the *effective radius* of an atomic group a value smaller than its *true radius*."

Starting in 1943, W. Kuhn and his gifted student, H. Kuhn, embarked on an extensive study of the behavior of flexible-chain molecules in a shear gradient. In their first publication[894] they cited evidence that short chains are free draining, with the frictional resistance to translation proportional to the chain length, whereas long chains behave like molecular coils with the solvent immobilized in its interior. It is strange that in this and later papers Kuhn never returned to the concept of the excluded volume effect, which he had been first to point out in 1934. As a result Kuhn and Kuhn concluded that although the free draining short chains should have $[\eta]$ proportional to M the impenetrable coils of long polymer chains would have $[\eta]$ proportional to $M^{1/2}$. In 1945 and 1946 Kuhn and Kuhn[895] published five papers that dealt with the behavior of flexible chains in a shear gradient, and although they considered at that time only the free draining case, which was later shown to have little relevance to real polymers, they introduced a number of important concepts. They pointed out that an extension of the end-to-end displacement of the coiled chain involves passages over energy barriers in the hindered rotations around bonds in the chain backbone and that the resistance to these transitions, characterized by an "internal viscosity" of the chain, will determine the hydrodynamic behavior of the chain in a shear gradient.* In considering hindered rotations

---

*Kuhn and Kuhn considered the "internal viscosity" of the chain to be independent of the sol-

around specific bonds in the chain they assumed, rather unrealistically, that a large portion of the chain would have to swing through the solution and that these rotations would rapidly slow down with an increasing distance of the bond from the chain end.* Yet they must have felt uneasy about this conclusion because they wrote a little later[†] that "simultaneous rotations (about two bonds) would make it possible for the middle part of the chain to rotate while portions of the chain to the left and right would remain stationary." This is the first formulation of a concept that became known later as a "crankshaftlike motion."

In a painstaking experimental study in which he used macroscopic models to simulate the hydrodynamic behavior of flexible chain molecules H. Kuhn[896] investigated the transition from free draining to impermeable coils. In his formulation $[\eta] = aZ_s/(1 + b\sqrt{Z_s})$, where Z_s was the number of statistical chain elements and a and b were coefficients, depending on the geometry of these elements. At the time when Kuhn carried out these experimental model studies in Switzerland Debye investigated the problem theoretically, using as his model for the molecular coil a sphere containing a uniform density of rigid beads.[897] Debye reasoned that the detailed distribution of the beads would have only a minor effect on the result, but Kirkwood and Riseman[898] felt that the solution of the hydrodynamic behavior of the "pearl necklace" chain model should be closer to physical reality. They showed that the ability of the solvent to flow through the molecular coil should decay with an increase in a dimensionless parameter $(f_o/\eta b)Z^{1/2}$, where f_o is the frictional coefficient of the beads, b, their spacing, Z their number in the chain, and η is the viscosity of the medium. For a sufficiently large value of this parameter the coil was supposed to behave like a sphere with a radius proportional to the end-to-end displacement of the chain. Because no account was taken of an excluded volume effect, $[\eta]$ was predicted, as by Kuhn, to be proportional to $M^{1/2}$ for sufficiently long chains. The steeper molecular weight dependence of intrinsic viscosity observed experimentally was believed by both Debye and Bueche and by Kirkwood and Riseman to be a consequence of a changing permeability of the molecular coil. Their neglect of the excluded volume effect and the consequent dependence of $[\eta]$ on the solvent power of the medium is surprising because Alfrey et al.[899] had shown in 1942 that the intrinsic viscosity decreases when a nonsolvent is added to a polymer solution and had quite clearly interpreted this observation as reflecting the dependence of the chain extension on the effective energetic interactions between chain segments.

After Flory formulated his theory of the manner in which the excluded volume effect modifies the expansion of a polymer chain[859,860] it was natural for him to explore the consequences of the excluded volume on the hydrody-

---

vent. Much later, Peterlin [895a] pointed out that conformational transitions will be resisted by the viscosity of the medium so that the internal viscosity should be the sum of a term depending on the energy barrier to conformational transition and a term proportional to the viscosity of the solvent.
*Helv. Chim. Acta, **29**, 619, 850–851 (1946).
[†]Helv. Chim. Acta, **29**, 857 (1946).

namic properties of flexible polymer chains. In a paper with Fox[900] he showed that with the use of reasonable values for the frictional coefficients of the chain elements the limit of coil impermeability should be attained with flexible polymers at moderate molecular weights. Thus the molecular weight dependence of [η] should reflect the expansion coefficient due to the excluded volume effect. One important consequence of this treatment was the prediction that [η] should be proportional to $M^{1/2}$ in θ-solvents and this was soon substantiated experimentally.[901] The use of intrinsic viscosities in θ-solvents became then a convenient tool for the estimation of "unperturbed dimensions" of polymer chains. It is interesting to note that Alfrey et al.[899] had written, without a full understanding of the import of their observation, that

> the point at which precipitation begins . . . should represent a mean value for any shape-dependent property. That solvent composition which is critical from the standpoint of solubility should correspond to a certain intrinsic viscosity no matter what the solvent and what the nonsolvent.

Another prediction of the Flory-Fox theory, that the ration of [η] in good solvents and in θ-solvents should yield α^3, proved to be somewhat inaccurate since the hydrodynamic radius expands with an increasing excluded volume somewhat less rapidly than the radius of gyration.[902,903]

Solutions of Helical Polypeptides

When Elliott and Ambrose[816] discovered, on the basis of spectroscopic evidence, that synthetic polypeptides existed in inert solvents in a state similar to that of the α-form of crystalline proteins they concluded that the chain molecules assumed a "folded" conformation stabilized by intramolecular hydrogen bonds. This conclusion was reinforced by the ease with which the chains could be oriented in a shear gradient, suggesting a rigid rodlike structure. However, after the nature of the α-form of crystalline fibrous proteins was clarified by Pauling et al.[734] the question naturally arose whether a structure with the characteristics of Pauling's α-helix could be observed in polypeptide solutions. Studies of this problem were stimulated by speculations that concerned the role of the α-helix in globular proteins, so that it was "of some importance to determine if this configuration is preserved in solution where the stabilizing forces of the crystalline state are no longer present."[904] This question could be tested experimentally after the development of techniques allowing the synthesis of high molecular weight polypeptides from α-amino acid N-carboxyanhydrides.[717] In fact, Doty et al.[905] found in 1954 that poly(γ-benzyl-L-glutamate) exhibited a molecular weight dependence of intrinsic viscosity which suggested random coils when the solvent was the strongly hydrogen bonding dichloroacetic acid, whereas in solvents such as chloroform or dioxane this dependence corresponded to that of rigid rodlike particles. A later paper from Doty's laboratory[906] established by light scattering that in solvents that favor the formation of rodlike polypeptides the radii of gyration correspond closely to those to be expected from Pauling's α-helix.* This discovery proved important not only because of the insight it eventually provided into the principles governing the formation of α-helical sequences in the structure of globular proteins but also because it made it possible for polymer chemists to investigate properties of solutions containing particles that could be described by the "wormlike chain" model with a high "persistence length" (see p. 124–125).

The helix-coil transition could be brought about by addition of a helix breaking cosolvent, by a change of temperature,[907] or, in the case of poly (L-glutamic acid), by the ionization of the carboxyl groups.[904] This transition could be observed most conveniently by a striking change in the optical activity, a phenomenon that excited particular interest because of a similar change observed during the denaturation of proteins. This, according to Cohen,[908] was the result of the elimination "of effects of specific main chain configurations," particularly "if modern concepts of helical chain configurations prevail." It was then attractive to speculate whether a synthetic polypeptide, which could undergo a complete helix-coil transition, could be used as a point of reference to estimate the "helix content" of globular proteins. This approach seemed to become possible when Moffitt[909] showed that the wavelength dependence of the optical activity ("optical rotatory dispersion," ORD) of asymmetric groups arranged in a helical array should differ in a characteristic manner from the ORD corresponding to a random distribution of these groups. Moffitt's analysis showed also that the ORD behavior of poly (benzyl L-glutamate) corresponds to right-handed helices.† This was important since Pauling had not been able to decide which sense of polypeptide α-helices composed of L-amino acid residues would be more stable.

A theory of helix-coil transition was first attempted by Schellman,[910] who showed that the helical form should increase in stability with increasing chain length because the formation of the first turn of the helix, a relatively improbable event, would be followed by a much more probable extension of the helix. Experimental results agreed with this prediction.[907] Later, Zimm and Bragg[911] formulated a more complete statistical-mechanical theory in which they took account of the possible alternation of helical and disordered sequences and this allowed an interpretation of the sharpness of the helix-coil transition in terms of the ease with which the helix could be "nucleated." They pointed out that this phenomenon has an importance transcending polymer science because "we have here the novelty of a rather sharp transition in a one-dimensional system."

---

*It is little known that the first demonstration that helical polymer chains can exist in solution is due to Rundle and Baldwin.[906a] They studied the amylose-iodine complex, which was known to consist of a helical polysaccharide with the iodine stacked up in the cavity of the helix. The flow dichroism observed for a solution of this complex showed that this structure remains intact on dissolution.

†This conclusion agreed with that based on the crystal structure of hemoglobin (p.   ).

While the equilibrium in helix-coil transition of various polypeptides was attained with extreme rapidity, a slow change in optical activity, which extended under some conditions over many hours, was discovered in Katchalski's laboratory[912] with solutions of poly(L-proline). This phenomenon was eventually shown[913,914] to be due to a transition between a left-handed and a right-handed polypeptide helix. The two forms differed very little in energy, so that one or the other represented the stable state, depending on the solvent. The slowness of the transition was shown to be due to the high energy barrier for a *cis-trans* isomerization around the amide bond; acids, which reduced its partial double bond character, therefore acted as catalysts for the mutarotation.

Polypeptides were the first rodlike polymers for which the formation of a liquid crystalline phase was recorded. Elliott and Ambrose[915] observed the formation of spherulites in solutions at concentrations above 15% as early as 1950, that is, before the structure of the α-helix had been proposed. Later, Robinson and his collaborators[916] subjected this liquid crystal formation to a detailed study. They found that an anisotropic phase was in equilibrium with an only slightly more dilute isotropic phase, with the phase transition moving to lower solution concentrations as the chain length of the polypeptide was increased. The anisotropic phase, although more concentrated, was much less viscous, and this suggested to Robinson "that in the birefringent solution layers are formed which can glide past one another with relative ease." The spherulites in the birefringent phase had periodicities of the order of 10^{-3} cm and enormous optical activities of several hundred degrees per mm. These findings were interpreted as suggesting a structure in which the orientation of the rodlike particles was rotated by a small angle in successive layers giving rise to a "superhelix."

The formation of an anisotropic phase in solutions of rodlike particles had first been observed with tobacco mosaic virus.[280] In considering this phenomenon, Langmuir[917] wrote that "the rodshaped molecules are presumably held in position in the hexagonal lattice by repulsive forces resulting from negative charges on the protein molecules." In an elaborate analysis of the phenomenon Onsager[918] also considered "results which may arise from electrostatic repulsion of highly anisotropic particles." Isihara[918a] confirmed Onsager's predictions of solutions of rodshaped particles on the basis of their large excluded volume without mentioning an electrical charge, but he suggested that his conclusions agreed with Langmuir's ideas. It was only when Flory[919] formulated a theory of such solutions on the basis of a lattice model that he stated clearly that the formation of the anisotropic liquid phase is expected even when the phase separation is athermal. He predicted that for an axial ratio x of the solute particles the concentrations of the isotropic and anisotropic phases in equilibrium would be $8/x$ and $12.5/x$. Flory was unaware of Robinson's work on anisotropic polypeptide solutions, published during the same year, or the earlier observations of Elliott and Ambrose. Robinson's second paper provided data on equilibria between the isotropic and anisotropic phase in remarkable agreement with Flory's prediction without any indication that he was aware of that theoretical development.

Polyelectrolytes

Kern's publications in 1938 and 1939 (pp. 158–159) contained a remarkable phenomenological survey of the great variety of effects observed with chain molecules that carry a large number of ionizable groups. Ten years later Fuoss, who for many years had been investigating the properties of electrolyte solutions, suggested the term "polyelectrolyte" for such highly charged polymers.[920] Considering the motivations for the study of such substances, he wrote that although "we may never learn the detailed sequence of amino acids in a protein molecule,* we may synthesize a molecule which will be its electrical analogue." Fuoss's main interest, however, was an extension of the physical chemistry of electrolytes. He pointed out that an interpretation of the behavior of poly(acrylic acid) and its salts was complicated by the weak acidity of carboxyl groups and he started, therefore, a program of study of "strong polyelectrolytes" obtained by the quaternization of poly(vinyl pyridine).[921] Dilute solutions of these substances will have "positive charges in clusters separated by solvent which contains only negative ions,"[920] and Fuoss pointed out that this unusual charge distribution should lead to characteristic effects on ionic conductivity, on the shape of the polymer chain and its dependence on concentration, on the ionic activity, and on osmotic pressure.

At the same time that Fuoss outlined his view of the scope of polyelectrolyte research, Kuhn et al.[922] formulated a detailed theory of the expansion of flexible chains that carry a large number of evenly spaced ionizable groups. This, it was argued, would distort the probability distribution of end-to-end distances derived earlier by Kuhn[370] because the *a priori* probability of each state of the chain would have to be weighted by a factor involving its electrostatic energy. As a result the chain would be expanded with increasing charge density.

The calculation of the crucial quantity, the electrostatic energy of the chain as a function of the chain length and the spacing of the charges, was carried out by Kuhn et al. with two assumptions, which were shown later to be unfounded: It was assumed implicitly that the distribution of shapes which an uncharged chain would assume for any given end-to-end displacement would also hold for the charged chain. In addition, it was stated explicitly that in a highly dilute system for which experimental data could be obtained, the counterions would be far removed from the polyion, so that their shielding of the repulsion between the charges of the polyion could be neglected. As a result of these assumptions the theory of Kuhn et al. led to a large overestimate of polymer

---

*This achievement was actually realized by Sanger et al.[727,728] only five years later.

expansion and to the prediction that long chains would approach full expansion, even at relatively low charge densities.

The dramatic change in the shape of polycarboxylic acid molecules when their degree of ionization is altered, as reflected in the large changes in the viscosity of their solutions, was found independently by Kuhn and Katchalsky to lead to corresponding changes in the swelling of such polymers when they were crosslinked.[923] Fibers made from these materials would extend and contract when exposed to alkaline and acid solutions, and it was shown by Kuhn and Hargitay[924] that the work that could be derived in this way per unit weight of the polymer was comparable to that performed by muscle. Because it was obvious that muscular contraction could not be due to rapid changes in the acidity of the living tissue, Kuhn and Hargitay wrote that

> it is possible that the stabilization of the extended state of resting muscle is not caused by the transition from an undissociated to a dissociated acid but that differences between the electrochemical behavior of potassium and sodium are utilized.

Kuhn never gave up his conviction that muscular contraction reflects a corresponding contraction of flexible biological macromolecules, even when studies of muscular contraction by electron microscopy rendered this concept untenable.* Katchalsky's approach was more cautious. Rather than stressing the analogy to biological systems, he tried to use the behavior of crosslinked polymeric acids to demonstrate the principles of the direct conversion of chemical into mechanical energy and to find the efficiency which can be attained in such "mechanochemical" processes.[925]

Kern had drawn attention to the fact that the apparent acidity of the carboxyl groups of poly(acrylic acid) decreases sharply with an increasing charge density of the polyion and this phenomenon promised to furnish a convenient approach to the study of its electrostatic potential. A theory of ionization equilibria of "large multivalent ions of colloidal solutes" was first formulated by Hartley and Roe[926] for spherical anionic micelles to which an indicator was adsorbed. In this case the "electrostatic free energy of ionization," ΔF_{el}^{ion}, which had to be added to the free energy of ionization of the free indicator, could be obtained from an extension of the Debye-Hückel theory, and a similar approach was then generally taken for the interpretation of the titration curves of globular proteins, which could be treated in first approximation as spheres with a uniformly distributed surface charge. The estimation of ΔF_{el}^{ion} for an ionizable flexible polymer chain obviously presented much greater uncertainties. Nevertheless, Katchalsky and Gillis[927] tried to derive a theory for the titration of a polycarboxylic acid based on the model of Kuhn et al. and concluded that the pK should be linear in the cubic root of the degree of dissociation. Somewhat earlier Overbeek[928] suggested a

---

*H. Huxley and J. Hanson, *Nature*, **173**, 973 (1954); H. E. Huxley, *J. Mol. Biol.*, **7**, 281 (1963).

solution based on a model in which the polyion was represented by a spherical charge cloud. His result had the serious deficiency that it implied a dependence of the pK on the chain length at constant charge density, although Kern[545] had already pointed out that titration curves of poly(acrylic acid) are independent of the degree of polymerization. It is particularly ironic that all the investigators of the early postwar period compared the results of their theories with the behavior of poly(methacrylic acid) (since the monomer was easily available[929]), not realizing that its behavior was particularly complicated. Only in 1957 did Arnold[930] point out that while the pK of poly(acrylic acid) is rising smoothly with increasing ionization, as expected on purely electrostatic grounds, the titration curve of poly(methacrylic acid) shows a transition at a critical charge density. A similar difference had been noted by Ferry et al.[931] as early as 1951 in the titration behavior of alternating maleic acid copolymers with vinyl methyl ether and styrene, and this had been interpreted as a result of the cohesion of the hydrophobic phenyl residues that resist an expansion of the styrene copolymer chain until a critical charge density is attained, but little notice seems to have been taken of that work. A similar titration behavior, due this time to a helix-coil transition, was later described by Wada[932] for poly(glutamic acid).

If a reasonable theory of the titration of a polycarboxylic acid was to be formulated the extent to which the field due to the ionic charges attached to the polymer chain was shielded by counterions was clearly of crucial importance. While I was working under Alfrey's direction on my doctoral dissertation we decided late in 1949 to try to derive the counterion distribution around a rodshaped polyion since a full extension of the chain would provide a lower limit to the attainable electrostatic potential. With the help of P. W. Berg we found[933] to our surprise that in a system that contained only polyions and their counterions, but no simple electrolyte, the distribution of the potential could be obtained analytically without use of the Debye-Hückel approximation that the electrical energy of the ions is small compared with their kinetic energy.* The corresponding distribution of counterions revealed that at the charge density of half-ionized poly(acrylic acid) a sizable fraction of the counterions would remain close to the polyion even at extreme dilution. Unknown to us, Lifson, working with Katchalsky on his thesis, had tackled the same question, stimulated by the discovery that a solution was known for a mathematically analogous problem, the distribution of electrons evaporated from a heated wire.[934] Their result,[935] equivalent but more elaborate than ours, disposed of Kuhn's assumption that it is possible to carry out experiments on polyelectrolytes under conditions in which the effect of counterions is negligible.

The necessity to use unrealistic assumptions in theories of polyion expansion in which the chain character of the polyion was taken explicitly into account made it appear more attractive to use models in which the polyion was repre-

---

*The Poisson-Boltzmann equation can be solved analytically for cylindrical symmetry but not for spherical symmetry.

sented by a continuous charge cloud. This approach was pioneered by Hermans and Overbeek[936] but their treatment was restricted to small electrostatic potentials; that is, to low charge densities or relatively high concentrations of salts added to the polyelectrolyte solution. A few years later it was shown by Kimball et al.[937] that in cases of greatest interest the excess of counterions over byions* in the volume occupied by the polyion compensates all but a negligible fraction of the ionic charge of the chain molecule. Flory[938] pointed out that under these conditions the driving force toward expansion of the coiled polyion may be treated more easily as resulting from osmotic effects of the excess concentration of mobile ions in the polyion domain than as a consequence of a tendency to reduce the electrostatic energy. With this approach the expansion coefficient α could be derived by the same procedure used for uncharged polymers, leading to the conclusion that $\alpha^5 - \alpha^3$ should be linear in $M^{1/2} i^2 / c_s$, where M is the molecular weight, i, the density of charges along the chain, and c_s, the concentration of an added uni-univalent salt.

In all the early work on polyion expansion solution viscosities were used to characterize the phenomenon. This proved difficult because the reduced viscosity, η/η_o, increased sharply as the system was being diluted, due to an expansion of the polyion, so that the data could not be extrapolated to zero concentration. Fuoss[939] suggested that the reciprocal reduced viscosity should be linear in the square root of the polyelectrolyte concentration to yield an extrapolated value characterizing the polyion dimensions. This approach was popular for some time but its significance proved to be illusory when Eisenberg and Pouyet[940] showed that at low shear rates dilution of a polyelectrolyte first leads to an increase in the reduced viscosity but then to a sharp drop. Eventually Pals and Hermans[941] showed that linear plots of the reduced viscosity against concentration, allowing an extrapolation to the intrinsic viscosity, were obtained if the polyelectrolyte was not diluted by water but by a salt concentration such that an "effective counterion concentration" remained constant. Using dilutions with salt solutions and interpreting the intrinsic viscosities in terms of the expansion coefficient α of the polyion over its unperturbed dimensions, Flory and Osterheld[942] found that α was lower than predicted by Flory's theory. They concluded that the neglect of the reduced ionic activity in the polyion domain must be the source of the discrepancy.

Studies by Pals and Hermans[943] of the osmotic pressure of polyelectrolyte solutions which contain added salt revealed second virial coefficients only about half as large as predicted from the uneven distribution of mobile ions when using the Donnan theory without consideration of activity coefficients. The determinations of dialysis equilibria between polyelectrolyte and salt solutions became a favorite method for the study of mean activity coefficients. However, it was realized that a polyion would exert a much stronger effect on counterions than on ions that carry a charge of the same sign. Electrochemical measurements confirmed that it was mostly the counterions whose activity was

---

* Ions with a charge of the same sign as that of the polyion.

strongly depressed by the presence of polyions.[944] This demonstration should have made it obvious that the concept of an "ionic strength," which predicts the same behavior for bi-univalent electrolytes no matter whether the cation or the anion is the bivalent species, is inapplicable to polyelectrolyte solutions. Yet theories continued to be formulated in terms of the ionic strength although this meant little in practice because experiments were mostly carried out only with uni-univalent salts.

A new insight into the nature of counterion inactivation resulted from the observation by Huizenga et al.[945] that in the electrophoresis of partly neutralized poly(acrylic acid) part of the radioactive sodium ions used as a tracer traveled with the polyion. This led to the concept that some of the counterions are bound to specific sites of the polyion rather than being partly inactivated by the relatively long-range effect of its electrical field. This site-binding could be understood only as a consequence of the close spacing of discrete charges and therefore it illustrated the limitation of models in which the polyion was represented by a continuous charge density smeared over the volume occupied by the molecular coil. Not surprisingly, such site binding exhibited considerable ion specificity.[946,947]

In 1955 Hermans and Fujita[948] developed a theory of the electrophoretic mobility of a flexible chain polyion and reached the surprising conclusion that under frequently encountered conditions (with a large ratio of the hydrodynamic shielding length to the Debye-Hückel thickness of the ion atmosphere) the mobility should be independent of chain length. Thus a polyion behaving like an impermeable coil in its effect on solution viscosity and its ultracentrifuge sedimentation was expected to behave like a free draining coil in electrophoresis. This prediction was fully confirmed by experiment.[949] On the other hand, under conditions in which a polyion is highly extended it would be expected to move more easily parallel to its extension than perpendicularly to it. The resulting anisotropy of conductance was actually demonstrated in polyelectrolyte solutions in which the polyions were oriented by shear.[950] Because the counterion atmosphere is highly polarizable, a polyion can also be oriented by a high-frequency alternating field. When the field is cut off the birefringence that results from the orientation decays, and the rate of this decay was used by Benoit[951] to obtain the rotational diffusion constant of tobacco mosaic virus. Orientation of a polyion in an alternating field also leads to an anisotropic conductance.[952]

I have already commented on the modification of polyion behavior, which may result from the hydrophobic forces between nonpolar chain substituents that resist an expansion of the polymer backbone, first demonstrated by Ferry et al.[931] An extreme case of this phenomenon was described by Strauss et al.,[952a] who showed that poly(vinyl pyridine) molecules quaternized with long-chain alkyl halides behave in solution like compact spherical particles, in spite of the high charge density of the polyion. The forces responsible for this shape are analogous to those that lead to the formation of soap micelles and polymers of this type have therefore been referred to as "polysoaps."

Characterization of Molecular Weight Distributions

Fractional precipitation of a polymer was reported as early as 1920 by Duclaux and Wollman,[953] who claimed to have been the first to observe that cellulose nitrate may be separated into fractions that differ greatly in their solution viscosities. In the spirit of the time they concluded that these fractions contain micelles of different size and assigned no significance to differences in their osmotic pressure because "as is known, osmotic pressure of colloidal solutions does not yield the true molecular weight." Later, when the nature of chain molecules was understood, the use of fractional precipitation to derive a molecular weight distribution was pioneered by Schulz[421] who compared the result of such an experiment with the distribution expected on theoretical grounds.

However, fractionations produced by the addition to a dilute polymer solution of increments of a nonsolvent (or by incremental changes of temperature), each time followed by the separation of the precipitate, were extremely time consuming and this soon led to efforts to develop a more rapid procedure. Desreux[954] devised a method in which a column was filled with an inert carrier coated with a thin polymer film. Fractions were separated by extraction using a solvent-nonsolvent mixture with a gradually changing composition or a solvent with a gradually rising temperature. A more elaborate technique was described by Baker and Williams,[955] who used columns with a temperature gradient so that the flow of a polymer solution led to repeated precipitation and dissolution. Both methods lent themselves to automation but they still required the characterization of the molecular weights of the polymer contained in the fractions.

Ever since the work of Signer and Gross[343] it was understood that ultracentrifuge sedimentation could, in principle, provide data for the molecular weight distribution of a polymer. Two effects were shown to lead to an apparent narrowing of the polydispersity; the superposition of diffusion on sedimentation and the concentration dependence of the sedimentation constants. Williams and his collaborators[956] showed, however, that both distortions could be taken care of by an extrapolation of the data to infinite dilution and infinite time to yield the distribution of sedimentation constants. These could then be translated to molecular weight distributions by a suitable calibration. This procedure was found to be the most powerful method for the characterization of polydispersity, being particularly valuable[957] in establishing the narrow chain-length distributions obtained by Szwarc's method of anionic polymerization (see p. 186).

Still, the great expense of an ultracentrifuge and the time required for the experiment and the subsequent calculations lent urgency to efforts to devise a simple technique by which polymer polydispersity could be characterized rapidly. In 1945 Morey and Tamblyn[958] suggested that following light scattering during the gradual precipitation of a polymer solution by the addition of a nonsolvent might provide "a means of 'weighing' the precipitate without the actual physical steps of filtration, washing, drying and weighing." If the nonsolvent concentration at which a polymer precipitated could be expressed as a function of its concentration and chain length the "turbidimetric titration curves" could be translated into molecular weight distributions. The method depended on optimistic assumptions; for example, that the size of the precipitated particles is independent of the polymer chain length and remains constant during the progress of the precipitation. Nevertheless it was able to resolve a mixture of three poly (methyl methacrylate) fractions[959] and enjoyed considerable popularity for some time.

An important new development was initiated when Porath and Flodin[960] introduced a technique they called "gel filtration." They filled a column with dextran, which had been crosslinked in the swollen state and showed that the passage of a solute through the column was slower for small molecules that could penetrate the gel particles than for large species which were restricted to the interstitial spaces. This suggested a powerful chromatographic method for the characterization of polymer polydisperisty and in 1960 Vaughn[961] tried to apply crosslinked polystyrene to the chromatography of linear polystyrenes. The attempt was not successful because only relatively short chains could be resolved. Success was eventually achieved after the realization that the polymer beads to be used for this application should be crosslinked in the presence of an inert diluent to obtain a structure with desirable characteristics. Voorn and Hermans[962] had shown that "a network is not uniquely defined by the number of crosslinks. One must distinguish between what may be called different topologies," and this hint led Lloyd and Alfrey[963] to the preparation of what was later called "macroporous resins" by heavy crosslinking in the presence of a diluent. These rigid structures were shown to sorb to an almost equal extent solvents with vastly different affinities for the polymer and became the basis for the development of gel permeation chromatography.[964]

25. Polymers in Bulk

During the period discussed in the preceding part of this book foundations were laid for an understanding of the relation between the molecular and bulk properties of polymers. Crystallography was applied to clarify the structure of some chain molecules and the "fringed micelle" model was proposed for semicrystalline polymeric materials. A first attempt was made to relate the crystal structure and spectral data of chain molecules to the tensile strength and the modulus of fibers composed of these molecules. The first studies of glass transition and the first formulations of theories of rubber elasticity also date to that period. Investigations along all these lines registered important advances during the last two decades covered in the present account.

The Glass Transition

The phenomenon of the transition of mobile fluids to rigid glasses was studied on inorganic systems and on organic molecules of low molecular weight well before the investigation of the analogous phenomena in high polymers. In 1922 Simon[965] discovered that the heat capacity of glycerol is almost doubled within a narrow temperature range which coincides with a transition from a brittle glass to a viscous liquid. Somewhat later Tamann and Kohlhaas[966] found that this transition is characterized by an increase in the thermal expansion coefficient. Simon[967] realized that a glass is in a nonequilibrium state and found that below the glass transition temperature* the heat capacity of the crystalline and amorphous states are nearly equal. According to measurements of various low molecular weight materials, glass transition occurred typically when the viscosity of a fluid reached a value of 10^{13} poises.[968]

The first study of glass transition in a polymer was carried out in 1936 by Ferry and Parks,[969] who used a polyisobutene sample with a number-average molecular weight of 4900. They found that the midpoint in the transition of the heat capacity and the thermal expansion coefficient (197°K) corresponded to a

---

*This was designated in German as *Einfriertemperatur*, that is, the temperature below which the configuration of the system is frozen.

viscosity of about 10^{13} poises, as observed for glass transitions of small molecules. They felt that "the steric hindrance of the long clumsy chain molecules and the variety of molecular species present undoubtedly account for the failure to crystallize".* Jenckel and Ueberreiter[970] measured the glass transition temperature (T_g) for polystyrenes of varying chain lengths and found that T_g approaches an asymptotic limit for long chains. This was interpreted by Ueberreiter[971] as indicating that only sections of the chain comprising 50–100 monomer residues have to be activated for flow. The glass transition of polystyrene was later studied in great detail by Fox and Flory,[972] who found that T_g, linear in the reciprocal of the chain length, was governed not by the viscosity but by the free volume of the polymer. Although T_g should, in principle, decrease as the cooling rate of the melt is reduced, Fox and Flory doubted that major changes could be observed within a reasonable range of experimental time.

In special cases the freezing of structures at the glass transition may be observed spectroscopically. This was first reported by Zhurkov and Levin,[973] who found that IR spectra of poly(vinyl alcohol) indicated a gradual dissociation of hydrogen bonds as the temperature was raised above T_g, but that the hydrogen-bonding equilibrium was frozen in the glassy state.

In early considerations of the phenomenon of glass transition it was assumed that it was due merely to the fact that the experimental time was insufficient for thermodynamic equilibrium to be attained and that this equilibrium should be characterized by an extrapolation of thermodynamic parameters from temperatures above T_g. This concept, however, led to some absurd conclusions, such as a lower entropy for the amorphous glass than a crystal of the same material at the same temperature.[973a] To avoid this difficulty Gibbs and DiMarzio[974] suggested that freezing the configurational entropy of systems composed of chain molecules should lead to a second-order transition, that is, a change in the temperature coefficients of extensive thermodynamic parameters, *even in the equilibrium state*. This would then correspond to the state toward which the nonequilibrium glass would gradually relax.

Polymer Crystallinity

In 1942 Bunn[975] published a paper in which he considered the probable conformations which a polymer molecule would assume in the crystalline state. Assuming staggered conformations of backbones comprised of carbon atoms and a regular repetition of internal angles of rotations, he specified repeat periods to be expected for some probable structures. One of these structures

---

*They seem to have been unaware of the evidence (see p. 78) that polymer polydispersity is no hindrance to crystallization. The crystallization of polyisobutene on stretching samples of much higher molecular weight than used by Ferry and Parks was reported only two years after their paper was published.

was generated by an alternation of trans (*t*) and gauche (*g*) bonds (all rotated in the same direction) which led to a helix containing three residues of a vinyl monomer per turn; this was characterized by a repeat of 6.2 Å. (Strangely, although Bunn wrote that this helical structure would be favored "when side substituents are present on the carbon chain" he never mentioned the need for stereoregularity.) When Natta and his collaborators, unaware of Bunn's analysis, reported repeat periods of 6.5Å and 6.7 Å for isotactic polypropylene and polystyrene respectively, Bunn and Howells[976] realized immediately that the chains of these polymers existed in the crystal as $(tg)_3$ threefold helices, with the repeat period extended over the predicted value due to an expansion of the bond angles.

Bunn and Holmes[977] considered the extent to which the efficiency of the packing of polymer chains in the crystal might affect their conformation. They concluded that the effect of these intermolecular interactions was likely to be minor, implying that the polymer would assume in the crystal a conformation close to that corresponding to the lowest potential energy of the isolated chain. Occasionally an exception was found to this generalization; thus isotactic poly(*p*-fluorostyrene) has a different conformation from that of isotactic polystyrene,[978] a finding that could be accounted for only by assuming that the difference in the packing of the two chains leads to a change in the preferred chain conformation. Still in the majority of cases the potential energy minimum of the isolated chain is a good indication of the conformation it assumes in the crystal and this can be predicted from maps in which the potential energy of nonbonded atom interactions is represented as a function of the two repeating internal angles of rotation of a stereoregular vinyl polymer chain. This method which was later used extensively, was applied first by Natta et al.[979] to isotactic and syndiotactic polypropylene. It may be noted that the potential energy map for the syndiotactic chain predicted that both the helical $(ttgg)_n$ and the linear all *trans* structures were possible; only the first was known at the time, the second was yet to be discovered.

Because even polymers that are able to crystallize generally contain an amorphous portion, various methods had to be developed to characterize the "degree of crystallinity." This was done on the assumption that a semicrystalline polymer consists of well defined crystalline and amorphous fractions. Thus the measurement of any property will yield the relative amounts of the two phases if this property for the crystalline and the amorphous state is known. This approach was used for polyethylene by using volume[980] and heat capacity[981] as the properties characterizing the transition from the crystalline to the amorphous state. Both methods revealed that melting occurs over an extended temperature range but that a sharp break presumably represents the melting of the most perfect crystallites. Hermans and Weidinger[982] estimated the degree of crystallinity of cellulose from the ratio of the integrated intensities of X-rays scattered in the sharp diffraction peaks and the amorphous halo. This method was greatly improved by Krimm and Tobolsky,[983] who used the scattering from a polyethylene melt as a basis of reference for estimating the amorphous content of semicrystalline samples. Their results were similar to those obtained by dilatometry. In the special case of cellulose, accessibility to isotopic hydrogen exchange was also used as a measure of the amorphous fraction.[822] In 1960 Ke[984] introduced differential thermal analysis, which later became the preferred method for the study of phase transitions.

The description of semicrystalline polymers as consisting of a crystalline and an amorphous fraction was, however, not the only way these materials could be viewed. Hosemann, in particular, was a champion of the interpretation of their X-ray diffraction pattern as reflecting distorted crystallites, and he defined "ideal paracrystals" in which the crystal lattice was subjected to statistical distortions.[984a] Hosemann and Bonart[984b] then evaluated the magnitudes of these distortions in a number of polymer fibers.

As early as 1931 Garner et al.[985] found that the enthalpy and entropy of fusion of paraffin hydrocarbons are linear in their chain length; it followed then that the melting point $T_m = \Delta H_m/\Delta S_m$ would approach a limit for long paraffin chains, estimated as 135°C. Later Powell et al.[986] presented arguments that such an approach to an asymptotic limit should be observed for melting points of all polymer homologous series. The fact that high-pressure polyethylene had melting points some 20°C lower than expected for a long paraffinic chain could be attributed to the presence of branches.[987] A phenomenon that required a theoretical interpretation was the breadth of the temperature range within which polymers were observed to melt. Frith and Tuckett[988] formulated a theory based on the "fringed micelle" model of semicrystalline polymers in which a chain molecule would pass alternately through crystalline and amorphous regions. As the crystals extended their length the enthalpy change would be proportional to the number of monomer residues added to the crystal, but the entropy change would decrease more rapidly due to constraints on the conformation of the amorphous portions of the chains connecting the crystallites; thus T_m would fall with increasing crystallization and the degree of crystallinity would be unable to exceed a limiting value.

In 1949 Flory[989] developed an elaborate statistical-mechanical theory of polymer crystallization that emphasized the temperature at which the last crystallite would disappear as "the melting point" of the polymer. He showed how this melting point would be depressed by an amount depending on the mole fraction of "extraneous ingredients," specifically diluents or comonomer residues, which could not be accommodated in the polymer crystal lattice. With less conviction Flory included chain ends among features that would be unable to enter the polymer crystallites, arguing that

> those which choose to occupy sites in the crystallites will be obliged to occur in pairs in end-to-end juxtaposition. . . . hence the chain ends will naturally gravitate to the greater freedom of the amorphous region.

Most important, Flory's theory predicted that the heat of melting of polymer residues in perfect crystallites could be derived from the dependence of the melting point on the concentration of a diluent, the concentration of a comonomer (distributed at random in the copolymer chain), or the number-average chain length. A comparison with the heat of melting of a semicrystalline sample would then yield a "degree of crystallinity."

The predictions of this theory were tested by Flory et al.,[990] who found that the melting-point depressions observed with different diluents yielded the same heat of melting for a crystalline polymer. Values for the heat of melting derived from the dependence of T_m on copolymer composition were somewhat lower but considered less reliable. Edgar and Hill[708b] found, in agreement with Flory's prediction, that in copolymers the melting point depended only on the mole fraction, not the kind, of the comonomer which is unable to enter the polymer crystal lattice. Thus poly(ethylene terephthalate) could be modified by the introduction of substantial *weight fractions* of a relatively high molecular weight poly(ethylene glycol) to increase its hydrophilicity substantially with little sacrifice of the high melting point. As Coleman[991] reported, the preparation of such a copolymer with a combination of desirable properties was suggested to him by his work on the silk fibroin structure. We may recall (see p. 72–73) that Brill's[243] work on the fibroin crystal structure led to the conclusion that at least two protein components must be present because some of the amino acid residues were much too bulky to be contained in the unit cell. Much effort was expended over the years in attempts to separate these components—all without success. It was now shown[992] that fibroin contained crystallizable sequences of amino acid residues separated by sequences which contained the bulky units and were therefore amorphous.

The paper by Edgar and Hill made two additional important points. First, although the tetramethylene portion of an adipate residue has a length close to that of paraphenylene in terephthalic acid, an adipate unit cannot enter the poly(ethylene terephthalate) lattice; yet poly(ethylene terephthalamide) and poly(ethylene adipamide) are isomorphous and copolymers containing both their residues form a single crystalline phase. This shows that small steric differences can be tolerated in a crystal lattice, provided they are counteracted by strong intermolecular interactions such as hydrogen bonding. Second, whereas the melting of a copolymer depends on the mole fraction of the minor component, the glass transition point depends on its weight fraction. Thus when the comonomer has a high molecular weight, T_g can be depressed by a large amount while retaining a melting point close to that of the homopolymer.

A different type of crystallinity was discovered as a result of the observation[993] that as the length of the alkyl side chain was being extended in a homologous series of polyacrylates the glass transition temperature first decreased but increased again after passing through a minimum. This suggested to Alfrey that long side chains might pack into paraffinlike crystallites, although the chain backbone cannot be included in a crystal lattice. Such "side-chain crystallization" was demonstrated[994] for polymers of long alkyl acrylates and methacrylates by X-ray diffraction and a dilatometric determination of melting points. Surprisingly, the melting point was little affected when a long alkyl methacrylate was copolymerized with moderate proportions of methyl methacrylate.

A discovery made almost simultaneously by Keller,[985] Till,[996] and Fischer[997] in 1957 produced a profound impact on the concept of polymer crystallinity. Their electron diffraction experiments showed that in polyethylene crystal platelets obtained from dilute solution the axes of the polymer chains are perpendicular to the large crystal faces. This was a most remarkable observation because the thickness of the crystal platelets (about 100 Å) was only a fraction of the length of the polymer molecule, inducing Keller and O'Connor[998] to write, "As the parallel alignment of the chains is practically perfect, we are forced to conclude that the molecules must sharply fold on themselves."* The vigorous discussion that followed the presentation of the Keller-O'Connor paper at the September 1957 Discussion of the Faraday Society may be exemplified by Bunn's[999] comment that however surprising chain folding may be in solution grown crystals, "if it occurs in crystals grown from a melt, this is still more surprising because we suppose the melt to be a tangle of long-chain molecules."† It was soon found that chain folding is not a phenomenon restricted to polyethylene but is typical of solution-grown polymer crystals. Rånby et al.[1001] obtained beautiful electron microscope pictures showing how such crystals grow by a screw dislocation mechanism. The observation of lamellae in polymers crystallized from the melt[997,1002] showed that principles similar to those operating in the crystallization from dilute solution may also be involved in the crystallization in bulk. The nature of chain folding, whether it involves a sharp bend of the chain with an "adjacent re-entry" into the crystal or whether the chain meanders for some distance on the crystal surface before reentering the ordered structure, became a subject of acrimonious debate only beyond the period covered in this book.

To appreciate the full impact of the discovery of solution-grown polymer crystals it should be realized that the formation of well formed crystals from polymer molecules was generally considered unlikely before the papers of Till, Keller, and Fischer appeared in print. Although Staudinger and Signer[1003] reported as early as 1929 that they had obtained macroscopic crystals of polyoxymethylene (50 μm long) by the addition of sulfuric acid to a concentration formaldehyde solution, this report seems to have been generally unknown. Thus Flory[1004] expressed a widely held belief when he wrote that "macroscopic single crystals seldom, if ever, are formed from high polymers (certain proteins excepted)." Once single crystals and chain folding were discovered the puzzling observation of long periodicities unrelated to the dimension of the unit cell of the lattice[1005,1006] found a ready explanation.

---

*E. W. Fischer has told me that his paper which suggested chain folding encountered such ridicule that he feared to lose his job. He felt secure only when news arrived that Keller had made similar observations.

†It is often pointed out that the pioneering electron microscopic work on ultrathin polymer films, which carried out by Storks in 1938[1000] already suggested that gutta-percha chains extended perpendicular to a film with a thickness of 100 Å so that the chains had to fold. This report may have failed to excite more interest because Storks also reported that chains of poly(ethylene sebacate) lie parallel to the film surface, so that no general principle seemed to be involved.

These periodicities were first interpreted as due to micelles, but the uniformity of their dimensions indicated by the sharpness of the low angle X-ray diffraction was difficult to account for. Ironically, Zahn and Winter[1006] noted that the polymer chains were much longer than this long periodicity and wrote that "it is difficult to imagine that the chains are folded so as to fit into a 100Å micelle." Zahn and Winter observed also that the long period in their polyurethane fiber increased after heating from 75 to 132Å, a striking analogy to the later observation[998] that the thickness of solution-grown single crystals increases with a rise in the crystallization temperature.

The principle of chain folding also made it easier to comprehend the formation of crystal structures like that of polycaprolactam in which nearest neighbor chains lie antiparallel to each other.[1002] It also explained a feature of polymer spherulite formation that earlier had seemed hard to understand. Spherulite formation in semicrystalline polymers was described first by Bunn and Alcock,[1007] who deduced from the spherulite birefringence that the polymer chains were oriented at right angles to the radius of the spherulite. This conclusion was confirmed by Herbst,[1008] who used X-ray diffraction from small portions of the spherulites. He concluded that crystalline needles radiating from the center of the spherulite "are built up of lamellae oriented perpendicular to the axis of the needle with the chain molecules parallel to the plane of the lamellae." This artificial description changed to a much more convincing model when the needlelike polymer crystal could be thought of as containing folded polymer chains.[1009]

A pioneering paper on the kinetics of crystallization was published by Wood and Bekkedahl[1010] in 1946. Using a dilatometric technique with *Hevea* rubber, they found that the rate of crystallization passes through a maximum at $-25°C$, being negligible at -50 and $0°C$. They also found that the melting range moves to higher temperatures as the crystallization temperature is raised.

This characteristic dependence of the rate of crystallization on temperature is typical of phase transitions that require a nucleation step. Becker[1011] analyzed such a process for the decomposition of a solid solution and concluded that the activation energy should be the sum of two terms, one representing the free energy for the formation of a nucleus of the critical size, the other, the activation energy for diffusion across the phase boundary. Because the size of the critical nucleus decreases with an increasing distance of the temperature from the melting point, it was predicted that the nucleation rate would pass through a maximum. Mandelkern et al.[1012] developed a specific theory for the rate of crystallization of chain molecules and concluded that not only the rate of nucleation but also the rate of spherulite growth should pass through a maximum as the temperature is varied. They predicted that the maximum rate of crystallization should be observed between 0.8 and 0.9 T_m, which agreed with the observation[1013] that poly(ethylene terephthalate) crystallized most rapidly at 0.85 T_m. It should be realized that these theories applied to crystallization with homogeneous nucleation. Gent[1014] showed that small additions of nucleating agents lead to large increases in the rate of rubber crystallization and

Price[1015] pointed out that a constant number of spherulites during polymer crystallization is indicative of heterogeneous nucleation.

A separate determination of the rate of nuclei formation and of spherulite growth as a function of temperature was first carried out by Flory and McIntyre.[1016] They studied a polymer close to its melting point, where the temperature dependence of these rates is particularly pronounced. As expected, both nucleation and spherulite growth accelerated with decreasing temperature; a drop of 5°C increased the nucleation rate by a factor of 10^4, the rate of spherulite growth by a factor of 10^3. The similarity of the two activation energies was interpreted as suggesting that the growth of spherulites is governed by a secondary nucleation.

Rheology

In 1948 Alfrey[1017] published the monograph "Mechanical Behavior of High Polymers," the first comprehensive summary of what was known at the time about the response of polymers, in bulk as well as in dilute or concentrated solution, to an applied stress. Six years later the fourth volume of Stuart's[1018] *Die Physik der Hochpolymeren* appeared with the subtitle "Theory and Molecular Interpretation of Technical Properties of Polymeric Materials," which indicated to what extent research in this area was influenced by the needs of the rapidly expanding polymer industry.

The first study of the steady-state melt viscosity on the molecular weight of a polymer was published by Flory[1019] in 1940. He used polyesters with weight-average chain lengths* up to about 1000 and found by capillary viscosimetry that the logarithm of viscosity was linear in the square root of the polymer chain length.† He suggested that melt viscosities might provide a simple method for the determination of the molecular weight of polymers. Around the same time Kauzmann and Eyring[1020] analyzed the dependence of the activation enthalpy for viscous flow, $\Delta H_\eta^\ddagger$, on the chain length of normal paraffins. For many compounds $\Delta H_\eta^\ddagger$ had been found to be one quarter of the heat of vaporization, but although this rule applied to the first members of the homologous series of paraffins $\Delta H_\eta^\ddagger$ seemed to approach an asymptotic limit as the chain was extended. This was interpreted as suggesting that "as the chain becomes sufficiently large, it moves in units of a fixed size independent of the size of the molecule." The independence of $\Delta H_\eta^\ddagger$ on chain length of polymers of high molecular weight was later strikingly demonstrated by Fox and Flory[1021] on polystyrene and polyisobutene. These authors[1022] also discovered that doubly logarithmic plots of melt viscosity against chain length exhibit an abrupt increase in slope beyond a critical length of the polymer molecule, a phenomenon interpreted by Bueche[1023] as reflecting the length of a chain molecule at which entanglements of chains become important in limiting their mobility.

---

*Number of atoms in the chain backbone.
†This dependence was not substantiated in later work on other polymers.

During the second world war and in the years immediately following, the theory of rubber elasticity was reconsidered by Wall,[1024] Flory and Rehner,[1025], James and Guth,[1026] and Treloar.[1027]* These treatments led all to a dependence of the retractive force, f on the number v of subchains between crosslinkages given by $f = vkT(\alpha - \alpha^{-2})$, where α is the relative extension. It was generally assumed that the crosslinks are firmly embedded in the rubber sample and that their displacements are proportional to those of the dimensions of the test specimen. It was also understood that chain entanglements would contribute to the retractive force and that the modulus of a rubber without crosslinkages could not simply be related to the length of the chain, as in Kuhn's formulation. Finally, it was pointed out that the presence of chain ends that are not incorporated into the network would reduce the modulus below the value expected if this feature is not taken into account. Originally, the crosslinkages were treated as fixed but James and Guth[1028] pointed out later that they would be expected to undergo Brownian motion; treating them as fixed should not lead to a change in the form of the functional relation between the retractive force and the extension but would result in an underestimate of the retractive force.

In all the treatments listed above the dependence of the retractive force f on the extention α involved the single parameter v. However, in 1940 Mooney[1028a] formulated a general theory of highly elastic substances and concluded that their description should involve two characteristic parameters, a modulus of rigidity and a "coefficient of asymmetry." Experiments by Rivlin[1028b] appeared to agree with Mooney's analysis.

That systems containing polymers may have to be described as exhibiting a superposition of viscosity and elasticity has been understood qualitatively for a long time. The concept of viscoelasticity goes back more than a hundred years to J. C. Maxwell, and as early as 1920 Hess[1029] reported that when the shearing stress on a starch solution was stopped the flow of the liquid was seen to reverse. Because no similar effect was observed with a glycerol solution of the same viscosity, it was concluded that "this can only be interpreted by the existence of elastic forces." Quantitative studies of viscoelastic phenomena were first reported in the late 1930s. Jenckel and Ueberreiter[1030] measured the creep and creep recovery of polystyrene fibers and demonstrated a delayed elasticity in both processes. They found that the permanent set decreases with an increasing molecular weight of the polymer. Creep recovery was later studied in more detail on plasticized poly(vinyl chloride) by Leaderman[1031] who found that the results agreed with the Boltzmann superposition principle: Thus, if the stress on a specimen was removed after a time t' the strain observed after a time t was equal to that expected if the stress had been applied for all of t with an equal negative stress acting for a time $t - t'$.

Dynamic testing in which the sample is subjected to periodic stresses was

---

*All these workers seem to have been unaware of Kuhn's 1938 publication[518a] in which he treated for the first time a vulcanized rubber network.

used first by Philippoff[1032] to demonstrate the viscoelasticity of a cellulose acetate solution. As noted before (see p. 152), this method was applied in 1939 by Aleksandrov and Lazurkin[521a] to the study of the viscoelasticity of rubber. They assumed that their samples were characterized by two discrete mechanisms of deformation, an "ordinary" and a "highly elastic." They wrote that "an increase in the frequency of the applied force has the same effect as a decrease in temperature" and that "various relaxation processes, such as electrical and mechanical, are marked by a unity in their physical nature." Both principles were to play an important role in the development of polymer rheology.

During the year in which Aleksandrov and Lazurkin published the results of their experimental work two important papers appeared that dealt with theoretical aspects of viscoelectricity. In Kuhn's[1033] paper rubber is characterized by different "cohesion mechanisms" *(Zusammenhaltsmechanismen)* with widely different relaxation times. Kuhn suggested that because the modulus of a rubber is typically five orders of magnitude smaller than that of a crystalline substance one may assume that most contributions to the modulus are relaxed so fast that they are not observed in conventional measurements. He conjectured, without adducing experimental data, that the spectrum of relaxation times of a rubber may extend over ten orders of magnitude. A second paper, by Bennowitz and Rötger,[1034] was not concerned with polymers but contained a detailed analysis of the behavior of materials with a continuous distribution of relaxation times in creep, stress relaxation, and dynamic testing.

Conclusions similar to those formulated by Kuhn were arrived at by Fuoss and Kirkwood[1035] to explain the breadth of the dielectric dispersion observed in polymers; in their words "a linear polymer must exhibit a distribution of relaxation times as a necessary consequence of its structure." A decade later Rouse[1036] developed a theory for this distribution. In his model the molecular chain was represented by beads connected by elastic springs and the motion of the beads corresponded to a series of vibrational modes, each associated with a relaxation time. Although the model described only isolated chains and therefore related directly only to the behavior of polymer molecules in highly dilute solution, it illustrated convincingly the physical causes for the breadth of relaxation times observed in bulk polymers.

At any given temperature only part of the relaxation spectrum is directly observable. However, with the assumption "that all relaxation mechanisms have the same temperature dependence" Ferry[1037] devised a method by which data obtained over a range of temperatures T could be combined to produce the complete spectrum of relaxation times at a given reference temperature T_0. He suggested that the dependence of any relaxation time τ_i on temperature can be obtained from the steady-state viscosities η after correction for the temperature-dependence of the density ρ from $\tau_i(T) = a_T \tau_i(T_0)$, where the "shift factor" a_T is given by $a_T = (\rho_0 T_0 / \rho T)[\eta(T)/\eta(T_0)]$. Using this approach, Tobolsky and McLoughlin[1038] were able to combine data obtained at temperatures from -70 to $+100°C$ to construct the spectrum of relaxation times of polyisobutene at 25°C, extending from microseconds to hundreds of hours. The assumption, already implicit in the comments of Aleksandrov and Lazurkin, that the shift factor obtained from mechanical testing should also describe the temperature-dependence of relaxation processes in dielectric measurements was confirmed in a study by Ferry and Fitzgerald[1039] on plasticized poly(vinyl chloride). Finally, because a_T primarily reflected the temperature-dependence of the segmental frictional coefficient, it seemed plausible that it would be related to the distance of a temperature from the glass transition point. Williams, Landel, and Ferry,[1040] in fact, found that a_T is a universal function of $T - T_g$, both for polymers and for substances of low molecular weight and their "WLF equation" has been central to further development of the field.

In summarizing extensive work on the distribution of relaxation times, Tobolsky and Catsiff[1041] pointed out that four distinct regions of the distribution function may be distinguished. The very long relaxation times of the glassy state are associated with moduli independent of chain length. In a transition region the modulus, still chain length-independent, changes rapidly with time, decreasing about a thousandfold. This region is followed by a "rubbery plateau" in which the modulus, governed by the chain length between entanglements, changes very slowly with time. Finally, in the "flow region" the modulus decreases with time rapidly to zero.

A pioneering study of normal and tangential stresses during the shearing of a polyethylene melt in a disc and cone rheometer was reported by Pollett[1042] in 1955. The complex changes in the rheological properties of the polymer during the shearing were interpreted as a result of "structural changes in the material" which were found to be reversible. This investigation opened up a whole new field of study aiming at the characterization of the rheological behavior of polymers under processing conditions.

Epilogue

Although by 1960 the science of high polymers had attained an impressive stature, there is no indication that its progress has slowed down in recent years. This book is not meant to cover advances realized during the last quarter century, but a few examples should serve to illustrate the health of the polymer science enterprise.

In the area of synthesis we have seen the preparation of polymers with previously undreamed of thermal stability. It is particularly impressive that much of the pioneering work in this field was carried out by C. S. Marvel after he had reached the normal retirement age. A series of rodlike aromatic polyamides that form liquid crystalline phases and yield fibers of outstanding strength was developed in the DuPont laboratories under the leadership of P. W. Morgan. The preparation of polymers for medical use has become an important component of polymer research, and an impressive success story in this speciality resulted from a careful study of the properties of hydrophilic gels which led to the development of soft contact lenses by O. Wichterle and his associates. R. B. Merrifield has shown that with the use of crosslinked polymers as inert supports the stepwise build-up of polypeptide hormones which contain a large number of amino acids in a predetermined sequence is technically feasable.* P. Rempp and his associates have devised methods for constructing model networks that contain isotopically labeled subchains which can serve to check assumptions that form the basis of theories of rubber elasticity.

As for the physical chemistry of polymers, advances in spectroscopic techniques opened up possibilities which could hardly have been imagined in 1960. At that time high resolution NMR spectroscopy had just begun to be used for the characterization of polymer stereoregularity in cases that were experimentally particularly simple. Advances in instrumentation and theory made it possible not only to resolve long stereoisomeric sequences but also to carry out sequential analyses of copolymers, to study conformational equilibria, the dynamics of conformational transitions, and to obtain highly resolved NMR spectra on solid samples. The introduction of neutron scattering as a new experimental tool enabled investigators to confirm predictions, such as the

---

*Merrifield was awarded in 1984 the Nobel Prize in Chemistry for this work.

shrinkage with increasing concentration of flexible polymer chains in good solvents, which had previously not seemed accessible to experimental verification. The use of gel permeation chromatography, a relatively unsophisticated development, simplified enormously the determination of molecular weight distributions. On the theoretical side techniques for predicting chain conformations in the crystalline state and in solution were greatly refined. The description of the diffusion of polymer chains in melts as a "reptation" process stimulated experimental studies of extremely minute diffusion coefficients. The investigation of conductive polymers became a subject of widespread interest and led to some remarkable results.

In 1960 four journals were devoted exclusively to investigations in the polymer field: the *Journal of Polymer Science*, *Makromolekulare Chemie*, *Vysokomolekularniye Soedineniya*, and *Polymer*. Since then a number of new journals have been added, among which are *Polymer Engineering and Science* (1961), *European Polymer Journal* (1965), *Journal of Macromolecular Science* (1967), *Macromolecules* (1968), and *Polymer Journal* (1969). Research centers devoted to polymer science have been established in many countries, and as this book goes to press a "*Max Planck Institut für makromolare Chemie*" is to be inaugurated in Mainz. Finally, the award of the Nobel Prize in chemistry for 1963 to K. Ziegler and G. Natta and for 1974 to P. J. Flory acknowledged the outstanding contributions of these scientists to all of chemistry. If ever an author in the years to come writes a sequel to this book he will have a rich story to relate.

References

1. J. Dalton, *Mem. Lit. Phil. Soc. Manchester*, (2), **1**, 286 (1805); *New System of Chemical Philosophy*, Manchester, 1808, pp. 219–220; Alembic Club Reprints No. 2, Edinburgh, 1923, pp. 30–33.
2. M. Gay-Lussac, *Mem. Soc. Arcueil*, **II**, 207–234 (1809) Alembic Club Reprints No. 4, Edinburgh, 1923, pp. 8–24.
3. J. Dalton, *New System of Chemical Philosophy*, Manchester, 1808, p. 71; Part II, Manchester, 1810, pp. 555–559; Alembic Club Reprints No. 4, Edinburgh, 1923, pp. 5 and 25–27.
4. A. Avogadro, *J. Phys. Chim., L'Histoire Naturelle et des Arts*, **73**, 58 (1811); Alembic Club Reprints No. 4, Edinburgh, 1923, pp. 28–51.
5. A. M. Ampère, *Ann. Chim. Phys.*, **90**, 45 (1814).
6a. M. Faraday, *Phil. Mag.*, **66**, 180 (1826).
6b. *Faraday's Diary*, T. Martin, Ed., Bell, London, 1932.
7. J. Berzelius, *Jahresber. Fortsch. Phys. Wissensch.*, (a) **6**, 98 (1827); (b) **11**, 44 (1832).
8. J. Berzelius, *Jahresber. Fortschr. Phys. Wissensch.*, **12**, 63 (1833).
9. H. Hennell, *Phil. Mag.*, **68**, 354 (1826).
10. I. Beilstein, 316, 327; P. Claesson, *J. Prakt. Chem. N. F.*, **19**, 260 (1879).
11. P. E. M. Berthellot, *Compt. Rend.*, **63**, 479, 515 (1866); *Bull. Soc. Chim. France. [2]*, **6**, 268 (1866); *Ann.*, **141**, 173 (1867).
11a. R. N. Pease, *J. Am. Chem. Soc.*, **52**, 1158 (1930).
11b. W. E. Vaughan, *J. Am. Chem. Soc.*, **54**, 3863 (1932).
12. J. Dumas, *Compt. Rend.*, **8**, 609 (1839); *Ann.*, **32**, 101 (1839).
13. J. Berzelius, *Ann. Chim. Phys. [2]*, **67**, 303 (1838); *Compt. Rend.*, **6**, 629 (1838).
14. J. Dumas, *Compt. Rend.*, **8**, 621 (1839); *Ann.*, **32**, 118 (1839).
15. F. Wöhler and J. v. Liebig, *Ann.*, **3**, 249 (1982).
16. J. Berzelius, *Ann.*, **3**, 282 (1832).
17. J. Schiel, *Ann.*, **43**, 107 (1842).
18. H. Kopp, *Ann.*, **41**, 169 (1842).
19. J. Redtenbacher and J. v. Liebig, *Ann.*, **38**, 113 (1841).
20. J. B. Dumas and J. S. Stas, *Compt. Rend.*, **11**, 991 (1840); *Ann. Chim. Phys. [3]*, **1**, 1 (1841).
21. C. Kolbe, *Ann.*, **109**, 257 (1859).

22. A. Wurtz, *Compt. Rend.*, **46**, 1228 (1858).
23. J. Wislicenus, *Ann.*, **125**, 41 (1863).
24. A. Kekulé, *Ann.*, **106**; (a) 153−155; (b) 140−146 (1858).
25. A. Crum Brown, *J. Chem. Soc.*, **18**, 230 (1865).
26. E. Frankland and B. F. Duppa, *Ann.*, **142** (1867).
27. A. Kekulé, *Lehrbuch der organischen Chemie. Erster Band*, Erlangen, 1861, pp. 152, 156−157.
28. L. Pasteur, *Ann. chim. phys. [3]*, **24**, 442 (1848); **28**, 56 (1850); **31**, 67 (1851); *Leçons de chimie professées en 1860*, Societé chimique de Paris, Paris, 1861, pp. 1−48; Alembic Club Reprints, No. 14, Edinburgh, 1905; R. J. Dubos, *Pasteur Free Lance of Science*, Little, Brown, Boston, 1950, pp. 90−115.
29. J. Wislicenus, *Ann.*, **166**, 47 (1873).
30. English translations of van't Hoff's paper are to be found in (a) G. M. Richardson, *The Foundations of Stereochemistry*, American Book, 1901; (b) J. E. Marsh, *Chemistry in Space*, Oxford, 1891. German translations are in (c) F. Herrmann, *Die Lagerung der Atome im Raume*, Braunschweig, 1877; (d) E. Cohen, *J. H. van't Hoff, sein Leben und Wirken*, Akad. Verlagsges., Leipzig, 1912, pp. 73−82.
31. A. Kekulé, *Z. Chem.*, **3**, 643 (1860); **4**, 257 (1861); **6**, 9 (1863).
32. J. Wislicenus, *Ber.*, **7**, 297 (1874).
33. J. Wislicenus, *Ann.*, **248**, 281 (1888).
34. Ref. 30d, pp. 104−113.
35. M. Berthelot, *Compt. Rend.*, **82**, 441 (1876); *Ann. Chim. Phys.* [5], **9**, 53 (1876).
36. J. H. van't Hoff, *Ber.*, **9**, 5, 1339 (1876); *Bull. Soc. Chim. France*, [2], **25**, 135 (1876).
37. H. Kolbe, *J. Prakt. Chem. N. F.*, **15**, 473 (1877).
38. E. Simon, *Ann.*, **31**, 265 (1839).
39. J. Blyth and A. W. Hofmann, *Mem. Chem. Soc.*, **2**, 334 (1845); *Ann.*, **53**, 289 (1845).
40. C. Gerhardt and A. A. T. Cahours, *Ann. Chim. Phys.* [3], **1**, 96 (1841); *Ann.*, **38**, 96 (1841).
41. E. Kopp, *Compt. Rend.*, **21**, 1378 (1845).
42. M. Berthelot, *Compt. Rend.*, **63**, 518 (1866); *Ann. Chim. Phys.* [4], **12**, 159 (1867); *Ann.*, **141**, 377 (1867).
43. M. Berthelot, *Bull. Soc. Chim. France,* [2], **6**, 294 (1866).
44. G. Wagner, *Ber.*, **11**, 1258 (1968).
45. E. Erlenmeyer, *Ann.*, **137**, 353 (1866).
46. M. Berthelot, *Leçons de chimie professées en 1864 et 1865*, Société chimique de Paris, Paris, 1866, pp. 18−65, 148−167.
47. M. Berthelot, *Compt. Rend.*, **36**, 425 (1853); *Ann. Chim. Phys.*, **38**, 38 (1853); **39**, 5 (1853).
48. A. Bauer, *Ann. Chim. Phys.* [3], **63**, 461 (1861).
49. M. Berthelot, *Ann. Chim. Phys.* [4], **6**, 349 (1865).
50. M. Berthelot, *Ann. Chim. Phys.* [3], **53**, 158 (1858).
51. A. L. Lavoisier and P. S. Laplace, *Histoire de l'Academie Royale des Sciences, Memoires*, 355 (read June 18, 1783), cited by J. R. Partington, *A History of Chemistry*, Vol. III, Macmillan, London, 1962, p. 428.

51a. K. H. Meyer and Y. Go, *Helv. Chim. Acta*, **17**, 1083 (1934).
51b. G. Gee, *Trans. Faraday Soc.*, **48**, 515 (1952).
52. M. Berthelot, *Bull. Soc. Chim. France*, **11**, 4 (1869).
52a. V. Goryainov and A. Butlerov, *Ann.*, **169**, 146 (1873).
52b. D. Berthelot and H. Gaudechon, *Compt. Rend.*, **150**, 1169 (1900).
52c. E. Bamberger and F. Tschirner, *Ber.*, **33**, 955 (1900).
52d. H. v. Pechmann, *Ber.*, **31**, 2640 (1899).
53. V. Regnault, *Ann. Chim. Phys.* [2], **69**, 152 (1838).
54. V. Sawitsch, *Ann.*, **119**, 183 (1861).
55. A. W. Hofmann, *Ann.*, **115**, 271 (1860).
56. E. Baumann, *Ann.*, **163**, 312 (1872).
57. I. Ostromislensky, *J. Russ. Phys. Chem. Soc.*, **44**, 204 (1911); *Chem. Zentralbl.*, **I**, 1980 (1912); *Chem. Abstr.*, **6**, 1538 (1912).
58. G. Williams, *Phil. Trans.*, **9**, 254 (1860).
59. G. Williams, *Proc. Roy. Soc. London*, **10**, 516 (1860).
60. G. Bouchardat, *Compt. Rend.*, **80**, 1446 (1875).
61. G. Bouchardat, *Compt. Rend.*, **89**, 1117 (1879).
62. O. Wallach, *Ann.*, **238**, 88 (1887).
63. F. E. Matthews and E. H. Strange, Br. Pat. 24,790, (October 25, 1910); *Chem. Abstr.*, **6**, 1542 (1912).
64. C. Harries, (a) *Ann.*, **383**, 184 (1911); (b) U.S. Pat. 1,058,056, (April 8, 1913).
65. F. Coutourier, *Ann. Chim. Phys.* [6], **26**, 491 (1892).
66. I. Kondakow, *J. Prakt. Chem.* [2], **62**, 66 (1900); [2], **64**, 109 (1901).
67. S. V. Lebedev, *J. Russ. Phys. Chem. Soc.*, **42**, 949 (1910); *Chem. Zentralbl.*, **II**, 1744 (1910).
68. S. V. Lebedev and N. A. Skavronskaya, *J. Russ. Soc. Phys. Chem.*, **43**, 1124 (1911); *Chem. Abstr.*, **6**, 855 (1912); *chem. Zentralbl.*, **I**, 1440 (1912).
68a. Ger. Pat. 255 786, January 12, 1912 (to BASF).
68b. Ref. 102a, p. 336 (original French translation).
68c. J. Redtenbacher, *Ann.*, **47**, 113 (1843).
68d. J. Wislicenus, *Ann.*, **192**, 106 (1878).
69. R. Fittig, *Ann.*, **188**, 47, 55 (1877); *Ann.*, **200**, 70 (1880).
70. C. Kolbe, *J. Prakt. Chem.* (2), **25**, 372 (1882).
71. C. Moureu, *Ann. Chim. Phys.* [7], **2**, 175 (1894).
72. A. Butlerov, *Ann.*, **180**, 247 (1876); **189**, 44 (1877).
72a. C. Engler, *Ber.*, **30**, 2358 (1897).
72b. A. Kronstein, *Ber.*, **35**, 4150, 4153 (1902).
73. S. Cannizzaro, *Nuovo Cim.*, **7**, 321 (1858); English translation, *Alembic Club Reprints*, Vol. 18, Edinburgh, 1910.
74. A. W. Williamson, *J. Chem. Soc.*, **22**, 433 (1869).
75. R. Anschütz, *A. Kekulé*, Vol. 2, Verlag Chemie, 1929, p. 903; A. Kekulé, *Nature*, **18**, 210 (1878).

76. R. Clausius, *Ann. Phys.*, **100**, 353 (1857).
77. A. Krönig, *Ann. Phys.*, **99**, 315 (1856).
78. J. P. Joule, *Phil. Mag.*, **14**, 211 (1857).
79. V. Regnault, *Ann. Chim. Phys. [3]*, **5**, 52 (1842).
80. J. D. van der Waals, "Over de continuiteit van den gasen vloeistoftoestand. *Academisch proefschrift*, A. W. Sijthoff, Leiden, 1873; German translation, "Die Continuität des gas-förmigen und Flüssigen Zustandes," 2nd ed., J. A. Barth, Leipzig, 1899.
81. J. H. Jeans, *J. Chem. Soc.*, **123**, 3398 (1923).
82. R. Watson, *Phil. Trans. Roy. Soc. London*, **61**, 213 (1771).
83. C. Blagden, *Phil. Trans. Roy. Soc. London*, **78**, 277 (1788).
84. L. C. deCoppet, *Ann. Chim. Phys.* [4], **25**, 502 (1872).
85. F. M. Raoult, *Ann. Chim. Phys.* (a) [5], **20**, 217 (1880); (b) [5], **28**, 133 (1883); (c) *[6]*, **2**, 66 (1884); (d) *[6]*, **8**, 289 (1886).
86. C. M. Guldberg, *Compt. Rend.*, **70**, 1349 (1870).
87. F. M. Raoult, *Compt. Rend.*, **103**, 1125, 1887; **104**, 976, 1430 (1887); *Ann. Chim. Phys. [6]*, **15**, 375 (1888).
88. J. H. van't Hoff, (a) *Z. Phys. Chem.*, **1**, (1887); (b) English translation, *Alembic Club Reprints*, No. 19, Edinburgh, 1928.
89. M. Planck, *Z. Phys. Chem.*, **1**, 537 (1887).
90. E. Beckmann, *Z. Phys. Chem.*, **2**, 638, 715 (1888).
91. A. W. Porter, *Proc. Roy. Soc. London*, **80**, 457 (1908).
92. J. E. Trevor, *J. Phys. Chem.*, **12**, 141 (1908).
93. E. G. M. Törnqvist, in *The Polymer Chemistry of Synthetic Elastomers*, Chapter 2, J. P. Kennedy and E. G. M. Törnquist, Eds., Wiley-Interscience, 1968.
94. C. M. de la Condamine, *Mem. Acad. Roy. Sci.*, **17**, 319 (1751).
95. M. Faraday, *Quart. J. Sci. Arts, Roy. Inst. Great Britain*, **21**, 19 (1826).
96. F. C. Himly, *Ann.*, **27**, 40 (1838).
97. W. A. Tilden, *J. Chem. Soc.*, **45**, 410 (1884).
98. W. Euler, *Ber.*, **30**, 1989 (1897); *J. Prakt. Chem.*, **57**, 131 (1898).
99. V. N. Ipatiev and V. Wittorf, *J. Prakt. Chem.*, **55**, 1 (1897).
100. C. Harries, *Ber.*, **35**, 3256, 4429 (1902); **36**, 1937 (1903).
101. C. Harries, *Ber.*, **37**, 2708 (1904); **38**, 1195 (1905).
101a. C. Harries, *Angew. Chem.*, **20**, 1265 (1907).
102. S. S. Pickles, *J. Chem. Soc.*, **97**, 1085 (1910).
102a. A. Dubosc and A. Luttringer, *Le caoutchouc, sa chimie nouvelle, ses syntheses*, Librairie Cillard, Paris, 1913; English translation, "Rubber, Its Production, Chemistry and Synthesis," E. W. Lewis, Ed., Griffin, London, 1918.
102b. S. Lebedev, *J. Russ. Phys. Chem. Soc.*, **45**, 1249 (1913); *Chem. Zentralbl.*, **I**, 1402 (1914).
102c. C. Harries, *Ann.*, **406**, 173 (1914).
103. C. Goodyear "Gum-Elastic and Its Variations with a Detailed Account of Its Applications and Uses, Vol. 1, New Haven, 1855; (a) p. 23; (b) p 118
104. J. Gough, *Mem. Lit. Phil. Soc. Manchester* [2], **1**, 288 (1805).

105. W. Thomson, *Quart. J. Math.*, **1**, 57 (1857).
106. J. P. Joule, *Phil. Trans. Roy. Soc. London*, **149**, 91 (1859).
107. J. Gay-Lussac and J. Thénard, *Recherches Physico-Chimiques*, Vol. 2, pp. 268–350 (1811).
108. Nasse, *J. Chem.*, **4**, 111 (1812).
109. C. Kirchhoff, *J. Chem.*, **14**, 385, 389 (1815).
110. J. J. Colin and H. Gaultier de Claubry, *Ann. Chim. Phys.*, **90**, 239 (1814).
111. H. T. Brown and J. Heron, *J. Chem. Soc.*, **35**, 596 (1879).
112. J. B. Biot and J. F. Persoz, *Ann. Chim. Phys.* [2].: (a) **52**, 72 (1852); (b) **53**, 73 (1833).
113. F. Musculus, *Ann. Chim. Phys.* [3], **60**, 203 (1860); [4], **6**, 117 (1865).
114. C. O'Sullivan, *J. Chem. Soc.*, **25**, 579 (1872); **29**, 478 (1876).
114a. A. P. Dubrunfaut, *Ann. Chim. Phys.* [3], **21**, 178 (1847).
115. C. O'Sullivan, *J. Chem. Soc.*, **30**, 125 (1876).
116. H. T. Brown and G. H. Morris, *J. Chem. Soc.*, **47**, 527 (1885).
117. H. T. Brown and G. H. Morris, *J. Chem. Soc.*, **55**, 462 (1889).
118. H. T. Brown and J. H. Millar, *J. Chem. Soc.*, **75**, 286 (1899).
119. L. Maquenne and E. Roux, *Ann. Chim. Phys.* [8], **9**, 179 (1906).
120. A. Payen, *Compt. Rend.*, **8**, 51 (1839).
121. H. Braconnot, *Ann. Chim. Phys.* [2], **12**, 172 (1819).
122. J. Pelouze, *Compt. Rend.*, **48**, 327 (1859).
123. *Jahresber. Fortschr. Chem.*, pp. 529–541 (1859).
124. E. Frémy, *Compt. Rend.*, **48**, 361 (1859).
125. K. Hess, *Chemie der Zellulose*, Akad. Verlagsges. Leipzig, 1928.
126. C. F. Cross, E. J. Bevan, and C. Beadle, *Cellulose*, Longmans, Green, London, 1895.
127. B. Tollens, *Handbuch der Kohlenhydrate*: (a) Vol. II, Breslau, 1895; (b) 3rd ed., Leipzig, 1914, p. 564.
128. E. Schulze, *Ber.*, **24**, 2277 (1891).
129. E. Flechsig, *Z. Physiol. Chem.*, **7**, 523 (1882).
130. C. F. Cross, E. T. Bevan, and C. Beadle, *Ber.*, **26**, 1090 (1893).
131. Z. H. Skraup and J. König, *Ber.*, **34**, 1115 (1901); *Monatsh.*, **22**, 1011 (1901).
132. H. Ost, *Ann.*, **398**, 313 (1913).
133. C. F. Cross, E. T. Bevan, and C. Beadle, *J. Chem. Soc.*, **79**, 366 (1901).
134. A. G. Green and A. G. Perkin, *J. Chem. Soc.*, **89**, 811 (1906).
135. A. Girard, *Compt. Rend.*, **81**, 1105 (1875); *Ann. Chim. Phys.* [5], **24**, 356 (1881).
136. A. Nastukoff, *Ber.*, **33**, 2237 (1900).
137. J. S. Fruton, *Molecules and Life*, Wiley, New York, 1972.
138. P. R. Srinivasan, J. F. Fruton, and J. T. Edsall, Eds., "The Origins of Modern Biochemistry. A Retrospect on Proteins", Ann. N.Y. Acad. Sci., Vol. 325.
139. J. v. Liebig, *Ann.*, **39**, 129 (1841).
140. O. Loewi, *Arch. Exper. Path. Pharm.*, **48**, 303 (1902).
141. H. Kolbe, *J. Prakt. Chem. NF*, **14**, 268 (1876).

142. O. Loew and T. Bokorny, *Pflüger's Arch.*, **25**, 150 (1881).
143. H. Hlasiwetz and J. Habermann, *Ann.*, **169**, 150 (1873).
144. E. Fischer, *Ber.*, **39**, 530 (1906).
145. E. Fischer, *Ber.*, **40**, 1754 (1907).
146. O. Zinoffsky, *Z. Physiol. Chem.*, **10**, 16 (1886).
147. R. Zsigmondy, *Kolloidchemie*, 3rd. ed., Leipzig, 1920, p. 381.
148. J. v. Liebig, *Ann.*, **30**, 250 (1839).
149. E. Fischer and H. Thierfelder, *Ber.*, **27**, 2031 (1894).
150. E. Fischer, *Ber.*, **27**, 2985 (1894).
151. A. Wróblewski, *Z. Physiol. Chem.*, **24**, 173 (1898).
152. W. B. Hardy, *Proc. Roy. Soc. London*, **66**, 110 (1900).
153. L. Michaelis and B. Mostynski, *Biochem. Z.*, **24**, 79 (1910).
154. P. Rona and L. Michaelis, *Biochem. Z.*, **28**, 193 (1910).
155. F. G. Donnan, *Z. Elektrochem.*, **17**, 572 (1911).
156. L. Michaelis and P. Rona, *Biochem. Z.*, **27**, 38 (1910).
157. H. Chick and C. J. Martin, *J. Physiol.*, **45**, 261 (1912).
158. W. Ostwald, *Grundriss der Kolloidchemie*, Dresden, 1909.
159. H. Freundlich, *Kapillarchemie*, Akad. Verlag, Leipzig, 1909.
160. W. Pauli, *Colloid Chemistry of Proteins*, Blakiston's, Philadelphia, 1922.
161. T. Graham, *Phil. Trans. Roy. Soc. London*, **151**, 183 (1861).
162. J. R. Katz, in K. Hess, *Chemie der Zellulose*, Akad. Verlag, Dresden, 1928, pp. 605–608.
163. F. Krafft, *Ber.*, **29**, 1334 (1896).
164. F. Krafft, *Ber.*, **32**, 1608 (1899).
165. E. Paterno, *Z. Phys. Chem.*, **4**, 457 (1889).
166. E. W. Reid, *Z. Physiol.*, **31**, 438 (1904).
167. H. Rodewald and A. Kattein, *Z. Phys. Chem.*, **33**, 579 (1900).
168. B. Moore, *Biochem. J.*, **2**, 34 (1906).
169. J. Perrin, *Compt. Rend.*, **147**, 530, 594 (1908); *Ann. Chim. Phys.* [8], **18**, 5 (1909).
170. A. A. Noyes, *J. Am. Chem. Soc.*, **27**, 85 (1905).
171. J. H. Gladstone and W. Hibbert, *Phil. Mag.* [5] **28**, 38 (1889).
172. W. Nernst, *Z. Phys. Chem.*, **2**, 613 (1888).
173. J. Thovert, *Compt. Rend.*, **135**, 579 (1902).
174. R. O. Herzog, *Kolloid Z.*, **2**, 2 (1907).
175. A. Einstein, *Ann. Phys.* [4], **19**, 289 (1906); **34**, 591 (1911).
176. E. Hatschek, *Kolloid Z.*, **7**, 302 (1910).
177. W. Nernst, *Theoretical Chemistry*, Enke, Stuttgart, 1921, pp. 311, 488.
177a. J. K. Mitchell, *Philadelphia J. Med. Sci.*, **13**, 36 (1831); *J. Roy. Inst. London*, **2**, 101, 307 (1831).
177b. T. Graham, *Phil. Mag.*, **32**, 218 (1866).
178. W. Pauli, *Kolloid Z.*, **3**, 2 (1908).
179. W. Ostwald, *Kolloid Z.*, **2**, 264 (1907).

180. K. Landsteiner, *Kolloid Z.*, **3**, 221 (1908).
181. M. H. Fischer and W. Ostwald, *Pflüger's Arch. Physiol.*, **106**, 229 (1905); *J. Am. Med. Assoc.*, **46**, 423 (1906).
182. T. Graham, *Ann.*, **121**, 1 (1862).
183. W. B. Hardy, *Proc. Roy. Soc. London*, **66**, 95 (1899).
184. W. Ostwald, *Kolloid Z.*, **1**, 291 (1906).
185. C. O. Weber, *Kolloid Z.*, **1**, 33, 65 (1906).
186. P. Schidrowitz and H. A. Goldsbrough, *J. Soc. Chem. Ind.*, **28**, 3 (1909); *Kolloid Z.*, **4**, 226 (1909).
187. W. Woodruff, *The Rise of the British Rubber Industry*, Liverpool University Press, 1958.
188. C. Goodyear, *Gum-Elastic and Its Varieties with a Detailed Account of Its Applications and Uses*, New Haven, 1855.
189. R. F. Wolf, *India Rubber Man—the Story of Charles Goodyear*, Caxton Printers, Caldwell, Idaho, 1939.
190. C. Goodyear, U.S. Pat. 3633 (June 15, 1844).
191. T. Hancock, *Personal Narrative of the Origin and Progress of the Caoutchouc or India-Rubber Manufacture in England*, Longman, London, 1857.
192. H. Wolf and R. Wolf, *Rubber, A Story of Glory and Greed*, Covici, Friede, New York, 1936.
193. H. J. Stern, *Rubber Natural and Synthetic*, MacClaren, London, 1967, p. 277.
194. P. Schidrowitz and T. R. Dawson, *History of the Rubber Industry*, Heffer, Cambridge, 1952, pp. 71–72.
194a. H. J. Stern, *Rubber Age and Synthetics, (Gr. Brit.)*, **16**, 268 (1945), 295 (1946).
195. H. Braconnot, *Ann.*, **7**, 242, 245 (1833).
196. C. F. Schönbein, *Phil. Mag.*, **31**, 7 (1847).
197. D. R. Wiggam and W. E. Gloor, *Ind. Eng. Chem.*, **26**, 551 (1934).
198. J. W. Swan, *Br. Pat.*, 5978 (1883).
198a. H. deChardonnet, *Compt. Rend.*, **105**, 899 (1887); **108**, 981 (1889) Fr. Pat. 165,349 (November 17, 1884).
199. H. deChardonnet, Fr. Pat. 203,202 (Jan. 16, 1890).
200. V. Hottenroth, *Die Kunstseide*, Hirzel, Leipzig, 1926; English translation, "*Artificial Silk*," Pitman, London, 1928; R. O. Herzog, Ed., *Technologie der Textilfasern*, Vol. 7, Kunstseide, Springer, 1927.
201. J. W. Hyatt and I. J. Hyatt, *U.S. Pat.* 105,338 (June 12, 1870).
202. M. Kaufman, *The First Century of Plastics, Celluloid and Its Sequel*. The Plastics and Rubber Institute, London, 1963.
203. W. E. Gloor, in E. Ott, Ed., *Cellulose and Cellulose Derivatives*, Interscience, 1943, Chapter X.
204. C. F. Cross, E. J. Bevan, and C. Beadle, *Ber.*, **26**, 1090 (1898).
205. C. F. Cross, E. J. Bevan, and C. Beadle, U.S. Pat. 520,770 (June 5, 1894).
206. C. F. Cross and E. J. Bevan, *Ber.*, **34**, 1513 (1901).
207. M. Müller, Ger. Pat. 187,947 (1905).
208. A. v. Baeyer, *Ber.*, **5**, 1094 (1872).
209. E. ter Meer, *Ber.*, **7**, 1200 (1874).

210. R. Fabrinyi, *Ber.*, **11**, 282 (1878).
211. A. Claus and E. Trainer, *Ber.*, **19**, 3004 (1886).
212. W. Kleeberg, *Ann.*, *263*, 283 (1891).
213. D. Menasse, *Ber.*, **27**, 2409 (1894).
214. L. H. Baekeland, *Ind. Eng. Chem.*, **1**, 149 (1909).
215. L. H. Beakeland, U.S. Pat. 942,699 (1907).
216. L. H. Baekeland, *Ind. Eng. Chem.*, **1**, 202 (1909).
217. L. H. Baekeland, *Ind. Eng. Chem.*, **1**, 545 (1909); **3**, 932 (1911).
218. *West Indian Bull.*, **7**, 263 (1906); *Chem. Z.*, **2**, 813 (1906).
218a. M. Bogemann, "Historical account of the synthetic rubber research in Farbenfabriken, form. F. Bayer and Co. during 1907–1918," written in 1956 for the archives of Bayer A. G., Leverkusen.
218b. F. Hofmann, *Chem. Z.*, **60**, 693 (1936).
219. C. Harries, *Ber.*, **383**, 157 (1911).
220. (a) Ger. Pat. 250, 690 (September 6, 1912); (b) Ger. Pat. 254, 672 (December 12, 1912; to Farbenfabriken, formerly F. Bayer Co.).
221. G. S. Whitby and C. C. Davis, Eds., *Synthetic Rubber*, Chapters III and IV, Wiley, New York, 1954.
222. A. Talalay and M. Magat, *Synthetic Rubber from Alcohol*, Interscience, New York, 1942.
222a. G. S. Whitby and M. Katz, *Ind. Eng. Chem.*, **25**, 1204, 1338 (1933).
223. Ref. 102a, p. 28 of original French edition.
224. H. Staudinger, *Schweiz. Chem. Z.*, 1–5, 28–33, 60–64 (1919); *Das Wissenschaftliche Werk von Herman Staudinger*, M. Staudinger, H. Hopff, and H. Kuhn, Eds., Hüthig, Basel, 1969, Vol. I, pp. 22, 40.
225. Ger. Pat. 511,145 (to I.G. Farbenindustrie, October 27, 1931).
226. E. Tschunkur and W. Bock, U.S. Pat., 1,938,731 (December 12, 1933); Ger. Pat. 570,980 (February 27, 1933).
226a. J. Borken, *The Crime and Punishment of I. G. Farben*, The Free Press, New York, 1978, pp. 26–27.
227. F. A. Howard, *Buna Rubber, the Birth of an Industry*, Van Nostrand, New York, 1947.
227a. Br. Pat. 401, 297 (to I. G. Farbenindustrie, November 3, 1933).
228. J. A. Nieuwland, W. S. Calcott, F. P. Downing, and A. S. Carter, *J. Am. Chem. Soc.*, **53**, 4197 (1931).
229. W. H. Carothers, I. Williams, A. M. Collins, and J. E. Kirby, *J. Am. Chem. Soc.*, **53**, 4203 (1931).
229a. J. W. Hill, *Proc. Welch Foundation Conf. Chem. Research*, **XX**, 232 (1976).
230. R. M. Thomas, J. E. Lightbown, W. J. Sparks, P. K. Frolich, and E. V. Murphree, *Ind. Eng. Chem.*, **32**, 1283 (1940).
231. W. J. Sparks and R. M. Thomas, U.S. Pat. 2,356,129 (August 22, 1944).
232. C. A. Thomas and W. H. Carmody, *J. Am. Chem. Soc.*, **54**, 2480 (1932).
233. H. Ambronn, *Kolloid Z.*, **20**, 173 (1917).
234. P. Scherrer, *Nachr. Ges. Wiss. Goettingen*, 98 (1918); *J. Chem. Soc.*, **116**, II 274 (1919); *Chem. Zentralbl.*, **I** 322 (1919).
235. R. O. Herzog, W. Jancke, and M. Polanyi, *Z. Phys.*, **3**, 343 (1920).

236. R. O. Herzog and W. Jancke, *Ber.*, **53**, 2162 (1920).
237. M. Polanyi, in *Fifty Years of X-ray Diffraction*, P. P. Ewald, Ed., A. Vosthoek, 1962.
238. K. Becker, R. O. Herzog, W. Jancke, and M. Polanyi, *Z. Phys.*, **5**, 61 (1921).
239. M. Polanyi, (a) *Z. Phys.*, **7**, (1921); (b) *Naturwiss.*, **9**, 288 (1921); (c) *Naturwiss.*, **9**, 337 (1921).
240. M. Polanyi and K. Weissenberg, *Z. Phys.*, **9**, 123 (1922).
241. P. Scherrer, in R. Zsigmondy, *Kolloidchemie*, 3rd. ed. Leipzig, 1920, pp. 407–408.
242. S. Nishikawa and S. Ono, *Proc. Tokyo Math. Phys. Soc.*, [2] **7**, 131 (1913); **9**, 888 (1915).
243. R. Brill, *Ann.*, **434**, 204 (1923).
243a. Ref. 266, Vol. II, p. 417 of German edition, p. 445 of English edition.
244. R. O. Herzog, *Naturwiss.*, **12**, 955 (1924).
245. R. O. Herzog and K. Weissenberg, *Kolloid Z.*, **37**, 23 (1925).
246. H. Mark, *Ber.*, **59**, 2982 (1926).
247. K. Weissenberg, *Ber.*, **59**, 1526 (1926).
248. J. R. Katz, *Kolloid Z.*, **36**, 300 (1925); **37**, 19 (1925).
248a. Ref. 266, p. 152 of German text, p. 162 of English text.
249. O. L. Sponsler and W. H. Dore, *Colloid. Symp. Monograph*, **4**, 174 (1926).
250. W. Charlton, W. N. Haworth and S. Peat, *J. Chem. Soc.*, 89 (1926).
251. E. L. Hirst, *J. Chem. Soc.*, 350 (1926).
252. H. Staudinger and M. Lüthy, *Helv. Chim. Acta*, **8**, 65 (1925).
253. J. Hengstenberg, *Ann. Phys.*, **84**, 245 (1927).
254. H. Staudinger, H. Johner, H. Signer, G. Mie, and J. Hengstenberg, *Z. Phys. Chem.*, **126**, 425 (1927).
255. H. Staudinger, M. Staudinger and E. Sauter, *Z. Phys Chem.*, **B33**, 403 (1937).
255a. G. Wegner, *Makromol. Chem.*, **154**, 35 (1971).
256. E. Sauter, *Z. Phys. Chem.* B, **18**, 417 (1932).
257. H. Lohmann, thesis, University of Freiburg, 1931. This thesis is included in H. Staudinger, *Die Hochmolecularen Organischen Verbindungen Kautschuk und Cellulose*, Springer, 1932. The pertinent passage is on pp. 294–295.
258. K. H. Meyer and H. Mark, *Ber*, B **61**, 593 (1928).
259. K. H. Meyer and H. Mark, *Ber*, B **61**, 1932 (1928).
260. K. H. Meyer and H. Mark, *Ber*, B **61**, 1936 (1928).
261. K. H. Meyer and H. Mark, *Ber*, B **61**, 1939 (1928).
262. R. O. Herzog, *J. Phys. Chem.*, **30**, 457 (1926).
263. K. H. Meyer and L. Misch, *Helv. Chim. Acta*, **20**, 232 (1937).
263a. K. Hess, C. Trogus, W. Osswald, and K. Dziengel, *Z. Phys. Chem. B*, **7**, 1 (1930).
264. H. Mark, *Scientia*, **51**, 405 (1932).
265. K. H. Meyer and W. Lotmar, *Helv. Chim. Acta*, **19**, 68 (1936).
266. K. H. Meyer and H. Mark, *Hochpolymere Chemie*, Akad. Verlagsges., Leipzig, 1940, II, p. 120; K. H. Meyer, *Natural and Synthetic High Polymers*, Interscience, New York, 1942, p. 129.
266a. Ref. 266, p. 95, of German text, English text, p. 102.
266b. R. Brill and F. Halle, *Naturwiss.*, **26**, 12 (1938).

266c. C. W. Bunn, *Trans. Faraday Soc.*, **35**, 484 (1939).
267. D. Crowfoot Hodgkin, in *The Origins of Modern Biochemistry*, P. R. Srinivasan, J. S. Fruton, and J. T. Edsall, Eds., *Anny. N.Y. Acad. Sci.*, **325**, 121 (1979).
268. W. T. Astbury and A. Street, *Phil. Trans. R. Soc. London A*, **230**, 75 (1931).
269. W. T. Astbury and H. J. Woods. *Phil. Trans. R. Soc. London A*, **232**, 333 (1933).
270. W. T. Astbury, *Trans. Faraday Soc.*, **34**, 378 (1938).
271. R. O. Herzog and H. W. Gonnell, *Ber.*, **58**, 2228 (1925).
272. J. R. Katz and O. Gerngross, *Naturwiss.*, **13**, 900 (1925).
273. O. Gerngross and J. R. Katz, *Kolloid Z.*, **39**, 181 (1926).
274. K. Herrmann, O. Gerngross, and W. Abitz, *Z. Phys. Chem. B* **10**, 371 (1930).
275. J. D. Bernal and D. Crowfoot, *Nature*, **153**, 794 (1934).
276. A. L. Patterson, *Z. Krist.*, **90**, 517, 543 (1935).
277. D. Crowfoot, *Proc. R. Soc. London A*, **164**, 508 (1938).
278. R. W. G. Wyckoff and R. B. Corey, *J. Biol. Chem.*, **116**, 51 (1936).
279. W. M. Stanley and T. F. Anderson, *J. Biol. Chem.*, **139**, 325 (1941).
280. J. D. Bernal and I. Fankuchen, *J. Gen. Physiol.*, **25**, 111 (1941).
281. H. Staudinger, *Arbeitserrinerungen*, A. Hüthig, Heidelberg, 1961. English translation "*From Organic Chemistry to Macromolecules,*" Wiley, New York, 1970, p. 77, 79.
282. A. Frey-Wyssling, *Mikroskopie*, **19**, 5 (1964), quoted by R. Olby, *The Path to the Double Helix*, University of Washington Press, 1974, p. 7.
283. H. Mark, *Naturwiss.*, **67**, 477 (1980).
284. R. O. Herzog, (a) *Ber.*, **58**, 1256 (1925); (b) *Helv. Chim. Acta*, **11**, 529 (1928).
285. R. O. Herzog and D. Krüger, *Naturwiss*, **13**, 1040 (1925).
286. R. O. Herzog and D. Krüger, *Kolloid Z.*, **39**, 256 (1926).
287. P. Karrer, *Helv. Chim. Acta*, **3**, 620 (1920).
288. P. Karrer and C. Nägeli, *Helv. Chim. Acta*, **4**, 263 (1921).
289. H. Pringsheim and A. Langhans, *Ber.*, **45**, 2533 (1922).
290. H. Pringsheim, *Ber.*, **59**, 3008 (1926).
291. K. Hess, W. Weltzien, and E. Messmer, *Ann.*, **435**, 1 (1924).
292. D. MacGillavry, *Rec. Trav. Chim.*, **48**, 18 492 (1929).
293. K. Hess, *Rec. Trav. Chim.*, **48**, 489,583 (1929).
293a. K. Freudenberg, *Ber.*, **54**, 767 (1921).
294. K. Hess and C. Trogus, *Ber.*, **61** 1982 (1928).
295. M. Bergmann and E. Knehe, *Ann.*, **45**, 1 (1926).
296. M. Bergmann, A. Miekeley and E. Kenn, *Ann.*, **445**, 17 (1925).
296a. K. H. Meyer, Angew. Chem., **41**, 935 (1928).
297. R. Pummerer, H. Nielson, and W. Gündel, *Ber.*, **60**, 2167 (1927).
298. (a) L. Lautenschläger, thesis, University Karlsruhe, 1913; (b) H. Staudinger and L. Lautenschläger, *Ann.*, **488**, 1 (1931).
299. H. Staudinger, *Ber.*, **53**, 1073 (1920).
300. C. Harries, *Untersuchungen über die natürlichen und synthetischen Kautschukarten*, Berlin, 1919, p. 48.
301. H. Staudinger and J. Fritschi, *Helv. Chim. Acta*, **5**, 785 (1922).

References

302. H. Staudinger, M. Brunner, F. Frey, P. Garbsch, R. Signer, and S. Wehrli, *Ber.*, **62**, 241 (1929).
303. E. Trommsdorff in H. Staudinger, *Die hochmolekularen organischen Verbindungen, Kautschuk und Cellulose*, Springer, Berlin, 1932, p. 333.
304. H. Staudinger, R. Signer, H. Johner, M. Lüthy, W. Kern, D. Russidis, and O. Schweitzer, *Ann*, **474**, 145 (1929).
305. H. Staudinger and R. Signer, *Helv. Chim. Acta*, **11**, 1047 (1928); *Z. Kryst.*, **70**, 193 (1929).
306. H. Staudinger, *Ber.*, **63**, 921 (1930).
307. H. Staudinger, *Die hochmolekularen organischen Verbindungen, Kautschuk und Cellulose*, Springer, Berlin, 1932, p. 115.
308. C. Priesner, "*H. Staudinger, H. Mark, und K. H. Meyer, Thesen zur Grösse und Struktur der Makromoleküle*, Verlag Chemie, 1980, pp. 87–94.
309. E. Fischer, *Ber.*, **46**, 3288 (1913).
310. (a) H. Staudinger and W. Heuer, *Z. Phys. Chem. A*, **171**, 129 (1934); (b) Ref. 307, pp. 35–36.
311. H. Staudinger, A. A. Ashdown, M. Brunner, H. A. Bruson, and S. Wehrli, *Helv. Chim. Acta*, **12**, 934 (1929).
312. Ref. 307, p. 114.
313. H. W. Carothers, *Chem. Revs.*, **8**, 353 (1931).
314. Ref. 307, p. 164.
315. H. Staudinger, E. Geiger, and E. Huber, *Ber.*, **62**, 263 (1929).
316. Kolloid Z, **53**, 1–78 (1930).
317. M. Dunkel, *Z. Phys. Chem. A*, **138**, 42 (1928).
318. K. H. Meyer, *Angew. Chem.*, **41**, 935 (1928).
319. H. Staudinger and V. Wiedersheim, *Ber.*, **62**, 2406 (1929).
320. H. Staudinger and W. Heuer, *Ber.*, **62**, 2933 (1929).
321. H. Staudinger and H. F. Bondy, *Ann.*, **468**, 1 (1929).
322. K. H. Meyer, *Trans. Faraday Soc.*, **32**, 116 (1936).
323. Ref. 266, p. 182 of German edition.
324. H. Morawetz, *Macromolecules in Solution*, 2nd ed., Wiley, New York, 1979, pp. 414–428.
325. J. Spěváček and B. Schneider, *Makromol. Chem.*, **176**, 3409 (1975).
326. T. Svedberg and K. Estrup, *Kolloid Z.*, **9**, 259 (1911).
327. S. Oden, *Kolloid Z.*, **18**, 33 (1916).
328. T. Svedberg and H. Rinde, *J. Am. Chem. Soc.*, **45**, 943 (1923).
329. R. Brown, *Phil. Mag.*, (a) [2], **4**, 161 (1828); (b) [2], **6**, 161 (1829).
330. *Phil. Mag.*, [2], **8**, 296 (1830).
331. M. Gouy, *Compt. Rend.*, **109**, 102 (1889).
332. A. Einstein, *Ann. Phys.*, (a) [4], **17**, 549 (1905); (b) **19**, 371 (1906).
333. J. Perrin, *Ann. Chim. Phys.*, [8], **18**, 5 (1909).
334. T. Svedberg and J. B. Nichols, *J. Am. Chem. Soc.*, **45**, 2910 (1923).
335. T. Svedberg and R. Fåhreus, *J. Am. Chem. Soc.*, **48**, 430 (1926).
336. T. Svedberg and J. B. Nichols, *J. Am. Chem. Soc.*, **49**, 2920 (1927).

337. T. Svedberg, Nobelvortrag 1927, *Kolloidchem. Beih.*, **26**, 230 (1928).
338. T. Svedberg and K. O. Pedersen, *The Ultracentrifuge*, Clarendon, Oxford, 1940, p. 4.
339. M. J. Thovert, *Ann. Chim. Phys.*,[7], **26**, 366 (1902); *Ann. Phys.*, [9], **2**, 369 (1914).
340. Ref. 337, pp. 253–273.
341. E. O. Kraemer and W. D. Lansing, *J. Am. Chem. Soc.*, **55**, 4319 (1933).
342. E. O. Kraemer and W. D. Lansing, *J. Phys. Chem.*, **39**, 153 (1935).
343. R. Signer and H. Gross, *Helv. Chim. Acta*, **17**, 59, 335, 726 (1934).
344. A. Einstein, *Ann. Phys.*, [4], (a) **19**, 289 (1906); (b) **34**, 581 (1911).
345. M. Bancelin, *Compt. Rend.*, **152**, 1382 (1911); *Kolloid Z.*, **9**, *154* (1911).
346. M. v. Smoluchowski, *Kolloid Z.*, **18**, 191 (1916).
346a. W. Kuhn, *Z. Phys. Chem.*, **161**, 1 (1932); *Kolloid Z.*, **62**, 269 (1933).
346b. R. Eisenschitz, *Z. Phys. Chem.*, **163**, 133 (1932).
347. R. Simha, *J. Phys. Chem.*, **44**, 25 (1940).
348. S. Axelrod, *Gummi Z.*, **19**, 1053 (1905); **20**, 105 (1905).
349. J. G. Fol, *Kolloid Z.*, **12**, 131 (1913).
350. F. Kirchhof, *Kolloid Z.*, **15**, 40 (1914).
351. W. Biltz, *Z. Phys. Chem.: (a)* **83**, 683 (1913); (b) **91**, 705 (1916).
352. W. Ostwald, *Kolloid Z.*, **12**, 213 (1913).
353. E. Hatschek, *Kolloid Z.*, **12**, 238 (1913).
354. K. Hess, E. Messmer, and N. Ljubitsch, *Ann.*, **444**, 287 (1925).
355. P. Karrer and E. v. Krauss, *Helv. Chim. Acta*, **12**, 1144 (1929).
356. K. Hess, C. Trogus, L. Akim, and I. Sakurada, *Ber.*, **64**, 408 (1931).
357. H. Fikentscher and H. Mark, *Kolloid Z.*, **49**, 138 (1939).
358. H. Staudinger and W. Heuer, **Ber.**, **63**, 222 (1930).
359. M. Staudinger, *Trends Biochem. Sci*, **3**, 123 (1978).
360. H. Staudinger, *Trans. Faraday Soc.*, **29**, 18 (1933).
361. H. Mark, *Kolloid Z.*, **53**, 32 (1930).
362. H. Mark, *Z. Elektrochem.*, **40**, 413 (1934).
363. H. Staudinger, *Z. Elektrochem.*, **40**, 434 (1934).
364. H. Staudinger, *Ber.*, **67**, 1242 (1934).
365. R. Signer and H. Gross, *Z. Phys. Chem. A*, **165**, 182 (1933).
366. G. W. Stewart, *Chem. Rev.*, **6**, 483 (1929).
367. A. Eucken and K. Weigert, *Z. Phys. Chem. B*, **23**, 265 (1933).
368. H. Staudinger, *Ber.*, **67**, 92 (1934).
369. Ref. 308, pp. 164–166.
369a. R. Signer, *Trans. Faraday Soc.*, **32**, 296 (1936).
369b. W. Haller, *Kolloid Z.*, **56**, 257 (1931).
370. W. Kuhn, *Kolloid Z.*, **68**, 2 (1934).
370a. H. Mark, in *Der feste Körper*, S. Hirzel, Leipzig, 1937, p. 103.
370b. R. Houwink, *J. Prakt. Chem.*, **157**, 15 (1940).
370c. H. Staudinger and H. Warth, *J. Prakt. Chem.*, **155**, 261 (1940).
370d. H. Staudinger and H. Hellfritz, *Makromol. Chem.*, **7**, 274 (1951).

370e. B. Vollmert, private communication.
371. F. Eirich and H. Mark, *Ergeb. Exakt. Naturwiss.*, **15**, 1 (1936).
372. E. O. Kraemer, *Ind. Eng. Chem.*, **30**, 1200 (1938).
373. W. Ostwald, *Kolloid Z.*, **53**, 42 (1930).
374. J. W. Strutt (Lord Rayleigh), *Phil. Mag.*: (a) [4], **41**, 107 (1871); (b) [5], **47**, 375 (1899).
375. W. Mecklenburg, *Kolloid Z.*, **16**, 97 (1915).
375a. L. Arisz, *Kolloid Beih.*, **7**, 22 (1915).
375b. E. O. Kraemer and S. T. Dexter, *J. Phys. Chem.*, **31**, 764 (1927).
375c. C. V. Raman, *Indian J. Phys.*, **2**, 1 (1927).
375d. C. V. Raman, *Nature*, **120**, 158 (1927).
376. B. J. Holwerda, *Rec. Trav. Chim.*, **50**, 601 (1931).
377. P. Putzeys and J. Brosteaux, *Trans. Faraday Soc.*, **31**, 1314 (1935).
378. D. Gehman and J. E. Field, *Ind. Eng. Chem.*, **29**, 793 (1937).
379. A. V. Lourenço, (a) *Compt. Rend.*, **49**, 619 (1959); (b) *Ann. Chim. Phys.* [3], **67**, 257 (1863).
380. A. Wurtz, (a) *Compt. Rend.*, **49**, 813 (1859); (b) *Compt. Rend.*, **50**, 1195 (1860).
381. J. v. Braun and W. Sobecki, *Ber.*, **44**, 1918 (1911).
382. C. Gerhardt, *Ann.*, **87**, 149 (1853).
383. K. Kraut, *Ann.*, **150**, 1 (1869).
384. R. Anschütz, *Ann.*, 273, 73 (1893); R. Anschütz and G. Schröter, *Ann.*, **273**, 97 (1893).
385. D. Vorländer, *Ann.*, **280**, 167 (1894).
386. E. Roithner, *Monatsch. Chem.*, **15**, 665 (1894).
387. C. A. Bischoff and A. v. Hedenström, *Ber.*, **35**, 3431 (1902).
388. A. Einhorn, *Ann.*, **300**, 135 (1898).
389. E. E. Blaise and L. Marcilly, *Bull. Soc. Chim. France*, [3], **31**, 308 (1904).
390. T. Curtius, *Ber.*, **37**, 1284 (1904).
391. P. J. Flory, *Principles of Polymer Chemistry*, Chapter I, Cornell University Press, Ithaca, New York, 1953.
392. (a) E. K. Bolton, *Ind. Eng. Chem.*, **34**, 53 (1942); (b) W. H. Carothers, "Early history of polyamide fibers," memorandum to A. P. Tanberg, February 19, 1936.
393. Ref. 307, p. 149.
394. W. H. Carothers, *J. Am. Chem. Soc.*, **51**, 2548 (1929).
395. A. Bayer, *Ber.*, **18**, 2277 (1885).
396. H. Sachse, *Z. Phys. Chem.*, **10**, 203 (1892); **11**, 185 (1893).
397. E. Mohr, *Z. Prakt. Chem. N.F.*, **98**, 315 (1918).
398. M. Stoll and G. Stoll-Comte, *Helv. Chim. Acta*, **13**, 1185 (1930).
398a. W. H. Carothers and A. Arvin, *J. Am. Chem. Soc.*, **51**, 2560 (1929).
399. W. H. Carothers and F. J. van Natta, **J. Am. Chem. Soc.**, **52**, 314 (1930).
400. W. H. Carothers, J. A. Arvin, and G. L. Dorough, *J. Am. Chem. Soc.*, **52**, 3292 (1930).
401. M. Stoll and A. Rouvé, *Helv. Chim. Acta*, **18**, 1087 (1935).
402. E. W. Spanagel and W. H. Carothers, *J. Am. Chem. Soc.*, **57**, 929 (1935).
403. W. H. Carothers and G. A. Arvin, *J. Am. Chem. Soc.*, **51**, 2560 (1929).
404. H. Staudinger, *Ber.*, **59**, 3019 (1926).

405. W. H. Carothers, J. W. Hill, G. E. Kirby, and R. A. Jacobson, *J. Am. Chem. Soc.*, **52**, 5279 (1930).
406. W. H. Carothers and G. L. Dorough, *J. Am. Chem. Soc.*, **52**, 711 (1930).
407. W. H. Carothers and J. W. Hill, *J. Am. Chem. Soc.*, **54**, 1559, 1579 (1932).
408. W. H. Carothers and J. W. Hill, *J. Am. Chem. Soc.*, **54**, 1566 (1932).
408a. P. Schlack, U.S. Pat. 2,241,321 (1941).
409. U.S. Pat. 2,130,947 and 2,130,948 (W. H. Carothers to DuPont, September 20, 1938).
410. Videotape interview "Reflections by Two Eminent Chemists, Drs. H. Mark and P. J. Flory," 1982, Copyright by the American Chemical Society.
411. W. H. Carothers, *Trans. Faraday Soc.*, **32**, 44 (1936).
412. P. J. Flory, *J. Am. Chem. Soc.*, **58**, 1877 (1936).
413. H. Dostal, *Monatsh. Chem.*, **70**, 324 (1937); H. Mark, *Trans. Faraday Soc.*, **32**, 51 (1936).
414. P. J. Flory, *J. Am. Chem. Soc.*, **59**, 466 (1937); **61**, 3334 (1939).
414a. P. J. Flory, *J. Am. Chem. Soc.*, **63**, 3083 (1941).
415. R. H. Kienle and A. G. Hovey, *J. Am. Chem. Soc.*, **51**, 509 (1929).
415a. W. H. Stockmayer, *J. Chem. Phys.*, **11**, 45 (1943).
415b. C. Friedel and J. M. Crafts, *Compt. Rend.*, **56**, 592 (1863).
415c. F. S. Kipping, *Proc. R. Soc. London A*, **159**, 139 (1937).
415d. J. A. Meads and F. S. Kipping, *J. Chem. Soc.*, **105**, 679 (1914).
415e. F. S. Kipping, *J. Chem. Soc.*, 2108 (1912).
415f. F. S. Kipping, *J. Chem. Soc.*, 2125 (1912).
415g. F. S. Kipping and A. G. Murray, *J. Chem. Soc.*, 142 (1928).
415h. H. A. Liebhafsky, *Silicones Under the Monogram*, Wiley-Interscience, 1978.
415i. E. G. Rochow and W. F. Gillam, *J. Am. Chem. Soc.*, **63**, 798 (1941); U.S. Pat. 2,258,218 (E.G. Rochow to General Electric Co. October 7, 1941).
415j. C. Combes, *Compt. Rend.*, **122**, 531 (1896).
415k. E. G. Rochow, *J. Am. Chem. Soc.*, **67**, 963 (1945); U.S. Pat. 2,380,995; 2,380,996; 2,380,997 (E. G. Rochow to General Electric Co., September 1941.
415l. F. C. Harrison, H. L. A. Tarr, and H. Hibbert, *Can. J. Res.*, **3**, 449 (1930); H. Hibbert, R. S. Tipson, and F. Brauns, *Can. J. Res.*, **4**, 221 (1931).
415m. G. T. Cori and C. F. Cori, *J. Biol. Chem.*, **135**, 733 (1940).
415n. C. S. Hanes, *Proc. R. Soc. B*, **128**, 421; *B*, 129, 174 (1940).
415o. E. Husemann, B. Fritz, B. Lippert, and B. Pfannemüller, *Makromol. Chem.*, **26**, 199 (1958).
416. G. S. Whitby and M. Katz, *J. Am. Chem. Soc.*, **50**, 1160 (1928).
417. M. Berthelot, Ref. 46, pp. 32–33.
418. H. Staudinger, M. Brunner, K. Frey, P. Garbsch, R. Signer, and S. Wehrli, *Ber.*, **62**, 241 (1929).
419. H. Staudinger and H. W. Kohlschütter, *Ber.*, **64**, 2019 (1931).
419a. H. Stobbe and G. Posnjak, *Ann.*, **371**, 259 (1909).
419b. H. Stobbe, *Ann.*, **409**, 1 (1915).
419c. W. O. Herrmann, *Vom Ringen mit den Molekülen*, Econ-Verlag, Düsseldorf, 1963.
419d. W. O. Herrmann and W. Haehnel, *Ber.*, **60**, 1658 (1927).

419e. H. Staudinger, K. Frey, and W. Starck, *Ber.*, **60**, 1782 (1927).
419f. H. v. Pechmann and O. Röhm, *Ber.*, **34**, 427 (1903).
420. H. Staudinger and W. Frost, *Z. Phys. Chem. B*, **29**, 235 (1935).
421. G. V. Schulz, *Z. Phys. Chem. B*, **30**, 379 (1935).
422. H. Dostal and H. Mark, *Z. Phys. Chem. B*, **29**, 299 (1935).
423. H. Mark and R. Raff, *Z. Phys. Chem. B*, **31**, 275 (1936).
424. H. Dostal and H. Mark, *Trans. Faraday Soc.*, **32**, 54 (1936).
425. G. Gee and E. K. Rideal, *Trans. Faraday Soc.*, **31**, 969 (1935).
426. G. Gee, *Trans. Faraday Soc.*, **32**, 656 (1936).
427. C. E. H. Bawn, *Trans. Faraday Soc.*, **32**, 178 (1936).
428. H. Staudinger and A. Steinhofer, *Ann.*, **517**, 35 (1935).
429. R. Signer and J. Weiler, *Helv. Chim. Acta*, **15**, 649 (1932).
430. G. V. Schulz and E. Husemann, *Z. Phys. Chem. B*, **34**, 187 (1935).
431. G. V. Schulz and E. Husemann, *Z. Phys. Chem. B*, **36**, 183 (1937).
432. H. S. Taylor and W. H. Jones, *J. Am. Chem. Soc.*, **52**, 1111 (1930).
433. H. W. Starkweather and G. B. Taylor, *J. Am. Chem. Soc.*, **52**, 4708 (1930).
434. H. S. Taylor and A. A. Vernon, *J. Am. Chem. Soc.*, **53**, 2527 (1931).
435. C. Moureu and C. Dufraisse, *Bull. Soc. Chim. France* [4], **35**, 1564 (1924).
436a. K. K. Jeu and H. N. Alyea, *J. Am. Chem. Soc.*, **55**, 575 (1933).
436b. W. Chalmers, *J. Am. Chem. Soc.*, **56**, 912 (1934).
436c. J. B. Conant and C. O. Tongberg, *J. Am. Chem. Soc.*, **52**, 1659 (1930).
436d. J. B. Conant and W. R. Peterson, *J. Am. Chem. Soc.*, **54**, 628 (1932).
437. H. Wieland and G. Rasuvayev, *Ann.*, **480**, 157 (1930).
438. D. H. Hay and W. A. Waters, *Chem. Revs.*, **21**, 169 (1937).
439. P. J. Flory, *J. Am. Chem. Soc.*, **59**, 241 (1937).
440. G. V. Schulz, *Z. Phys. Chem. B*, **43**, 25 (1939).
441. G. V. Schulz and A. Dinglinger, *Z. Phys. Chem. B*, **43**, 47 (1939).
442. R. G. W. Norrish and E. F. Brookman, *Proc. R. Soc. London, A*, **171**, 147 (1939).
443. G. V. Schulz and F. Blaschke, *Z. Phys. Chem. B*, **50**, 305 (1941).
444. R. G. W. Norrish and R. R. Smith, *Nature*, **150**, 336 (1942).
445. G. V. Schulz and G. Harboth, *Makromol. Chem.*, **1**, 106 (1947).
445a. E. Trommsdorff, H. Kohle, and P. Legally, *Makromol. Chem.*, **1**, 169 (1947).
445b. M. W. Perrin, *Research*, **6**, 111 (1953); see also "Polyethylene—the first fifty years," Plastics and Rubber International, **8**, 127 (1983).
445c. E. W. Fawcett and R. O. Gibson, *J. Chem. Soc.*, 386 (1934).
445d. E. W. Fawcett, *Trans. Faraday Soc.*, **32**, 119 (1936).
445e. E. W. Fawcett, R. O. Gibson, M. W. Perrin, J. G. Patton, and E. G. Williams (to ICI), Br. Pat. 471,590 (September 6, 1937).
445f. W. Reppe, *Polyvinylpyrrolidon*, Verlag Chemie, Weinheim, 1954.
446. T. Wagner-Jauregg, *Ber.*, **63**, 3213 (1930).
446a. H. Staudinger and E. Husemann, *Ber.*, **68**, 1618 (1935); H. Staudinger, W. Heuer, and E. Husemann, *Trans. Faraday Soc.*, **32**, 323 (1936).

446b. R. G. W. Norrish and E. F. Brookman, *Proc. R. Soc. A*, **163**, 205 (1937).
446c. R. Hill, J. R. Lewis and J. L. Simonson, *Trans. Faraday Soc.*, **35**, 1073 (1939).
446d. U.S. Pat. 2,100,900 (1937).
446e. H. Staudinger and J. Schneiders, *Ann.*, **541**, 151 (1939).
446f. C. S. Marvel, G. D. Jones, T. W. Mastin, and G. L. Schertz, *J. Am. Chem. Soc.*, **64**, 2356 (1942).
446g. H. Dostal, *Monatsh. Chem.*, **69**, 424 (1936).
446h. F. T. Wall, *J. Am. Chem. Soc.*, **63**, 1862 (1941).
446i. H. Staudinger and B. Ritzenthaler, *Ber.*, **68**, 455 (1935).
446j. D. S. Frederick, H. D. Cogan, and C. S. Marvel, *J. Am. Chem. Soc.*, **56**, 1815 (1934); M. Hunt and C. S. Marvel, *J. Am. Chem. Soc.*, **57**, 1691 (1935).
447. K. Ziegler and K. Bahr, *Ber.*, **61**, 253 (1928).
448. K. Ziegler and H. Kleiner, *Ann.*, **473**, 57 (1929).
449. K. Ziegler, F. Dersch, and H. Wollthan, *Ann.*, **511**, 13 (1934).
450. K. Ziegler and L. Jakob, *Ann.*, **511**, 45 (1934).
451. K. Ziegler, L. Jakob, H. Wollthan, and A. Wenz, *Ann.*, **511**, 64 (1934).
452. F. Auerbach and H. Barschall, *Chem. Abst.*, **2**, 1125 (1908).
453. H. Staudinger, H. Johner, M. Lüthy, and R. Signer, *Ann.*, **474**, 155 (1929).
454. (a) Ref. 307, p. 256; (b) H. Staudinger and W. Kern, *Ber.*, **66**, 1863 (1933).
455. Ref. 307, p. 288-290.
456. (a) H. Leuchs, *Ber.*, **39**, 857 (1906); (b) H. Leuchs and W. Geiger, *Ber.*, **41**, 1721 (1908).
457. F. Wessely and F. Sigmund, *Z. Physiol. Chem.*, **159**, 102 (1926).
458. K. H. Meyer and Y. Go, *Helv. Chim. Acta*, **17**, 1488 (1934).
459. P. J. Flory, *J. Am. Chem. Soc.*, **62**, 1561 (1940).
460. E. Abderhalden, *Naturwiss.*, **12**, 716 (1924).
461. M. Bergmann, *Naturwiss.*, **13**, 1045 (1925).
462. E. J. Cohn and J. B. Conant, *Proc. Natl. Acad. Sci. U.S.A.*, **12**, 433 (1926).
463. T. Svedberg, *Kolloid Z.*, **51**, 10 (1930).
464. (a) T. Svedberg and F. F. Heyroth, *J. Am. Chem. Soc.*, **51**, 550 (1929); (b) T. Svedberg and B. Sjögren, *J. Am. Chem. Soc.*, **52**, 279 (1930).
465. T. Svedberg and N. B. Lewis, *J. Am. Chem. Soc.*, **50**, 525 (1928).
466. T. Svedberg, *Proc. R. Soc. London A*, **170**, 40 (1939).
467. S. P. L. Sørensen, *Kolloid Z.*, **53**, 102, 308 (1930).
468. E. Abderhalden and H. Brockman, *Biochem. Z.*, **211**, 395 (1929).
469. A. Tiselius, *Trans. Faraday Soc.*, **33**, 524 (1927); *Kolloid Z.*, **85**, 129 (1928).
470. A. Tiselius, *Biochem. J.*, **31**, 1464 (1937).
471. L. G. Longsworth, R. K. Cannan, and D. A. MacInnes, *J. Am. Chem. Soc.*, **62**, 2580 (1940).
472. H. B. Vickery and T. B. Osborne, *Physiol. Revs.*, **8**, 393 (1928).
473. E. Waldschmidt-Leitz and E. Simons, *Z. Physiol. Chem.*, **156**, 114 (1926).
474. A. L. v. Muralt and J. T. Edsall, *J. Biol. Chem.*, **89**, 315, 351 (1930).
475. G. Boehm and R. Signer, *Helv. Chim. Acta*, **14**, 1370 (1931).

476. D. M. Wrinch, *Proc. R. Soc. A*, **161**, 505 (1936).
477. M. Spiegel-Adolf, *Biochem. Z.*, **170**, 126 (1926).
478. A. E. Mirsky and M. L. Anson, *J. Phys. Chem.*, **35**, 185 (1931).
479. M. L. Anson and A. E. Mirsky, *J. Gen. Physiol.* **14**, 725 (1931).
480. W. C. M. Lewis, *Chem. Revs.*, **8**, 81 (1931).
481. H. Wu, *Chinese J. Physiol.*, **5**, 321 (1931); Chem Abstr., **26**, 740 (1932).
482. A. E. Mirsky and L. Pauling, *Proc. Natl. Acad. Sci., U.S.*, **22**, 439 (1936).
483. Nobel Lectures, Chemistry, Elsevier, Amsterdam, 1964.
484. J. B. Sumner, *J. Biol. Chem.*, **69**, 435 (1926).
485. R. M. Willstätter, *Naturwiss.*, **15**, 585 (1927).
486. J. H. Northrop, *J. Gen. Physiol.*, **13**, 739 (1930).
487. J. H. Northrop, *J. Gen. Physiol.*, **14**, 313 (1931).
488. M. L. Anson and A. E. Mirsky, *J. Gen. Physiol.*, **17**, 393 (1934).
488a. J. S. Fruton, in *Of Oxygen, Fuels and Living Matter*, Chapter 5, Part 2, G. Semenza, Ed., Wiley, New York, 1982.
489. M. Bergmann and J. S. Fruton, *Advan. Enzymol.*, **1**, 63 (1941).
490. M. Bergmann, *Chem. Revs.*, **22**, 423 (1938).
491. K. Landsteiner, *Die Spezifizität der serologischen Reaktionen*, Springer, Berlin, 1933; *The Specificity of Serological Reactions*, Thomas, Springfield, Illinois, 1936; updated edition, Dover, New York, 1945.
492. K. Landsteiner and H. Lampl, *Biochem. Z.*, **86**, 343 (1918).
493. J. R. Marrack and F. C. Smith, *Nature*, **128**, 1037 (1931).
494. F. Haurowitz and F. Breinl, *Z. Physiol. Chem.*, **214**, 111 (1933).
495. M. Heidelberger, *Physiol. Rev.*, **7**, 207 (1927); *Chem. Revs.*, **3**, 403 (1927).
496. K. Landsteiner, *Science*, **73**, 403 (1931).
497. F. Haurowitz, *Z. Physiol. Chem.*, **245**, 23 (1936).
498. M. Heidelberger and F. E. Kendall, *J. Exp. Med.*, **61**, 559 (1935); M. Heidelberger, *Chem. Revs.*, **24**, 323 (1939).
499. W. C. Boyd and S. B. Hooker, *Proc. Soc. Exp. Biol. Med.*, **39**, 491 (1938).
500. F. Breinl and F. Haurowitz, *Z. Physiol. Chem.*, **192**, 45 (1930).
501. L. Pauling, *J. Am. Chem. Soc.*, **62**, 2643 (1940).
501a. L. Pauling and D. H. Campbell, *Science*, **95**, 440 (1942).
502. H. Freundlich and A. E. Hauser, *Kolloid Z.*, **36**, Suppl. 15 (1925).
503. W. Ostwald, *Kolloid Z.*, **40**, 58 (1926).
504. A. E. Hauser, *Ind. Eng. Chem.*, **21**, 249 (1929).
505. L. Hock (a) *Kolloid Z.*, **35**, 40 (1924); (b) *Z. Elektrochem.*, **31**, 404 (1925).
506. H. Feuchter, *Kautschuk*, (a) **2**, 171, 197 (1926); (b) **3**, 98, 122 (1927).
507. H. Mark and E. Valkó, *Kautschuk*, **6**, 210 (1930).
508. E. Wöhlisch, *Verh. Phys. Med. Ges. Wüzzburg N. F.*, **51**, 53 (1926).
509. K. H. Meyer, *Biochem. Z.*, **214**, 253 (1929).
510. H. Fikentscher and H. Mark, *Kautschuk*, **6**, 2 (1930).
511. G. S. Whitby, *J. Phys. Chem.*, **36**, 198 (1932).

512. H. Staudinger, *Helv. Chim. Acta*, **13**, 1324 (1930); Ref. 307, pp. 121–122.
513. H. L. Bender, H. F. Wakefield, and H. A. Hoffman, *Chem. Rev.*, **15**, 123 (1934).
514. K. H. Meyer, G. v. Susich, and E. Valkó, *Kolloid Z.*, **59**, 208 (1932).
515. H. Staudinger, *Kolloid Z.*, **60**, 296 (1932).
516. W. F. Busse, *J. Phys. Chem.*, **36**, 2862 (1932).
517. E. Guth and H. Mark, *Monatsh. Chem.*, **65**, 93 (1934).
518. W. Kuhn, *Kolloid Z.*, **76**, 258 (1936).
518a. W. Kuhn, *Angew, Chem.*, **51**, 640 (1938).
519. K. H. Meyer and C. Ferri, *Helv. Chim. Acta*, **18**, 570 (1935).
520. R. L. Anthony, R. H. Caston, and E. Guth, *J. Phys. Chem.*, **46**, 826 (1942).
520a. S. E. Bresler and Ya. I. Frenkel, *J. Exp. Theor. Phys. (U.S.S.R.)*, **9**, 1094 (1939); English translation, *Rubber Chem. Tech.*, **16**, 1 (1943).
521. C. W. Bunn, *Proc. R. Soc. A*, **180**, 82 (1942).
521a. A. P. Aleksandrov and Yu. S. Lazurkin, *J. Technol. Phys. U.S.S.R.*, **9**, 1249 (1939); English Translation *Rubber Chem. Technol.*, **13**, 886, 898 (1940).
522. J. Duclaux and E. Wollman, *Compt. Rend.*, **152**, 1580 (1911).
523. W. A. Caspari, *J. Chem. Soc.*, **105**, 2139 (1914).
524. E. Posnjak, *Kolloidchem. Beih.*, **3**, 417 (1912).
525. W. Ostwald and K. Mündler, *Kolloid Z.*, **24**, 7 (1919).
526. G. S. Adair, *Proc. R. Soc. A*, **120**, 573 (1928).
527. W. Haller, *Kolloid Z.*, **49**, 74 (1929).
528. G. V. Schulz, *Z. Phys. Chem. A*, **158**, 237 (1932).
529. K. H. Meyer and P. Lühdemann, *Helv. Chim. Acta*, **18**, 307 (1935).
530. E. Hückel, *Z. Elektrochem.*, **42**, 753 (1936).
531. C. G. Boissonas, *Helv. Chim. Acta*, **20**, 768 (1937).
532. E. A. Guggenheim, *Trans. Faraday Soc.*, 33, 151 (1937).
533. R. H. Fowler and G. S. Rushbrooke, *Trans. Faraday Soc.*, **33**, 1272 (1937).
534. K. H. Meyer, *Helv. Chim. Acta*, **23**, 1063 (1940).
535. M. L. Huggins, (a) *J. Phys. Chem.*, **46**, 151 (1942); (b) *Ann. N.Y. Acad. Sci.*, **43**, 1 (1942); (c) *J. Am. Chem. Soc.*, **64**, 1712 (1942).
536. P. J. Flory, *J. Chem. Phys.*, **10**, 51 (1942).
537. J. Loeb, *Proteins and the Theory of Colloid Behavior*, McGraw-Hill, New York, 1922.
538. E. J. Cohn, *Physiol. Rev.*, **5**, 349 (1925).
539. W. Pauli, *Biochem. Z.*, **202**, 337 (1928).
540. J. Loeb, *J. Gen. Physiol.*, **3**, 827 (1921); **4**, 73, 97 (1922).
541. H. R. Procter, *J. Chem. Soc.*, **105**, 313 (1914); H. R. Procter and J. A. Wilson, *J. Chem. Soc.*, **109**, 307 (1916).
542. E. Hammersten, *Biochem. Z.*, **144**, 383 (1923).
543. A. W. Thomas and H. A. Murray, Jr., *J. Phys. Chem.*, **32**, 676 (1928).
544. D. R. Briggs, *J. Phys. Chem.*, **38**, 867 (1934).
545. W. Kern, *Z. Phys. Chem. A*, **181**, 249, 283 (1938); **184**, 197 (1939).
546. H. Green, *J. Chem. Soc.*, **93**, 2049 (1908).

546a. *Chemie Kunststoffe Aktuel*, **34**, (2); also personal communication.
546b. File DII 15.12, Staudinger Archiv, Deutsches Museum, Munich.
547. K. Nozaki and P. D. Bartlett, *J. Am. Chem. Soc.*, **68**, 1686 (1946).
548. G. V. Schulz, *Naturwiss.*, **27**, 659 (1939).
549. (a) F. M. Lewis and M. S. Matheson, *J. Am. Chem. Soc.*, **71**, 747 (1949); (b) C. G. Overberger, P. Fram, and T. Alfrey, Jr., *J. Polym. Sci.*, **6**, 539 (1951).
550. C. C. Price and B. E. Tate, *J. Am. Chem. Soc.*, **65**, 517 (1943).
550a. W. Kern and H. Kämmerer, *Makromol. Chem.*, **2**, 127 (1948).
551. J. C. Bevington, H. W. Melville, and R. P. Taylor, *J. Polym. Sci.*, **12**, 449; **14**, 476 (1954).
552. (a) H. Logemann (to Farbenfabriken Bayer) Ger. Pat. 891,025 (March 18, 1939); (b) J. Monheim (to Farbwerke Hoechst) Ger. Pat. 926,091 (July 21, 1942); (c) A. Hardt and W. Kern (to Farbwerke Hoechst) Ger. Pat. 863,996 (November 11, 1943).
553. W. Kern, *Makromol. Chem.*, **1**, 209 (1948).
554. R. G. R. Bacon, *Trans. Faraday Soc.*, **42**, 140 (1946).
555. J. H. Baxendale, M. G. Evans, and G. S. Park, *Trans. Faraday Soc.*, **42**, 155 (1946).
555a. F. Haber and J. Weiss, *Proc. R. Soc. (London)*, **A 147**, 332 (1939).
556. L. B. Morgan, *Trans. Faraday Soc.*, **42**, 169 (1946).
557. H. W. Melville, *Proc. R. Soc. A*, **163**, 511 (1937).
558. A. Berthoud and H. Bellenot, *Helv. Chim. Acta*, **7**, 307 (1924).
559. T. T. Jones and H. W. Melville, *Proc. R. Soc. A*, **175**, 392 (1940).
560. F. Paneth and W. Hofeditz, *Ber.*, **62**, 1335 (1929).
561. G. M. Burnett and H. W. Melville, *Nature*, **156**, 661 (1945); *Proc. Roy. Soc. A*, **189**, 956 (1947).
562. C. H. Bamford and M. J. S. Dewar, *Proc. R. Soc. A*, **192**, 309 (1947).
563. G. M. Burnett, *Trans. Faraday Soc.*, **46**, 772 (1950).
564. N. Grassie and H. W. Melville, *Proc. R. Soc.*, **207**, 285 (1951).
565. T. G. Majury and H. W. Melville, *Proc. R. Soc.*, **205**, 496 (1951).
566. J. Matheson, E. E. Auer, E. B. Bevilacqua, and E. J. Hart, *J. Am. Chem. Soc.*, **71**, 497 (1951).
567. S. E. Bresler, E. N. Kazbekov, and E. M. Saminskii, *Vysokomol. Soedin.*, **1**, 132 (1959).
568. T. Alfrey, Jr., and G. Goldfinder, *J. Chem. Phys.*, **12**, 205 (1944).
569. F. R. Mayo and F. M. Lewis, *J. Am. Chem. Soc.*, **66**, 1594 (1944).
570. (a) F. R. Mayo and C. Walling, *Chem. Revs.*, **46**, 191 (1950); (b) T. Alfrey, Jr., J. J. Bohrer, and H. Mark, *Copolymerization*, Interscience, New York, 1952.
571. F. M. Lewis, F. R. Mayo, and W. F. Hulse, *J. Am. Chem. Soc.*, **67**, 1701 (1945).
572. T. Alfrey, Jr., and C. C. Price, *J. Polym. Sci.*, **2**, 101 (1947).
573. C. Walling, *J. Am. Chem.Soc.*, **71**, 1930 (1949).
574. A. C. Cuthbertson, G. Gee, and E. K. Rideal, *Proc. R. Soc. A*, **170**, 300 (1939).
575. H. Suess, K. Pilch, and H. Rudorfer, *Z. Phys. Chem. A*, **179**, 361 (1937); H. Suess and A. Springer, *Z. Phys. Chem. A*, **181**, 81 (1937).
576. G. V. Schulz, A. Dinglinger, and F. Husemann, *Z. Phys. Chem. B*, **43**, 385 (1939).
577. J. W. Breitenbach and A. Maschin, *Z. Phys. Chem. A*, **187**, 175 (1940).

578. F. R. Mayo, *J. Am. Chem. Soc.*, **65**, 2324 (1945).
579. J. W. Livingston and J. T. Cox, Jr., Chapter 7, in *Synthetic Rubber*, G. S. Whitby, C. C. Davis, and R. F. Dunbrook, Eds., Wiley, New York, 1954.
580. C. S. Marvel, private communication.
581. H. Logemann and G. Pampus, *Kautsch. Gummi Kunstst*, **23**, 479 (1970).
582. K. Meisenburg, I. Dennstedt and E. Zancker, Ger. Pat. 751,604 (March 3, 1937).
583. H. Wollthan and W. Becker, Ger. Pat. 753,991 (1937).
584. R. Snyder, J. M. Steward, R. E. Allen, and R. T. Dearborn, *J. Am. Chem. Soc.*, **68**, 1422 (1946).
585. M. S. Kharasch, A. T. Read, and F. R. Mayo, *Chem. Ind.*, **57**, 752 (1938); F. R. Mayo and C. Walling, *Chem. Revs.*, **27**, 351 (1940).
586. R. M. Houtz and H. Adkins, *J. Am. Chem. Soc.*, **55**, 1609 (1933).
587. O. L. Wheeler, S. L. Ernst, and R. N. Crozier, *J. Polym. Sci.*, **8**, 409 (1952); O. L. Wheeler, E. Lavin, and R. N. Crozier, *J. Polym. Sci.*, **9**, 157 (1952).
588. H. Staudinger and H. Warth, *J. Prakt. Chem. NF*, **155**, 261 (1940).
589. T. G. Fox and S. Gratch, *Ann. N.Y. Acad. Sci.*, **57**, 367 (1953).
590. F. E. Matthews, Fr. Pat. 459,134 (October 28, 1913).
590a. U.S. Pat. 1,613,673 (I. Ostromislensky to Naugatuck Co. January 11, 1927).
590b. U.S. Pat. 2,694,692 (J. L. Amos, J. L. McCurdy, and O. R. McIntire to Dow Chemical Co., November 16, 1954).
590c. R. G. R. Bacon, C. R. Morrison-Jones, and K. D. Errington, *Rubber Tech. Conference*, London, Preprint 56 (1938); *Chem. Abstr.*, **32**, 8201); R. G. R. Bacon and E. H. Farmer, *Rubber Chem. Techn.*, **12**, 200 (1939).
590d. P. Compagnon and J. LeBras, *Compt. Rend.*, **212**, 616 (1941).
591. R. C. Carlin and N. E. Shakespeare, *J. Am. Chem. Soc.*, **68**, 876 (1946).
592. D. Bandel, Thesis, Polytechnic Institute of Brooklyn, 1950; quoted in Ref. 570b, p. 159, and by E. H. Immergut and H. Mark, *Makromol. Chem.*, **18/19**, 322 (1956).
593. G. Smets and M. Claesen, *J. Polym. Sci.*, **8**, 289 (1952).
594. (a) E. E. Schneider, *Discuss. Faraday Soc.*, **19**, 158 (1951); (b) E. J. Lawton, R. S. Powell, and J. S. Balwit, *J. Polym. Sci.*, **32**, 277 (1958).
595. A. Chapiro, *J. Chim. Phys.*, **53**, 805 (1956); L. A. Wall and D. W. Brown, Jr., *J. Res. Nat. Bur. Stand.*, **57**, 131 (1956); J. C. Bevington and D. E. Eaves, *Nature*, **178**, 1112 (1956).
596. A. Chapiro, *J. Polym. Sci.*, **29**, 321 (1958).
597. A. Chapiro, *J. Polym. Sci.*, **34**, 481 (1959).
598. U. Henning, E. M. Möslein, B. Arreguin, and F. Lynen, *Biochem. Z.*, **333**, 534 (1961).
599. J. C. W. Crawford, Br. Pat. 427,494 (to ICI, October 25, 1933).
600. O. Röhm and E. Trommsdorff, Ger. Pat. 735,284 (to Röhm and Haas, December 31, 1934).
601. J. H. Baxendale, M. G. Evans, and J. K. Kilham, *J. Polym. Sci.*, **1**, 466 (1946).
602. W. P. Hohenstein and H. Mark, *J. Polym. Sci.*, **1**, 549 (1946).
603. W. D. Harkins, (a) *J. Chem. Phys.*, **13**, 381 (1945); **14**, 47 (1946); (b) *J. Am. Chem. Soc.*, **69**, 1428 (1947).
604. W. V. Smith and R. H. Ewart, *J. Chem. Phys.*, **16**, 592 (1948).

605. W. V. Smith, *J. Am. Chem. Soc.*, **70**, 3695 (1948).
606. J. W. Vanderhoff, J. F. Vitkuske, E. B. Bradford, and T. Alfrey, Jr., *J. Polym. Sci*, **20**, 225 (1954).
607. T. Alfrey, E. B. Bradford, J. W. Vanderhoff, and G. Oster, *J. Opt. Soc. Am.*, **44**, 603 (1954).
608. H. Morawetz and T. A. Fadner, *Makromol. Chem.*, **34**, 162 (1959); *J. Polym. Sci.*, **45**, 475 (1960).
609. H. Morawetz and I. D. Rubin, *J. Polym. Sci.*, **57**, 669 (1962).
610. D. M. White, *J. Am. Chem. Soc.*, **82**, 5678 (1960).
611. S. Okamura, K. Hayashi, and Y. Kitanishi, *J. Polym. Sci.*, **58**, 927 (1962); S. Okamura, K. Hayashi, and M. Nishii, *J. Polym. Sci.*, **60**, 526 (1962).
612. F. Bohlmann and E. Inhoffen, *Ber.*, **89**, 1276 (1956).
613. H. S. Taylor and A. V. Tobolsky, *J. Am. Chem. Soc.*, **67**, 2063 (1945).
614. R. B. Mesrobian and A V. Tobolsky, *J. Am. Chem. Soc.*, **67**, 785 (1945); *J. Polym. Sci.*, **2**, 463 (1947).
615. R. D. Snow and F. E. Frey, *Ind. Eng. Chem.*, **30**, 176 (1938); *J. Am. Chem. Soc.*, **65**, 2417 (1943).
616. F. S. Dainton and K. J. Ivin, *Nature*, **162**, 705 (1948).
617. S. L. Madorsky, S. Straus, D. Thompson, and L. Williamson, *J. Res. Natl. Bur. Stand.*, **42**, 499 (1948).
618. L. A. Wall, *J. Res. Natl. Bur. Stand.*, **41**, 315 (1948).
619. N. Grassie and H. W. Melville, *Proc. R. Soc. A*, **199**, 1, 14, 24 (1949).
620. R. Simha, L. A. Wall, and P. J. Blatz, *J. Polym. Sci.*, **5**, 615 (1950); R. Simha, *Ann. N.Y. Acad. Sci.*, **14**, 151 (1952).
621. A. Charlesby, *Nature*, **171**, 167 (1953); A. Charlesby and M. Ross, *Nature*, **171**, 1153 (1953).
622. E. J. Lawton, A. M. Bueche, and J. S. Balwit, *Nature*, **176**, 76 (1953).
623. A. A. Miller, E. J. Lawton, and J. S. Balwit, *J. Polym. Sci.*, **14**, 503 (1954); *J. Phys. Chem.*, **60**, 599 (1956).
624. L. A. Wall, *J. Polym. Sci.*, **17**, 141 (1955).
625. M. Dole, C. E. Keeling, and G. Rose, *J. Am. Chem. Soc.*, **76**, 4304 (1953).
626. P. Alexander and A. Charlesby, *Proc. R. Soc. A*, **230**, 136 (1955).
627. K. Ziegler, H. Grimm, and R. Willer, *Ann.*, **542**, 90 (1939).
628. H. Staudinger and F. Breusch, *Ber.*, **62**, 442 (1929).
629. H. Staudinger and H. A. Bruson, *Ann.*, **447**, 110 (1926).
630. P. Walden, *Ber.*, **35**, 2018 (1902).
631. A. Baeyer, *Ber.*, **38**, 569 (1905).
632. A. Hantsch, *Ber.*, **54**, 2573, 2613 (1921).
633. H. Meerwein and F. Montfort, *Ann.*, **435**, 203 (1924).
634. F. C. Whitmore, *Ind. Eng. Chem.*, **26**, 94 (1934).
635. W. H. Hunter and R. V. Yohe, *J. Am. Chem. Soc.*, **55**, 1248 (1933).
636. C. C. Price, *Ann. N.Y. Acad. Sci.*, **44**, 351 (1944).
637. M. Polanyi, *Nature*, **158**, 94 (1946); A. G. Evans and M. Polanyi, *J. Chem. Soc.*, 252, (1947).

638. A. G. Evans and G. W. Meadows, *Trans. Faraday Soc.*, **46**, 327 (1950).
639. C. M. Fontana and G. A. Kidder, *J. Am. Chem. Soc.*, **70**, 3745 (1948).
639a. A. Wohl and E. Wertyporoch, *Ber.*, **64**, 1357 (1931); E. Wertyporoch, *Ber.*, **64**, 1369.
640. P. H. Plesch, *J. Chem. Soc.*, 1659 (1953).
641. J. P. Kennedy and R. M. Thomas, *J. Polym. Sci.*, **45**, 227 (1960).
642. Ger. Pat. 504,730 (F. Hofmann and M. Otto, March 24, 1933).
643. D. C. Pepper, *Nature*, **169**, 828 (1952).
644. P. H. Plesch, ed., *Cationic Polymerization*, Heffer, Cambridge, 1953.
645. P. H. Plesch, *J. Chem. Soc.*, 1653 (1953).
646. C. G. Overberger and G. F. Endres, *J. Am .Chem. Soc.*, **75**, 6349 (1953); *J. Polym. Sci.*, **16**, 283 (1955).
647. C. M. Fontana, G. A. Kidder, and R. J. Harold, *Ind. Eng. Chem.*, **44**, 1688 (1952).
648. Z. Zlámal, L. Ambrož, and K. Veselý, *J. Polym. Sci.*, **24**, 285 (1957).
649. R. M. Thomas, W. J. Sparks, P. K. Frolich, M. Otto, and M. Muellar-Conradi, *J. Am. Chem. Soc.*, **62**, 276 (1940).
650. R. G. Beaman, *J. Am. Chem. Soc.*, **70**, 3115 (1948).
651. W. C. E. Higginson and N. S. Wooding, *J. Chem. Soc.*, 760, 774 (1952).
652. Y. Landler, *Compt. Rend.*, **230**, 539 (1950).
653. C. Walling, E. R. Briggs, W. Cummings, and F. R. Mayo, *J. Am. Chem. Soc.*, **72**, 48 (1950).
654. K. Ziegler, E. Eimers, H. Hechelhammer, and H. Wilms, *Ann.*, **567**, 43 (1950).
655. F. W. Staveley, *Ind. Eng. Chem.*, **48**, 778 (1956).
656. H. Hsieh, D. J. Kelley, and A. V. Tobolsky, *J. Polym. Sci.*, **26**, 240 (1957).
657. H. Morita and A. V. Tobolsky, *J. Am. Chem. Soc.*, **79**, 5853 (1957).
658. R. S. Stearns and L. E. Forman, *J. Polym. Sci.*, **51**, 381 (1959).
659. T. G. Fox, B. S. Garrett, W. E. Goode, S. Gratch, J. F. Kincaid, A. Spell, and J. D. Stroupe, *J. Am. Chem. Soc.*, **80**, 1768 (1958); J. D. Stroupe and R. E. Hughes, *ibid.*, **80**, 2341 (1958).
660. N. D. Scott, J. F. Walker, and V. N. Hansley, *J. Am. Chem. Soc.*, **58**, 2442 (1936).
661. U.S. Pat. 2,146,447 (N. D. Scott to Du Pont, February 7, 1939).
662. D. E. Paul, D. Lipkin, and S. I. Weissman, *J. Am. Chem. Soc.*, **78**, 116 (1956).
663. M. Szwarc, in "*Anionic Polymerization, Kinetics, Mechanisms and Synthesis*," J. E. McGrath, Ed., *Am. Chem. Soc. Symp. Ser. No. 166*, Washington, D. C., 1981, p. 1.
664. M. Szwarc, M. Levy, and R. Milkovich, *J. Am. Chem. Soc.*, **78**, 2656 (1956).
665. A. Abkin and S. Medvedev, *Trans. Faraday Soc.*, **32**, 286 (1936).
666. Belg. Pat. 627,652 (G. Holden and R. Milkovich to Shell International Research Maatschappij, 1963).
667. U.S. Pat. 2,677,700 (D. R. Jackson and L. G. Lundsted to Wyandotte Chem. Co., 1954); U.S. Pat. 2,674,619 (L. G. Lundsted to Wyandotte Chem. Co., 1954).
668. M. Szwarc, *Nature*, **178**, 1168 (1956).
669. A. F. Siriani, D. J. Worsfeld, and S. Bywater, *Trans. Faraday Soc.*, **55**, 2124 (1959).
670. K. Ziegler, *Angew. Chem.*, **76**, 3 (1964).
671. K. Ziegler, E. Holzkamp, H. Breil, and H. Martin, *Angew. Chem.*, **67**, 541 (1955).

672. K. Ziegler and H.-G. Gellert, *Ann.*, **567**, 179,185 (1950); *Angew. Chem.*, **64**, 323 (1952).
673. H. Martin, *Über Reaktionen von Olefinen mit aluminiumorganischen Verbindungen*, Ph.D. thesis, University of Aachen, 1952.
674. E. Holzkamp, *Gelenkte Polymerisation des Äthylen*, Ph.D. thesis, University of Aachen, 1953.
675. H. Breil, *Über metallorganische Mischkatalysatoren*, Ph.D. thesis, University of Aachen, 1955.
676. Ger. Pat. 874215 (M. Fischer to I. G. Farbenindustrie, April 20, 1953).
677. Ger. Pat. 973626 (K. Ziegler, H. Breil, E. Holzkamp, and H. Martin, April 14, 1960).
678. P. Pino, in *Giulio Natta, Present Significance of his Scientific Contribution*, Editrice di Chimica, Milano, 1982, pp. 21–54; *Atti Accad. Naz. Lincei Mem.* (8), **68**, 229 (1980). (contains a list of Natta's publications).
679. G. Natta, P. Pino, P. Corradini, F. Danusso, E. Mantico, G. Mazzanti, and G. Moraglio, *J. Am. Chem. Soc.*, **77**, 1708 (1955).
680. (a) F. M. McMillan, *The Chain Straighteners*, Macmillan, London, 1979, p. 115; (b) P. J. Flory, personal communication.
681. C. W. Bunn and H. S. Peiser, *Nature*, **159**, 161 (1947).
682. M. L. Huggins, *J. Am. Chem. Soc.*, **66**, 1991 (1944).
683. C. E. Schildknecht, S. T. Gross, H. R. Davidson, J. M. Lambert, and A. O. Zoss, *Ind. Eng. Chem.*, **48**, 2104 (1948).
684. P. J. Flory, *Principles of Polymer Chemistry*, Cornell University Press, 1953, pp. 237–238.
685. G. Natta, I. W. Bassi and P. Corradini, *Makromol. Chem.*, **18-19**, 455 (1955).
686. A. A. Morton, *Ind. Eng. Chem.*, **42**, 1488 (1950).
687. J. L. R. Williams, J. Van Den Berghe, J. L. Dulmage, and K. R. Dunham, *J. Am. Chem. Soc.*, **78**, 1260 (1956).
688. H. R. Sailors and J. P. Hogan, in *History of Polymer Science and Technology*, R. B. Seymour, Ed., Dekker, New York, 1982, p. 313.
688a. Ref. 680a, p. 111.
689. Ref. 680, pp. 102–111.
689a. S. E. Horne, *Ind. Eng. Chem.*, **48**, 784 (1956).
690. C. C. Price, M. Osgan, R. E. Hughes, and C. Shambelan, *J. Am. Chem. Soc.*, **78**, 690 (1956).
691. G. Natta, *Angew. Chem.*, **68**, 393 (1956).
692. G. Natta and G. Mazzanti, *Tetrahedron*, **8**, 86 (1960).
693. D. S. Breslow and N. R. Newburg, *J. Am. Chem. Soc.*, **81**, 81 (1959).
694. G. Natta and F. Danusso, *Stereoregular Polymers and Stereospecific Polymerization*, Pergamon, Oxford, 1967.
695. G. Natta, M. Farina, and M. Peraldo, *Atti Accad. Nazl. Lincei Cl. Sci. Fis. Mat. Nat. Rend.*, **25**, 424 (1958).
696. G. Natta, M. Farina, M. Peraldo, P. Corradini, G. Bassan, and P. Ganis, *Atti Accad. Naz. Lincei Cl. Sci. Fis. Mat. Nat. Rend*, **28**, 442 (1960).
697. G. Natta, G. Dell'Asta, G. Mazzanti, U. Giannini, and S. Cesca, *Angew. Chem.*, **71**, 205 (1959).

698. G. Natta, G. Mazzanti, P. Longi, and F. Bernardini, *Chim. Ind. Milan*, **41**, 519 (1959).
699. O. Bayer, *Angew. Chem.*, **59A**, 257 (1947).
700. O. Bayer, E. Müller, S. Petersen, H.-F. Piepenbrink, and E. Windemuth, *Angew. Chem.*, **62**, 57 (1950).
701. O. Bayer and E. Müller, *Angew. Chem.*, **72**, 934 (1960).
702. U.S. Pat. 2,929,802 (M. Katz to DuPont, March 22, 1960).
703. J. R. Whinfield, *Chem. Ind.*, **62**, 354 (1943).
704. Br. Pat. 578,079 (J. R. Whinfield, and J. T. Dickson, June 14, 1946).
705. W. J. Reader *Imperial Chemical Industries. A History*, Vol. 2, Oxford University Press, 1975, p. 381.
706. J. R. Whinfield, *Nature*, **158**, 930 (1946).
707. A. Conix and R. Van Kerpel, *J. Polym. Sci.*, **40**, 521 (1959).
708. (a) O. B. Edgar and E. Ellery, *J. Chem. Soc.*, 2633 (1952); (b) O. B. Edgar and R. Hill, *J. Polym. Sci.*, **8**, 1 (1952).
709. U.S. Pat. 2,708,617 (E. E. Magat and D. R. Strachan to DuPont, May 17, 1955).
710. (a) Ger. Pat. 818,580 (W. Hechelhammer and M. Coenan to Farbenfabriken Bayer, October 25, 1951); (b) Ger. Pat. 919,610 (H.-G. Trieschmann and G. Wenner, October 28, 1954).
711. E. L. Witbecker and P. W. Morgan, *J. Polym. Sci.*, **40**, 289 (1959).
712. P. W. Morgan and S. L. Kwolek, *J. Polym. Sci.*, **40**, 299 (1959).
713. V. Shashoua and W. M. Eareckson III, *J. Polym. Sci.*, **40**, 343 (1959).
714. H. Meerwein, D. Delfs, and H. Morschel, *Angew. Chem.*, **72**, 927 (1960).
715. H. Meerwein, G. Heinz, P. Hofmann, E. Kroning, and E. Pfeil, *J. Prakt. Chem.*, **147**, 257 (1937); H. Meerwein, E. Battenberg, H. Gold, E. Pfeil, and G. Willfang, *ibid.*, **154**, 83 (1939).
715a. W. E. Hanford and R. M. Joyce, *J. Polym. Sci.*, **3**, 167 (1948).
715b. O. Wichterle, J. Šebenda, and J. Králiček, *Fortschr. Hochpol. Forsch.*, **2**, 578 (1961).
715c. E. Katchalski, I. Grossfeld, and M. Frankel, *J. Am. Chem. Soc.*, **70**, 2094 (1948).
716. R. B. Woodward and C. H. Schramm, *J. Am. Chem. Soc.*, **69**, 1551 (1947).
717. E. R. Blout, R. H. Karlson, P. Doty, and B. Hargitay, *J. Am. Chem. Soc.*, **76**, 4492 (1954).
718. E. Katchalski and M. Sela, *Adv. Protein Chem.*, **13**, 243 (1958).
718a. S. Sarid, A. Berger, and E. Katchalski, *J. Biol. Chem.*, **234**, 1740 (1959).
719. C. H. Bamford, A. Elliott, and W. E. Hanby, *Synthetic Polypeptides*, Academic, New York, 1956.
720. R. D. Lundberg and P. Doty, *J. Am. Chem. Soc.*, **79**, 3961 (1957).
721. D. G. H. Ballard, C. H. Bamford, and A. Elliott, *Makromol. Chem.*, **35**, 222 (1960).
722. A. J. P. Martin and R. L. M. Synge, *Adv. Protein Chem.*, **2**, 1 (1945).
723. S. Moore and W. H. Stein, *J. Biol. Chem.*, **192**, 663 (1951).
724. M. Bergmann and C. Niemann, *J. Biol. Chem.* (a) **115**, 77 (1936); (b) **118**, 309 (1937).
725. M. Bergmann and W. H. Stein, *J. Biol. Chem.*, **128**, 217 (1939); **134**, 627 (1940).
726. F. Sanger, *Biochem. J.*, **39**, 507 (1945).
727. F. Sanger and H. Tuppy, *Biochem. J.*, **49**, 463, 481 (1951).

References

728. F. Sanger and E. O. P. Thompson, *Biochem. J.*, **53**, 353, 366 (1953).
729. A. P. Ryle, F. Sanger, L. F. Smith, and R. Kitai, *Biochem. J.*, **60**, 541 (1955).
730. C. H. W. Hirs, S. Moore, and W. H. Stein, *J. Biol. Chem.*, **235**, 633 (1960); D. H. Sparkman, W. H. Stein, and S. Moore, *J. Biol. Chem.*, **235**, 648 (1960).
730a. F. Haurowitz, *Chemistry and Biology of Proteins*, Academic, New York, 1950, p. 116.
731. L. Pauling, H. A. Itano, S. J. Singer, and I. C. Wells, *Science*, **110**, 543 (1949).
732. V. M. Ingram, *Nature*, **180**, 326 (1957).
733. M. L. Huggins, *Chem. Revs.*, **32**, (1943).
734. L. Pauling, R. B. Corey and H. R. Branson, *Proc. Nat. Acad. Sci. USA*, **37**, 205 (1951).
735. F. C. Caldwell and C. Hinshelwood, *J. Chem. Soc.*, 3156 (1950).
736. F.W. Putnam, in *The Proteins*, Vol. I, Pt. B, H. Neurath and K. Bailey, Eds., Academic, New York, 1953, p. 807.
737. W. Kauzmann, *Adv. Protein Chem.*, **14**, 1 (1959).
738. F. H. White and C. B. Anfinsen, *Ann. N.Y. Acad. Sci.*, **81**, 515 (1959); F. H. White, *J. Biol. Chem.*, **235**, 383 (1960).
739. W. T. Astbury, *Nature*, **157**, 121 (1946).
740. W. L. Bragg, *Rep. Progr. Phys.*, **28**, 1 (1965).
741. W. L. Bragg, J. C. Kendrew, and M. F. Perutz, *Proc. R. Soc. London A*, **203**, 321 (1950).
742. J. Boyes-Watson, E. Davidson, and M. F. Perutz, *Proc. R. Soc. London*, *A*, **191**, 82 (1947).
743. W. L. Bragg, E. R. Howells, and M. F. Perutz, *Acta Cryst.*, **5**, 136 (1952).
744. J. D. Bernal as quoted by W. L. Bragg, Ref. 740.
745. J. C. Kendrew and M. F. Perutz, in "Hemoglobin," F. J. W. Roughton and J. C. Kendrew, Eds., Butterworths, London, 1949, p. 161–179.
746. D. W. Green, V. M. Ingram, and M. F. Perutz, *Proc. R. Soc. London*, *A*, **225**, 287 (1954).
747. M. M. Bluhm, G. Boda, H. M. Dintzis, and J. C. Kendrew, *Proc. R. Soc. London*, *A*, **246**, 369 (1958).
748. J. C. Kendrew, G. Boda, H. B. Dintzis, R. G. Parrish, H. Wyckoff, and D. C. Phillips, *Nature*, **181**, 662 (1958).
749. J. C. Kendrew, R. E. Dickerson, B. E. Strandberg, R. G. Hart, D. R. Davies, D. C. Phillips, and V. C. Shore, *Nature*, **185**, 422 (1960).
750. L. Pauling and R. B. Corey, *Proc. Nat. Acad. Sci. USA*, **37**, 241 (1951).
751. M. F. Perutz, M. G. Rossmann, A. F. Cullis, H. Muirhead, G. Will, and A. C. T. North, *Nature*, **185**, 416 (1960).
752. T. H. Morgan, in *Nobel Lectures in Molecular Biology 1933–1975*, D. Baltimore, Ed., Elsevier, Amsterdam, 1977.
753. H. J. Muller, *Scient. Monthly*, **44**, 210 (1937).
754. M. Delbrück, *Trans. Connecticut Acad. Sci.*, **38**, 173 (1949), reprinted in *"Phage and the Origin of Molecular Biology,"* J. Cairns, G. S. Stent, and J. D. Watson, Eds., Cold Spring Harbor Symp. Quant. Biology, 1966, p. 9.
755. N. W. Timoféaff-Ressovsky, K. G. Zimmer, and M. Delbrück, *Nachr. Ges. Wiss. Göttingen, Math. Phys. Kl. Fachgr. 6 1*, 189 (1935); see also K. G. Zimmer, *Physik. Z.*, **44**, 233 (1943) and *"Phage and the Origin of Molecular Biology"* (as in Ref. 754), p. 35.

756. R. Olby, *The Path to the Double Helix*, University of Washington Press, Seattle, 1974.
757. E. Schrödinger, *What is Life*, Oxford University Press, 1944.
758. E. Fischer, *Sitz. königl. preuss. Akad.*, **40** 990 (1916) (*Chem. Abstr.*, **11**, 1972 (1917)).
759. D. Wrinch, *Nature*, **134**, 978 (1934).
760. S. Wright, *Physiol. Revs.*, **25**, 643 (1945).
761. F. G. Fischer, *Naturwiss.*, **30**, 377 (1942).
762. J. M. Gulland, G. R. Barker, and D. O. Jordan, *Ann. Rev. Biochem.*, **14**, 175 (1945).
763. T. Caspersson and J. Schultz, *Nature*, **142**, 294 (1938).
764. A. Hollaender and C. W. Emmons, *Cold Spring Harbor Symp. Quant. Biol.*, **9**, 179 (1941).
765. J. L. Alloway, *J. Exp. Med.*, **55**, 91 (1932); **57**, 265 (1933).
766. O. T. Avery, C. M. MacLeod, and M. McCarty, *J. Exp. Med.*, **79**, 137 (1944).
767. M. H. F. Wilkins, Nobel Prize address, *Science*, **140**, 941 (1963).
768. A. Boivin and R. Vendrely, *Helv. Chim. Acta*, **29**, 1338 (1946).
769. A. Boivin, *Cold Spring Harbor Symp. Quant. Biol.*, **12**, 7 (1947).
770. E. Chargaff, *Experientia*, **6**, 201 (1950).
771. S. Zamenhof and E. Chargaff, *J. Biol. Chem.*, **187**, 1, (1950).
772. J. H. Northrop, *J. Gen. Physiol.*, **34**, 315 (1951).
773. A. D. Hershey and M. Chase, *J. Gen. Physiol.*, **36**, 39 (1952).
774. H. Fraenkel-Conrat, *J. Am. Chem. Soc.*, **78**, 882 (1956).
775. W. T. Astbury and F. O. Bell, *Nature*, **141**, 77 (1938).
776. W. T. Astbury, *Symp. Soc. Exp. Biol.*, **1**, 66 (1947).
777. F. Haurowitz, *Chemistry and Biology of Proteins*, Academic, New York, 1950, pp. 340–355.
778. L. Pauling and M. Delbrück, *Science*, **92**, 77 (1940).
779. J. D. Watson, *Science*, **140**, 17 (1963).
780. R. E. Franklin and R. G. Gosling, *Acta Cryst.*, **6**, 673, 678 (1953).
781. J. D. Watson and F. H. C. Crick, *Nature*, (a) **171**, 737; (b) **171**, 964 (1953).
782. L. Pauling and R. B. Corey, *Proc. Nat. Acad. Sci. USA*, 84 (1953).
783. J. D. Watson, *The Double Helix*, Atheneum, New York, 1968, pp. 167–169, 181–182.
784. R. C. Williams, *Biochim. Biophys. Acta*, **9**, 237 (1952); H. Kahler and B. J. Lloyd, Jr., *ibid.*, **10**, 355 (1953).
785. J. M. Gulland and D. O. Jordan, *Symp. Soc. Exp. Biol.*, **1**, 56 (1947).
786. F. H. C. Crick and J. D. Watson, *Proc. Roy. Soc. London A*, **223**, 80 (1954).
787. R. E. Franklin and R. G. Gosling, (a) *Nature*, **171**, 740 (1953); *Nature*, **172**, 156 (1953).
788. F. H. C. Crick, *Science*, **139**, 461 (1963).
789. J. D. Watson, *Science*, **140**, 17 (1963).
790. A. Klug, *Nature*, **219**, 818, 843 (1968).
791. R. E. Franklin, A. Klug, J. T. Finch, and K. C. Holmes, *Discuss. Faraday Soc.*, **25**, 197 (1958).
792. M. Meselson, F. W. Stahl, and J. Vinograd, *Proc. Nat. Acad. Sci. USA*, **43**, 581 (1957).
793. M. Meselson and F. W. Stahl, in *Phage and the Design of Molecular Biology* (Ref. 754), p. 246.

794. A. Kornberg, *Science*, **131**, 1503 (1960).
795. (a) G. W. Beadle, and E. L. Tatum, *Proc. Nat. Acad. Sci. USA*, **27**, 499 (1941); (b) G. W. Beadle, *Physiol. Rev.*, **25**, 643 (1945).
796. F. H. C. Crick, *Symp. Soc. Exp. Biol.*, **12**, 138 (1958).
797. M. Grunberg-Manago and S. Ochoa, *J. Am. Chem. Soc.*, **77**, 3165 (1955); M. Grunberg-Manago, P. J. Ortiz, and S. Ochoa, *Science*, **122**, 907 (1955).
798. S. Ochoa, in *Nobel Lectures in Molecular Biology*, D. Baltimore, Ed., Elsevier, North-Holland, New York, 1977, p. 107.
799. E. A. Carlson, *Genes, Radiation and Society. The Life Work of H. Muller*, Cornell University Press, Ithaca, 1981, p. 392.
800. G. Pantecorvo, *Symp. Soc. Exp. Biol.*, **12**, 1 (1958).
801. J. Lederberg, Nobel Prize lecture, *Science*, **131**, 269 (1959).
802. E. J. Meehan, *J. Polym. Sci.*, **1**, 175 (1946).
803. E. Vischer and E. Chargaff, *J. Biol. Chem.*, **176**, 703 (1948).
804. R. Thomas, *Biochim. Biophys. Acta*, **14**, 231 (1954).
805. G. Felsenfeld and A. Rich, *Biochim. Biophys. Acta*, **26**, 457 (1957).
806. P. D. Ross and J. M. Sturtevant, *Proc. Nat. Acad. Sci. U.S.A.*, **46**, 1360 (1960).
807. I. Tinoco, *J. Am. Chem. Soc.*, **82**, 4785 (1960).
808. T. J. Fox and A. E. Martin, *Proc. R. Soc. London A*, **175**, 208 (1940).
809. W. H. Thompson, *Proc. R. Soc. London A*, **184**, 3 (1945); *J. Chem. Soc.*, 289 (1947).
810. A. R. Kemp and H. Peters, *Ind. Eng. Chem. Anal. Ed.*, **15**, 453 (1943).
811. I. M. Kolthoff and T. S. Lee, *J. Polym. Sci.*, **2**, 206 (1947).
812. E. J. Hart and A. W. Meyer, *J. Am. Chem. Soc.*, **71**, 1980 (1949); R. R. Hampton, *Anal. Chem.*, **21**, 923 (1949).
813. G. B. B. M. Sutherland, *Discuss. Faraday Soc.*, **9**, 274 (1950).
814. E. J. Ambrose, A. Elliott, and R. B. Temple, *Nature*, **163**, 859 (1949).
815. C. H. Bamford, W. E. Hanby and F. Happey. *Proc. R. Soc. London A*, **205**, 30 (1951).
816. A. Elliott and E. J. Ambrose, *Nature*, **165**, 921 (1950); E. J. Ambrose and A. Elliott, *Proc. R. Soc. London A*, **205**, 43 (1951).
817. C. Y. Liang, S. Krimm, and G. B. B. M. Sutherland, *J. Chem. Phys.*, **25**, 543 (1956).
818. S. Krimm, C. Y. Liang, and G. B. B. M. Sutherland, *J. Chem. Phys.*, **25**, 549 (1956).
819. S. Krimm, *Adv. Polym. Sci.*, **2**, 51 (1960).
820. S. Krimm and C. Y. Liang, *J. Polym. Sci.*, **22**, 95 (1952).
820a. S. Krimm, A. R. Berens, V. L. Folt, and J. J. Shipman, *Chem. Ind.*, 433 (1959).
821. S. Krimm, C. Y. Liang, and G. B. B. M. Sutherland, *J. Polym. Sci.*, **22**, 227 (1952).
822. V. J. Frilette, J. Hanle, and H. Mark, *J. Am. Chem. Soc.*, **70**, 1107 (1948).
823. M. Tsuboi, *J. Polym. Sci.*, **25**, 159 (1953); J. Mann and H. J. Marrinan, *J. Polym. Sci.*, **32**, 357 (1958); C. Y. Liang and R. H. Marchessault, *J. Polym. Sci.*, **37**, 385 (1959).
824. C. Y. Liang and R. H. Marchessault, *J. Polym. Sci.*, **43**, 71 (1960).
825. R. E. Cook, F. S. Dainton, and K. J. Ivin, *J. Polym. Sci.*, **26**, 351 (1957).
826. W. J. Burlant and J. L. Parsons, *J. Polym. Sci.*, **22**, 249 (1956).
827. N. L. Alpert, *Phys. Rev.*, **72**, 637 (1947).

828. F. Grün, *Experientia*, **3**, 490 (1947).
829. L. V. Holroyd, R. S. Codington, B. A. Mrowca, and E. Guth, *J. Appl. Phys.*, **22**, 696 (1951).
830. C. W. Wilson, III and G. E. Pake, *J. Polym. Sci.*, **10**, 503 (1931).
831. H. S. Gutowsky and L. H. Meyer, *J. Chem. Phys.*, **21**, 2122 (1953).
832. N. Fuschillo, E. Rhian, and J. A. Sauer, *J. Polym. Sci.*, **25**, 381 (1957).
833. W. P. Slichter and D. W. McCall, *J. Polym Sci.*, **25**, 230 (1957).
834. H. S. Gutowsky and C. J. Hoffman, *J. Chem. Phys.*, **19**, 1259 (1951).
835. W. P. Slichter, *Fortschr. Hochpol. Forsch.*, **1**, 35 (1958).
836. R. E. Naylor, Jr., and S. W. Lasoski, Jr., *J. Polym. Sci.*, **44**, 1 (1960).
837. F. A. Bovey, G. V. D. Tiers, and F. Filipovich, *J. Polym. Sci.*, **38**, 73 (1959).
838. F. A. Bovey and G. V. D. Tiers, *J. Polym. Sci.*, **44**, 173 (1960).
839. B. D. Coleman, *J. Polym. Sci.*, **31**, 155 (1958).
840. H. Morawetz and E. Gaetjens, *J. Polym. Sci.*, **32**, 526 (1958).
841. P. J. Flory and F. S. Leutner, *J. Polym. Sci.*, **3**, 880 (1948); **5**, 267 (1950).
842. G. K. Fraenkel, J. M. Hirshon, and C. Walling, *J. Am. Chem. Soc.*, **76**, 3606 (1954).
843. N. M. Atherton, H. Melville, and D. H. Whiffen, *J. Polym. Sci.*, **34**, 199 (1959).
844. C. H. Bamford, A. D. Jenkins, D. J. E. Ingram, and M. C. R. Symons, *Nature*, **175**, 894 (1955); C. H. Bamford, A. D. Jenkins, M. C. R. Symons, and M. G. Townsend, *J. Polym. Sci.*, **34**, 181 (1959).
845. S. E. Bresler, E. N. Kazbekov, and E. M. Saminskii, *Vysokomol. Soedin.*, **1**, 132, 1374 (1955).
846. S. W. Benson and A. M. North, *J. Am. Chem. Soc.*, **81**, 1339 (1959).
847. E. E. Schneider, M. J. Day, and G. Stein, *Nature*, **168**, 645 (1951); E. E. Schneider, *Discuss. Faraday Soc.*, **19**, 158 (1955).
848. R. J. Abraham and D. H. Whiffen, *Trans. Faraday Soc.*, **54**, 1291 (1958).
849. E. J. Lawton, R. S. Powell, and J. S. Balwit, *J. Polym. Sci.*, **32**, 247 (1958).
850. S. Oka, *Proc. Phys. Nat. Soc. Jpn.*, **24**, 657 (1942).
851. H. Benoit (a) *J. Chim. Phys.*, **44**, 18 (1947); (b) *J. Polym. Sci.*, **3**, 376 (1948).
852. H. Kuhn, *J. Chem. Phys.*, **15**, 843 (1947).
853. W. J. Taylor, (a) *J. Chem. Phys.*, **15**, 412 (1947); (b) *J. Chem. Phys.*, **16**, 257 (1948).
854. K. S. Pitzer, *J. Chem. Phys.*, **8**, 711 (1940).
855. M. V. Volkenstein, *Dokl. Akad. Nauk SSSR*, **78**, 879 (1951); *Zhur. Fiz. Khim.*, **26**, 1072 (1952).
856. M. V. Volkenstein, *J. Polym. Sci.*, **29**, 441 (1958).
857. M. V. Volkenstein, *Configurational Statistics of Polymeric Chains*, Interscience-Wiley, New York, 1963.
858. O. B. Ptitsyn and Yu. A. Sharonov, *Zh. Tekh. Fiz.*, **27**, 2744, 2762 (1957).
859. P. J. Flory, *J. Chem. Phys.*, **17**, 303 (1949).
860. T. G. Fox, Jr. and P. J. Flory, *J. Phys. Colloid Chem.*, **53**, 197 (1949).
861. T. R. Grimley, *J. Chem. Phys.*, **21**, 185 (1952).
862. B. H. Zimm, W. H. Stockmayer and M. Fixman, *J. Chem. Phys.*, 1, 1716 (1952).
863. F. T. Wall, A. Hiller, Jr., and D. J. Wheeler, *J. Chem. Phys.*, **22**, 1056 (1954).

References

864. F. T. Wall and J. J. Erpenbeck, *J. Chem. Phys.*, **30**, 634 (1959).
865. O. Kratky and G. Porod, *Rec. Trav. Chim.*, **68**, 1106 (1949).
865a. S. E. Bresler and Ya. I. Frenkel, as quoted in L. D. Landau and E. M. Lifshitz, *Statistical Physics*, 2nd rev. ed., Pergamon, Oxford, 1958, pp. 475–479.
866. R. M. Fuoss and D. J. Mead, *J. Phys. Chem.*, **47**, 59 (1942).
867. B. H. Zimm and I. Meyerson, *J. Am. Chem. Soc.*, **68**, 911.
868. H. P. Frank and H. F. Mark, *J. Polym. Sci.*, **10**, 129 (1953); **17**, 1, (1955).
869. A. J. Staverman, *Rec. Trav. Chim.*, **71**, 623 (1952).
870. H. Morawetz, *J. Polym. Sci.*, **6**, 117 (1951).
871. N. H. Ray, *Trans. Faraday Soc.*, **48**, 809 (1952).
872. K. G. Schön and G. V. Schulz, *Z. Phys. Chem.*, **2**, 197 (1954).
873. G. V. Schulz and H. Marzolph, *Z. Elektrochem.*, **58**, 211 (1954).
874. P. J. Flory, *J. Chem. Phys.*, **13**, 453 (1945).
875. P. J. Flory and W. R. Krigbaum, *J. Chem. Phys.*, **18**, 1086 (1950).
876. W. R. Krigbaum and P. J. Flory, *J. Am. Chem. Soc.*, **75**, 1775 (1953).
877. P. Debye, *Ann. Phys.*, **46**, 809 (1915).
878. P. Debye, *Phys. Z.*, **31**, 419 (1930).
879. P. Debye, *J. Appl. Phys.*, **15**, 338 (1944); *J. Phys. Colloid Chem.*, **51**, 18 (1947).
880. A. Einstein, *Ann. Phys.*, **33**, 1275 (1910).
881. P. M. Doty, B. H. Zimm and H. Mark, *J. Chem. Phys.*, **12**, 144 (1944).
882. B. H. Zimm and P. M. Doty, *J. Chem. Phys.*, **12**, 203 (1944).
883. R. S. Stein and P. Doty, *J. Am .Chem. Soc.*, **68**, 159 (1946).
884. B. H. Zimm, *J. Chem. Phys.*, **16**, 1099 (1948).
885. R. H. Ewart, C. P. Roe, P. Debye, and J. R. McCartney, J. Chem. Phys., **14**, 687 (1946).
886. J. G. Kirkwood and R. J. Goldberg, *J. Chem. Phys.*, **18**, 54 (1950).
887. W. H. Stockmayer, *J. Chem. Phys.*, **18**, 58 (1950).
888. W. H. Stockmayer and H. E. Stanley, *J. Chem. Phys.*, **18**, 153 (1950).
889. W. R. Krigbaum and P. J. Flory, *J. Chem. Phys.*, **20**, 873 (1952).
890. W. Bushuk and H. Benoit, *Can. J. Chem.*, **36**, 1616 (1958).
891. H. Benoit and C. Wippler, *J. Chim. Phys.*, **57**, 524 (1960).
892. S. A. Rice, A. Wada, and E. P. Geidushek, *Discuss. Faraday Soc.*, **25**, 130, 152 (1958).
893. M. L. Huggins, *J. Phys. Chem.*, **43**, 439 (1939).
894. W. Kuhn and H. Kuhn, *Helv. Chim. Acta*, **26**, 1394 (1943).
895. W. Kuhn and H. Kuhn, *Helv. Chim. Acta*, **28**, 1533 (1945); **29**, 71,609,830,1095 (1946).
895a. A. Peterlin, *J. Polym. Sci. B*, **10**, 101 (1972).
896. H. Kuhn, Habilitationsschrift, Basel, 1946; *J. Colloid Sci.*, **5**, 331 (1950).
897. P. Debye and A. M. Bueche, *J. Chem. Phys.*, **16**, 573 (1948).
898. J. G. Kirkwood and J. Riseman, *J. Chem. Phys.*, **16**, 565 (1948).
899. T. Alfrey, A. Bartovics, and H. Mark, *J. Am. Chem. Soc.*, **64**, 1557 (1942).
900. P. J. Flory and T. G. Fox, Jr., *J. Am. Chem. Soc.*, **73**, 1904 (1951).
901. W. R. Krigbaum, L. Mandelkern and P. J. Flory, *J. Polym. Sci.*, **9**, 381 (1952).

902. W. R. Krigbaum and D. K. Carpenter, *J. Phys. Chem.*, **59**, 1166 (1955).
903. O. B. Ptitsyn and Yu. E. Eizner, *Zh. Tekhn. Fiz.*, **29**, 1113 (1959) [English translation *Soviet Phys. Tech. Phys.*, **4**, 1020 (1960)].
904. P. Doty, A. Wada, J. T. Yang, and E. R. Blout, *J. Polym. Sci.*, **23**, 851 (1957).
905. P. Doty, A. M. Holtzer, J. H. Bradbury, and E. R. Blout, *J. Am. Chem. Soc.*, **76**, 4493 (1954).
906. P. Doty, J. H. Bradbury, and A. M. Holtzer, *J. Am. Chem. Soc.*, **78**, 947 (1956).
906a. R. E. Rundle and R. R. Baldwin, *J. Am. Chem. Soc.*, **65**, 554 (1943).
907. P. Doty and T. J. Yang, *J. Am. Chem. Soc.*, **78**, 498 (1956).
908. C. Cohen, *Nature*, **175**, 129 (1955).
909. W. Moffitt, *J. Chem. Phys.*, **25**, 467 (1956).
910. J. A. Schellman, *Compt. rend. trav. lab. Carlsberg, Sér. chim.*, **29**, No. 15 (1955).
911. B. H. Zimm and J. K. Bragg, *J. Chem. Phys.*, **31**, 526 (1959).
912. J. Kurtz, A. Berger, and E. Katchalski, *Nature*, **178**, 1066 (1956).
913. W. F. Harrington and M. Sela, *Biochim. Biophys. Acta*, **27**, 24 (1958).
914. I. Z. Steinberg, W. F. Harrington, A. Berger, and E. Katchalski, *J. Am. Chem. Soc.*, **82**, 5263 (1960).
915. A. Elliott and E. J. Ambrose, *Discuss. Faraday Soc.*, **9**, 246 (1950).
916. C. Robinson, *Trans. Faraday Soc.*, **52**, 571 (1956); C. Robinson, J. C. Ward, and R. D. Beevers, *Discuss. Faraday Soc.*, **25**, 29 (1958).
917. I. Langmuir, *J. Chem. Phys.*, **6**, 873 (1938).
918. L. Onsager, *Phys. Rev.*, **62**, 558 (1942); *Ann. N.Y. Acad. Sci.*, **51**, 627 (1949).
918a. A. Isihara, *J. Chem. Phys.*, **19**, 1142 (1951).
919. P. J. Flory, *Proc. R. Soc. London A*, **234**, 73 (1956).
920. R. M. Fuoss, *Science*, **108**, 545 (1948).
921. R. M. Fuoss and U. P. Strauss, *J. Polym. Sci.*, **3**, 246 (1948).
922. W. Kuhn, O. Künzle, and A. Katchalsky, *Helv. Chim. Acta*, **31**, 1994 (1948).
923. W. Kuhn, *Experientia*, **5**, 318 (1949); A. Katchalsky, *Experientia*, **5**, 319 (1949).
924. W. Kuhn and B. Hargitay, *Experientia*, **7**, 1 (1951).
925. A. Katchalsky and M. Zwick, *J. Polym. Sci.*, **16**, 221 (1955).
926. G. S. Hartley and J. W. Roe, *Trans. Faraday Soc.*, **36**, 101 (1940).
927. A. Katchalsky and J. Gillis, *Rec. Trav. Chim.*, **68**, 879 (1949).
928. J. T. G. Overbeek, *Bull. Soc.Chim. Belges*, **57**, 252 (1948); R. Arnold and J. T. G. Overbeek, *Rev. Trav. Chim.*, **69**, 192 (1950).
929. H. Eisenberg, *Biophys. Chem.*, **7**, 3 (1977).
930. R. Arnold, *J. Colloid Sci.*, **12**, 549 (1957).
931. J. D. Ferry, D. C. Udy, F. C. Wu, G. E. Heckler, and D. B. Fordyce, *J. Colloid Sci.*, **6**, 429 (1951).
932. A. Wada, *Mol. Phys.*, **3**, 409 (1960).
933. T. Alfrey, P. W. Berg, and H. Morawetz, *J. Polym. Sci.*, **7**, 543 (1951).
934. F. Noether in *Die Differential und Integralgleichungen der Mechanik und Physik*, Vol. 2, P. Frank and R. v. Mises, Eds., Vieweg, Braunschweig, 1927, p. 319.
935. R. M. Fuoss, A. Katchalsky, and S. Lifson, *Proc. Nat. Acad. Sci. U.S.A.*, **37**, 579 (1951).

284 References

936. J. J. Hermans and J. T. G. Overbeek, *Rec. Trav. Chim.*, **67**, 761 (1948).
937. G. E. Kimball, M. Cutler, and H. Samelson, *J. Phys. Chem.*, **56**, 57 (1952).
938. P. J. Flory, *J. Chem. Phys.*, **21**, 162 (1953).
939. R. M. Fuoss, *J. Polym. Sci.*, **3**, 603 (1948).
940. H. Eisenberg and J. Pouyet, *J. Polym. Sci.*, **13**, 85 (1954).
941. D. T. F. Pals and J. J. Hermans, *J. Polym. Sci.*, **5**, 733 (1950).
942. P. J. Flory and J. E. Osterheld, *J. Phys. Chem.*, **58**, 653 (1954).
943. D. T. F. Pals and J. J. Hermans, *Rec. Trav. Chim.*, **71**, 458 (1952).
944. M. Nagasawa, M. Izumi and I. Kagawa, *J. Polym. Sci.*, **37**, 375 (1959).
945. J. R. Huizenga, P. F. Grieger and F. T. Wall, *J. Am. Chem. Soc.*, **72**, 2636 (1950).
946. U. P. Strauss and P. D. Ross, *J. Am. Chem. Soc.*, **81**, 5299 (1959).
947. H. Eisenberg and G. R. Mohan, *J. Phys. Chem.*, **63**, 671 (1959).
948. J. J. Hermans and H. Fujita, *Proc. Roy. Netherland Acad. Sci.*, **13**, 48, 182 (1955); J. J. Hermans, *J. Polym. Sci.*, **18**, 527 (1955).
949. M. Nagasawa, A. Soda, and I. Kagawa, *J. Polym. Sci.*, **31**, 439 (1958).
950. H. Schindewolf, *Z. Phys. Chem. Frankfurt*, **1**, 129 (1954).
951. H. Benoit, *J. Chim. Phys.*, **49**, 517 (1952).
952. M. Eigen and G. Schwarz, *J. Colloid Sci.*, **12**, 181 (1957).
952a. U. P. Strauss and N. L. Gershfeld, *J. Phys. Chem.*, **58**, 747 (1954); U. P. Strauss, N. L. Gershfeld, and E. H. Crook, *J. Phys. Chem.*, **60**, 577 (1956).
953. J. Duclaux and E. Wollman, *Bull. Soc. Chim.* France, **27**, 414 (1920).
954. V. Desreux, *Rec. Trav. Chim.*, **68**, 789 (1949).
955. C. A. Baker and R. J. P. Williams, *J. Chem. Soc.*, 2352 (1956).
956. J. W. Williams, R. L. Baldwin, W. M. Saunders and P. G. Squire, *J. Am. Chem. Soc.*, **74**, 1542 (1952); J. W. Williams and W. M. Saunders, *J. Phys. Chem.*, **58**, 854 (1954).
957. H. W. McCormick, *J. Polym. Sci.*, **36**, 341 (1959); **41**, 327 (1959).
958. D. R. Morey and J. W. Tamblyn, *J. Appl. Phys.*, **16**, 419 (1945).
959. I. Harris and R. G. J. Miller, *J. Polym. Sci.*, **7**, 377 (1951).
960. J. Porath and P. Flodin, *Nature*, **183**, 1657 (1959).
961. M. F. Vaughn, *Nature*, **188**, 55 (1960).
962. M. J. Voorn and J. J. Hermans, *J. Polym. Sci.*, **35**, 113 (1959).
963. W. G. Lloyd and T. Alfrey, Jr., *J. Polym. Sci.*, **62**, 301 (1962).
964. J. C. Moore, *J. Polym. Sci. A*, **2**, 835 (1964).
965. F. Simon, *Ann. Phys.*, **68**, 278 (1922).
966. G. Tamann and A. Kohlhaas, *Z. Anorg. Allg. Chem.*, **182**,
967. F. Simon, *Z. Anorg. Allg. Chem.*, **203**, 219 (1931).
968. G. Tamman and W. Hesse, *Z. Anorg. Allg. Chem.*, **156**, 256 (1926).
969. J. D. Ferry and G. S. Parks, *J. Chem. Phys.*, **4**, 70 (1936).
970. E. Jenckel and K. Ueberreiter, *Z. Phys. Chem. A*, **182**, 361 (1938).
971. K. Ueberreiter, *Kolloid Z.*, **102**, 272 (1943).
972. T. G. Fox, Jr. and P. J. Flory, *J. Appl. Phys.*, **21**, 581 (1950).
973. S. N. Zhurkov and B. Ya. Levin, *Dokl. Akad. Nauk SSSR*, **67**, 89 (1949).

973a. W. Kaufmann, *Chem. Revs.*, **43**, 219 (1948).
974. J. H. Gibbs and E. A. DiMarzio, *J. Chem. Phys.*, **28**, 373 (1958).
975. C. W. Bunn, *Proc. R. Soc. London*, **180**, 67 (1942).
976. C. W. Bunn and E. R. Howells, *J. Polym. Sci.*, **18**, 307 (1955).
977. C. W. Bunn and D. R. Holmes, *Discuss. Faraday Soc.*, **25**, 95 (1958).
978. G. Natta, P. Corradini, and I. W. Bassi, *Gazz. Chim. Ital.*, **89**, 784 (1959).
979. G. Natta, P. Corradini, and P. Ganis, *Makromol. Chem.*, **39**, 238 (1960).
980. E. Hunter and W. G. Oakes, *Trans. Faraday Soc.*, **41**, 49 (1945).
981. H. C. Raine, R. B. Richards, and H. Ryder, *Trans. Faraday Soc.*, **41**, 56 (1945).
982. P. H. Hermans and A. Weidinger, *J. Polym. Sci.*, **4**, 135 (1949); **5**, 565 (1950).
983. S. Krimm and A. V. Tobolsky, *J. Polym. Sci.*, **7**, 57 (1951).
984. B. Ke, *J. Polym. Sci.*, **42**, 15 (1960).
984a. R. Hosemann, *Z. Phys.*, **128**, 1 (1950).
984b. R. Hosemann and R. Bonart, *Kolloid Z.*, **152**, 53 (1957).
985. W. E. Garner, K. Van Bitter, and A. M. King, *J. Chem. Soc.*, 1533 (1931).
986. R. E. Powell, C. R. Clark, and H. Eyring, *J. Chem. Phys.*, **9**, 268 (1941).
987. R. B. Richards, *Trans. Faraday Soc.*, **41**, 127 (1944).
988. E. M. Frith and R. F. Tuckett, *Trans. Faraday Soc.*, **40**, 251 (1944).
989. P. J. Flory, *J. Chem. Phys.*, **17**, 223 (1949).
990. P. J. Flory, L. Mandelkern, and H. K. Hall, *J. Am. Chem.Soc.*, **73**, 2532 (1951).
991. D. Coleman, *J. Polym. Sci.*, **14**, 15 (1954).
992. D. Coleman and F. O. Howitt, *Proc. R. Soc. (London), A*, **180**, 166 (1947); B. Drucker and S. G. Smith, *Nature*, **165**, 197 (1950).
993. C. E. Rehberg and C. H. Fisher, *Ind. Eng. Chem.*, **40**, 1429 (1948).
994. S. A. Greenberg and T. Alfrey, *J. Am. Chem. Soc.*, **76**, 6280 (1954).
995. A. Keller, *Phil. Mag.*, [8], **2**, 1171 (1953).
996. P. H. Till, Jr., *J. Polym. Sci.*, **24**, 301 (1957).
997. E. W. Fischer, *Z. Naturforsch.*, **120**, 753 (1957).
998. A. Keller and A. O'Connor, *Discuss. Faraday Soc.*, **25**, 114 (1958).
999. C. W. Bunn, *Discuss. Faraday Soc.*, **25**, 208 (1958).
1000. K. H. Storks, *J. Am. Chem. Soc.*, **60**, 1753 (1938).
1001. B. G. Rånby, E. F. Moorhead, and N. M. Walter, *J. Polym. Sci.*, **44**, 349 (1960).
1002. R. Eppe, E. W. Fischer, and H. A. Stuart, *J. Polym. Sci.*, **34**, 721 (1959).
1003. H. Staudinger and R. Signer, *Z. Krist.*, **70**, 193 (1929).
1004. P. J. Flory, *Principles of Polymer Chemistry*, Cornell University Press, Ithaca, New York, 1953, p. 564.
1005. K. Hess and H. Kiesig, *Naturwiss.*, **31**, 171 (1943); *Kolloid Z.*, **130**, 10 (1953).
1006. H. Zahn and U. Winter, *Kolloid Z.*, **128**, 140 (1952).
1007. C. W. Bunn and T. C. Alcock, *Trans. Faraday Soc.*, **41**, 317 (1945).
1008. M. Herbst, *Z. Elektrochem.*, **54**, 318 (1950).
1009. A. Keller, *Makromol. Chem.*, **34**, 1 (1958).
1010. L. A. Wood and N. Bekkedahl, *J. Appl. Phys.*, **17**, 362 (1946).

1011. R. Becker, *Ann. Phys.*, **32**, 128 (1938).
1012. L. Mandelkern, F. A. Quinn, Jr. and P. J. Flory, *J. Appl. Phys.*, **25**, 830 (1954).
1013. W. H. Cobbs and R. L. Burton, *J. Polym. Sci.*, **10**, 235 (1953).
1014. A. H. Gent, *J. Polym. Sci.*, **18**, 321 (1955).
1015. F. P. Price, *J. Am. Chem. Soc.*, **74**, 311 (1952).
1016. P. J. Flory and A. D. McIntyre, *J. Polym. Sci.*, **18**, 592 (1955).
1017. T. Alfrey, Jr., *Mechanical Behavior of High Polymers*, Interscience, New York, 1948.
1018. H. A. Stuart, *Die Physik der Hochpolymeren*, Vol. 4, Springer, Berlin, 1952.
1019. P. J. Flory, *J. Am. Chem. Soc.*, **62**, 1057 (1940).
1020. W. Kauzmann and H. Eyring, *J. Am. Chem. Soc.*, **62**, 3113 (1940).
1021. T. G. Fox and P. J. Flory, *J. Am. Chem. Soc.*, **70**, 2384 (1948).
1022. T. G. Fox and P. J. Flory, *J. Phys. Colloid Chem.*, **55**, 221 (1951).
1023. F. Bueche, *J. Chem. Phys.*, **20**, 1959 (1952).
1024. F. T. Wall, *J. Chem. Phys.*, **10**, 485 (1942).
1025. P. J. Flory and J. Rehner, Jr., *J. Chem. Phys.*, **11**, 512 (1943).
1026. H. M. James and E. Guth, *J. Chem. Phys.*, **11**, 455 (1943).
1027. L. R. G. Treloar, *Trans. Faraday Soc.*, **39**, 36 (1943).
1028. H. M. James and E. Guth, *J. Chem. Phys.*, **15**, 669 (1947).
1028a. M. Mooney, *J. Appl. Phys.*, **11**, 582 (1940).
1028b. R. S. Rivlin, *J. Appl. Phys.*, **18**, 444 (1947).
1029. W. R. Hess, *Kolloid Z.*, **27**, 154 (1920).
1030. E. Jenckel and K. Ueberreiter, *Z. Phys. Chem. A*, **182**, 361 (1938).
1031. H. Leaderman, *Ind. Eng. Chem.*, **35**, 374 (1943).
1032. W. Philippoff, *Physik. Z.*, **35**, 884, 900 (1934).
1033. W. Kuhn, *Z. Phys. Chem. B*, **42**, 1 (1939).
1034. K. Bennowitz and H. Rötger, *Physik. Z.*, **40**, 416 (1939).
1035. R. M. Fuoss and J. G. Kirkwood, *J. Am. Chem. Soc.*, **63**, 385 (1941).
1036. P. E. Rouse, Jr., *J. Chem. Phys.*, **21**, 1272 (1953).
1037. J. D. Ferry, *J. Am. Chem. Soc.*, **72**, 3746 (1950).
1038. A. V. Tobolsky and J. R. McLoughlin, *J. Polym. Sci.*, **8**, 543 (1952).
1039. J. D. Ferry and E. R. Fitzgerald, *J. Colloid Sci.*, **8**, 224 (1953).
1040. M. L. Williams, R. F. Landel and J. D. Ferry, *J. Am. Chem. Soc.*, **77**, 3701 (1955).
1041. A. V. Tobolsky and E. Catsiff, *J. Polym. Sci.*, **19**, 111, (1956).
1042. W. F. V. Pollett, *Brit. J. Appl. Phys.*, **6**, 199 (1955).

Name Index

For authors not cited on a given page, the numbers of references citing their work are *italicized* in parentheses.

Abderhalden, E., 139, 140(*468*)
Abitz, W., 84
Abott, H., 186
Abraham, R.J., 220
Adair, G.S., 154
Adkins, H., 175
Akim, L., 105(*356*)
Alcock, T.C., 247
Alexander, P., 181(*626*)
Alexsandrov, A.P., 152, 250
Alfrey, T., Jr., 169(*549b*), 172, 173, 176, 178(*606*), 179, 230, 231, 236, 240, 245(*994*), 248
Allegra, G., 223
Allen, R.E., 174(*584*)
Alloway, J.L., 206
Alpert, N.L., 217
Alyea, H.N., 129
Ambronn, H., 70-72
Ambrose, E.J., 215, 216, 231, 233
Ambrož, L., 184(*648*)
Amos, J.L., 176
Ampère, A.M., 3, 4, 26
Anderson, T.F., 85(*279*)
Anfinsen, C.B., 200
Anschütz, R., 114
Anson, M.L., 142, 144(*488*)
Anthony, R.L., 151
Arisz, L., 111(*375a*)
Arnold, C., 6
Arnold, R., 235(*928*), 236
Arreguin, B., 177(*598*)

Arrhenius, S., 159
Arvin, J.A., 118
Ashdown, A.A., 94(*311*), 184(*311*)
Astbury, W.T., 83, 84, 197, 200, 208
Atherton, N.M., 220
Auer, E.E., 172(*566*)
Auerbach, F., 137
Avery, O.T., 206-208, 211
Avogadro, A., 3, 26-28
Axelrod, S., 104

Bacon, R.G.R., 170, 176
Baekeland, L.H., 58, 59
Baeyer, A., 58, 117, 182
Bahr, K., 135(*447*), 182(*447*)
Baker, C.A., 239
Baldwin, R.L., 239(*956*)
Baldwin, R.R., 232
Ballard, D.G.H., 196
Balwit, J.S., 176(*594b*), 180(*622*), 181(*623*), 220(*849*)
Bamberger, E., 21
Bamford, C.H., 172(*562*), 195, 196(*721*), 216(*815*), 220
Bancelin, M., 103
Bandel, D., 176
Barker, G.R., 206(*762*)
Barschall, H., 137
Bartlett, P.D., 169(*547*)
Bartovics, A., 230(*899*), 231(*899*)
Bassan, G., 192(*696*)
Bassi, I.W., 190(*685*), 191(*685*), 243(*978*)

Name Index

Battenberg, E., 194(*715*)
Bauer, A., 18
Bauer, W., 177
Baumann, E., 21, 22, 125
Bawn, C.E.H., 127
Baxendale, J.H., 170(*555*), 177(*601*)
Bayer, O., 193
Beadle, C., 39, 40, 57(*204, 205*)
Beadle, G.W., 212
Beaman, R.G., 184
Bearman, D., 142
Becker, K., 71(*238*)
Becker, R., 247
Becker, W., 71(*583*)
Beckmann, E., 30, 31
Beevers, R.D., 233(*916*)
Bekkedahl, N., 247
Bell, F.O., 208
Bellenot, H., 128, 171(*558*)
Bence Jones, H., 27
Bender, H.L., 150
Benedetti, E., 223
Bennowitz, K., 250
Benoit, H., 222, 228, 238
Benson, S.W., 220(*846*)
Benzer, S., 205
Berg, P.W., 236
Berger, A., 195(*718a*), 233(*912, 914*)
Bergmann, M., 87, 89, 139, 145(*489*), 162, 198
Bernal, J.D., 84, 85(*280*), 143, 198, 201, 233(*280*)
Bernardini, F., 192(*698*)
Berens, A.R., 216(*820a*)
Berthelot, D., 21
Berthelot, P.E.M., 6, 14, 17(*42*), 18-21, 124, 182
Bertoud, A., 128, 171(*558*)
Berzelius, J., 3, 5-8, 18, 117
Bevan, E.J., 39, 40, 57(*204, 205, 206*)
Bevilacqua, E.B., 172(*566*)
Bevington, J.C., 169-170, 176(*595*)
Bikales, N., 79
Biltz, W., 104, 105
Biot, J.B., 10, 37
Bischoff, C.A., 13, 115
Blackwell, J., 81
Blagden, C., 29
Blaise, E.E., 116
Blaschke, F., 131
Blatz, P.J., 180(*620*)
Blout, E.R., 195, 231(*717, 904, 905*), 232(*904*)
Bluhm, M.M., 202(*747, 748*)
Blyth, J., 16, 17, 22
Bock, W., 66-67(*226*)
Boda, G., 202(*747, 748*)

Boehm, G., 141
Bogemann, M., 64(*218a*)
Bohlmann, F., 179
Bohr, N., 203
Bohrer, J.J., 173(*570b*)
Boissonas, C.G., 155
Boivin, A., 207
Bokorny, T., 43(*142*)
Bolton, E.K., 116(*392a*)
Bonart, R., 244
Bondy, H.F., 96(*321*)
Borkin, J., 67(*226a*)
Bouchardat, G., 22, 33, 63
Boudelle, M., 81
Bovey, F.A., 218, 219
Boyd, W.C., 146
Boyes-Watson, J., 201(*742*)
Boyle, R., 28
Braconnot, H., 39, 56
Bradbury, J.H., 195(*906*), 231(*905, 906*)
Bradford, E.B., 178(*606*), 179(*607*)
Bragg, J.K., 232
Bragg, W.H., 83
Bragg, W.L., 201, 202
Branson, H.R., 199(*734*), 231(*734*)
Braun, E., 80
v. Braun, J., 114
Brauns, F., 123(*4151*)
Breil, H., 186(*671*), 188
Breinl, F., 145, 146
Breitenbach, J.W., 174
Bresler, S.E., 151, 172, 220, 225
Breslow, D.S., 191
Breusch, F., 182(*628*)
Briggs, D.R., 158
Briggs, E.R., 185
Brill, R., 72, 73, 75, 82, 139, 140, 245
Brockmann, H., 140(*468*)
Brodie, B.C., 19, 27
Brookman, E.F., 131, 134, 135, 172(*442*)
Brosteaux, J., 112
Brown, D.W., Jr., 176(*595*)
Brown, H.T., 37, 38
Brown, R., 100
Brunner, M., 92(*302*), 93(*302*), 94(*302, 311*), 124(*418*), 182(*302*), 189(*311*)
Bruson, H.A., 94(*311*), 182, 185(*629*), 189(*311*)
Bueche, A.M., 180(*622*), 230(*897*)
Bueche, F., 248
Bunn, C.W., 82, 190(*681*), 242, 243, 246, 247
Burlant, W.J., 217
Burnett, G.M., 172
Burton, R.L., 247(*1013*)

Bushuk, W., 228
Busse, W.F., 150
Butlerov, A., 12, 18, 20, 24, 182
Bywater, S., 186(*669*)

Cahours, A.A., 16, 17
Calcott, W.S., 67–68(*228*)
Caldwell, F.C., 200, 208, 209
Campbell, D.H., 147(*501a*)
Cannan, R.K., 141(*471*)
Cannizzaro, S., 26
Carlin, R.C., 176
Carlson, E.A., 213
Carmody, W.H., 65(*232*)
Carothers, W.H., 64, 68, 82(*229*), 94, 96, 113, 116–121, 127, 129
Carpenter, D.K., 231(*902*)
Carter, A.S., 67–68(*228*)
Caspari, W.A., 153
Caspersson, T., 206
Caston, R.H., 151(*520*)
Catsiff, E., 251
Cesca, S., 192(*697*)
Chalmers, W., 129, 135, 179
Chanzy, H., 81
Chapiro, A., 176, 177
de Chardonnet, H., 56, 57
Chargaff, E., 207, 208, 210, 211, 214
Charlesby, A., 180, 181(*626*)
Charlton, W., 76(*250*)
Chase, M., 208
Chick, H., 46
Claesen, M., 176
Claesson, P., 6(*10*)
Clark, C.R., 244(*986*)
Claus, A., 58
Clausius, R., 28
Cobbs, W.H., 247(*1013*)
Codington, R.S., 217(*827*)
Coenan, M., 194(*710a*)
Cogan, H.D., 137(*446j*)
Cohen, C., 232
Cohen, E., 12
Cohn, E.J., 139, 157
Coleman, B.D., 219
Coleman, D., 247
Colin, J.J., 37
Collins, A.M., 68(*229*), 82(*229*), 129(*229*)
Columbus, C., 32
Combes, C., 123
Compagnon, R., 176
Conant, J.B., 130, 132, 139
Conix, A., 194(*707*)

Cook, R.E., 217(*825*)
Corey, R.B., 85, 199(*734*) 202, 210, 231(*734*)
Cori, C.F., 123
Cori, G.T., 123
Corradini, P., 189(*679*), 190(*685*), 191(*685*), 192(*696*), 243(*978, 979*)
Cox, J.T., Jr., 174(*579*)
Crafts, J.M., 26, 122
Crawford, J.C.W., 177(*599*)
Crick, F.H.C., 200, 205, 209–212
Crook, E.H., 238(*952a*)
Cross, C.F., 39, 40, 57
Crowfoot Hodgkin, D., 84–85, 143, 198, 201, 202
Crozier, R.N., 175(*587*)
Crum Brown, A., 9
Cullis, A.F., 202(*751*)
Cummings, W., 185(*653*)
Curtius, T., 116
Cuthbertson, A.C., 173(*574*)
Cutler, M., 237(*937*)

Dainton, F.S., 180, 217(*825*)
D'Alelio, G.F., 6
Dalton, J., 3
Danusso, F., 189(*679*), 192
Davidson, E., 201(*742*)
Davidson, H.R., 190(*683*)
Davies, D.R., 202(*749*)
Davis, C.C., 64(*221*)
Dawson, T.R., 55(*194*)
Day, M.J., 220(*847*)
Dearborn, R.T., 174(*584*)
Debye, P., 71, 112, 226, 227, 228(*885*), 230, 235, 236, 238
de Coppet, L.C., 29
de la Condamine, C.M., 32, 33, 53
Delbrück, M., 198, 203–205, 209
Delfs, D., 194(*714*)
Dell'Asta, G., 192(*697*)
Dennstedt, I., 174(*582*)
Dersch, F., 135(*449*), 136(*449*), 182(*449*)
Desreux, V., 239
Dewar, M.J.S., 172(*562*)
Dexter, S.T., 111, 112
Dickerson, R.E., 202(*749*)
Dickhaüser, E.R., 125
Dickson, J.T., 194
Di Marzio, E.A., 242
Dinglinger, A., 131(*441*), 174(*576*)
Dintzis, H.M., 202(*747, 748*)
Dole, M., 181(*625*)
Donnan, F.G., 46, 108, 157, 237

Dore, W.H., 76-78, 80
Dorough, G.L., 118(406)
Dostal, H., 120(413), 127, 131, 134, 135, 172(446)
Doty, P., 195(717, 906), 196, 226, 227, 228(885), 230
Downing, E.P., 67-68(228)
Drucker, B., 245(999)
Dubosc, A., 35, 66
Dubrunfaut, A.P., 37
Duclaux, J., 153, 239
Dufraisse, C., 129
Dulmage, J.L., 190(687)
Dumas, J.B., 7, 8
Dunham, K.R., 190(687)
Dunkel, M., 95, 96
Dunlop, J.B., 55
Dunston, W.R., 63
Duppa, B.F., 9
Dziengel, K., 81(263)

Eareckson, W.M., 111, 194
Eaves, D.E., 176(595)
Edgar, O.B., 194(708), 245
Edsall, J.T., 42(138), 141, 142
Eigen, M., 238(952)
Eimers, E., 185(654)
Einhorn, A., 115
Einstein, A., 50, 51, 100, 101, 103, 104, 110, 227
Eirich, F., 85, 110(371), 162
Eisenberg, H., 236(929), 237, 238(947)
Eisenschitz, R., 104
Eizner, Yu.E., 231(903)
Ellery, E., 194(708a)
Elliott, A., 195(719), 196(721), 215(814), 216, 231, 233
Emmons, C.W., 206
Endres, G.F., 184(646)
Engler, C., 24, 25, 90, 129
Eppe, R., 246(1002), 247(1002)
Erlenmeyer, E., 17
Ernst, S.L., 175(587)
Erpenbeck, J.J., 224(864)
Errington, K.D., 176(590c)
Estrup, K., 99
Eucken, A., 108, 151(367)
Euler, W., 33(98)
Evans, A.G., 183
Evans, M.G., 170, 177(601)
Ewart, R.H., 178, 228
Eyring, H., 244(986), 248

Fabrinyi, R., 58

Fadner, T.A., 179, 220
Fåhreus, R., 101(335)
Fankuchen, I., 85(281), 233(281)
Faraday, M., 4, 5, 27, 33, 54
Farina, M., 192(695, 696), 217(695)
Farmer, E.H., 176(590c)
Fawcett, E.W., 132
Felsenfeld, G., 214
Ferri, C., 151
Ferry, J.D., 236, 238, 241, 242, 250, 251
Feuchter, H., 149
Field, J.E., 112
Fikentscher, H., 105, 134, 149, 178
Filipovich, F., 218(837), 219(837)
Finch, J.T., 211(791)
Fischer, E., 43-45, 58, 76, 93, 141, 205
Fischer, E.W., 246, 247(1002)
Fischer, M., 188
Fischer, M.H., 52(181)
Fisher, C.H., 245(933)
Fittig, R., 18, 24
Fitzgerald, E.R., 251
Fixman, M., 224(862)
Flechsig, E., 39
Flodin, P., 240
Flory, P.J., 96, 116, 120, 121, 127, 130, 131, 138, 156, 169, 170, 173, 175, 186, 189, 190, 222-226, 228(889), 230, 231, 233, 234, 237, 242, 244-246, 247(1012), 248, 249, 253
Fol, J.G., 104
Folt, V.L., 216(820a)
Fontana, C.M., 183
Fordyce, D.B., 236(931), 238(931)
Forman, L.E., 185(657)
Fourcroy, A.F., 43
Fowler, R.H., 155
Fox, T.G., 185(659), 219, 224, 230, 231, 242, 248
Fox, T.J., 215
Fraenkel, G.K., 220
Fraenkel-Conrat, H., 208
Fram, P., 169(549b)
Frank, H.P., 225(868), 228(868)
Frankel, M., 195(715c)
Frankland, E., 9, 27
Franklin, R.E., 209-211
Frederick, D.S., 135(446j)
Frémy, E., 39
Frenkel, Ya.I., 151, 225
Freudenberg, K., 80, 89
Freundlich, H., 47, 49, 148, 153
Frey, F., 92(302), 93(302), 94(302), 182(302)
Frey, F.E., 180
Frey, K., 124(418), 126(419e)

Frey-Wyssling, A., 86
Friedel, C., 26, 122
Frilette, V.J., 217(*822*), 244(*822*)
Frith, E.M., 244
Fritschi, J., 91
Fritz, B., 123(*415o*)
Frolich, P.K., 69(*230*), 184(*649*)
Frosch, C.J., 82
Frost, W., 126, 131, 177
Fruton, J.S., 42, 43(*137*), 45, 144, 145(*489*), 204
Fujita, H., 238
Fuller, C.S., 82
Fuoss, R.M., 225, 234, 236(*935*), 237, 250
Fuschillo, N., 218

Gaetjens, E., 219
Ganis, P., 243(*979*)
Garbsch, P., 92(*302*), 93(*302*), 94(*302*), 124(*418*), 182(*302*)
Gardner, K.H., 81
Garner, W.E., 244
Garrett, B.S., 185(*659*), 219(*659*)
Gaudechon, H., 21
Gaultier de Claubry, H., 37
Gay-Lussac, M., 3, 5, 28, 36, 38
Gee, G., 20, 127, 173(*574*)
Gehman, D., 112
Geidushek, E.P., 228(*892*)
Geiger, E., 137(*456b*)
Gellert, H.G., 186(*672*), 187(*672*), 190(*672*)
Gent, A.H., 247
Gerhardt, C., 16, 17, 114
Gerngross, O., 84
Gerrens, H., 178
Greshfeld, N.L., 238(*952a*)
Giannini, U., 192(*697*)
Gibbs, J.H., 242
Gibson, R.O., 132
Gillam, W.F., 123(*415i*)
Gillis, J., 235
Girard, A., 41(*135*)
Gladstone, J.H., 50
Gloor, W.E., 56(*197*), 57
Go, Y., 20, 138, 195
Gold, H., 194(*715*)
Goldberg, R.J., 228
Goldfinger, G., 172
Goldsbrough, H.A., 52
Gonnell, H.W., 84
Goode, W.E., 185(*659*), 219(*659*)
Goodyear, C., 35, 54, 55, 148
Goryainov, V., 20
Gosling, R.G., 210, 211

Gough, J., 35, 36, 149
Gouy, M., 100
Graham, T., 28, 47, 48, 51, 52
Grassie, N., 172(*564*), 180
Gratch, S., 185(*659*), 219(*659*)
Graulich, W., 174
Green, A.G., 40
Green, D.W., 202(*746*)
Green, H., 159
Greenberg, S.A., 245(*994*)
Grieger, P.F., 238(*945*)
Grimley, T.R., 224(*861*)
Grimm, H., 182(*627*)
Gross, H., 102, 105, 107, 108, 239
Gross, S.T., 190(*683*)
Grossfeld, I., 195(*715a*)
Grün, F., 217
Grunberg-Manago, M., 213
Guggenheim, E.A., 155
Guldberg, C.M., 30
Gulland, J.M., 206, 210
Gündel, W., 90(*297*)
Guth, E., 150, 151, 162, 217(*829*), 249
Gutowsky, H.S., 218

Haber, F., 65, 161, 170
Habermann, J., 43
Habgood, B.J., 133
Haehnel, W., 125
Haldane, J.S., 205
Hall, H.K., 245(*990*)
Halle, F., 82
Haller, W., 109, 154
Hammersten, E., 158
Hampton, R.R., 215(*812*)
Hanby, W.E., 195(*719*), 216(*815*)
Hancock, T., 54, 55
Hanes, C.S., 123
Hanford, W.E., 195(*715a*)
Hanle, J., 217(*822*), 244(*822*)
Hansley, V.N., 185(*660*)
Hanson, J., 235
Hantsch, A., 182, 184
Happey, F., 216(*815*)
Harboth, G., 131
Hardt, A., 170(*552c*)
Hardy, W.B., 45, 50, 52
Hargitai, B., 195(*717*), 231(*717*), 235
Harkins, W.D., 174(*603b*), 178
Harold, R.J., 184(*647*)
Harries, C., 19, 23, 33–35, 52, 63, 90, 182(*64*)
Harrington, W.F., 112, 233(*913, 914*)
Harris, I., 240(*959*)

Harrison, F.C., 123(*4151*)
Hart, E.J., 172(*566*), 215(*812*)
Hart, R.G., 202(*749*)
Hartley, G.S., 235
Hatschek, E., 51, 105
Haurowitz, F., 145, 146, 162, 199, 209
Hauser, A.E., 148
Haworth, W.N., 76(*250*)
Hay, D.H., 130(*438*)
Hayashi, K., 179(*611*)
Hayward, N., 54
Hechelhammer, H., 185(*654*), 194(*710a*)
Heckler, G,E., 236(*931*), 238(*931*)
v. Hedenström, A. 115
Heidelberger, M., 145, 146, 198
Heinz, G., 194(*715*)
Heisenberg, W., 15
Helferich, B., 6
Hellfritz, H., 110(*370d*)
Hengstenberg, J., 78, 79, 91(*254*), 92(*253, 254*), 94(*254*), 134
Hennell, H., 6, 7
Henning, U., 177(*598*)
Herbst, M., 247
Hérrissant, L.A.P., 53
Hermans, J.J., 237, 238, 240
Hermans, P.H., 243
Heron, J., 37, 38
Herrmann, F., 15
Herrmann, W.O., 125, 133, 161
Hershey, A.D., 208
Herzfeld, K.F., 128
Herzog, R.O., 50, 71, 75, 76, 80, 84, 87, 89, 95
Hess, K., 39, 81, 87–89, 105, 161, 246(*1005*)
Hess, W.R., 249
Hesse, W., 241(*968*)
Heuer, W., 93(*310*), 96(*320*), 105, 106
Heyroth, F.F., 140(*464a*)
Hibbert, H., 123(*4151*)
Hibbert, W., 50
Higginson, W.C.E., 185
Hill, J.W., 68, 119, 120(*229a, 408*)
Hill, R., 134, 194(*708b*), 245
Hiller, A., Jr., 224(*863*)
Himly, F.C., 33
Hinshelwood, C., 200, 208, 209
v.Hippel, P.H., 112
Hirs, C.H.W., 199(*730*)
Hirshon, J.M., 220(*842*)
Hirst, E.L., 76(*251*)
Hitler, A., 67, 162
Hlasiwetz, H., 43
Hofeditz, W., 172

Hoffman, C.J., 218(*834*)
Hoffman, H.A., 150(*513*)
Hofmann, A.W., 16, 17, 21, 22
Hofmann, F., 23, 63, 64, 65(*218b*), 126, 184(*642*)
Hofmann, P., 194(*715*)
Hogan, J.P., 186(*688*)
Hohenstein, W.P., 177(*602*)
Holden, G., 186(*666*)
Hollaender, A., 206(*791*)
Holmes, K.C., 211, 243
Holroyd, L.V., 217
Holscher, F., 170
Holtzer, A.M., 195(*906*), 231(*905, 906*)
Holwerda, B.J., 112
Holzkamp, E., 186(*671*), 187, 188
Hooke, R., 56
Hooker, S.B., 146
Horne, S.E., 191
Hörz, H., 15
Hosemann, R., 244
Hottenroth, V., 57
Houtz, R.M., 175
Houwink, R., 110
Hovey, A.G., 121
Howard, F.A., 68
Howells, E.R., 201(*743*), 243
Howitt, F.O., 245(*992*)
Hsieh, H., 185(*656*)
Huber, E., 94(*315*)
Hückel, E., 155, 156, 235, 236, 238
Huggins, M.L., 156, 190, 199, 223, 225, 226, 229
Hughes, R.E., 185(*659*), 191(*690*)
Huizenga, J.R., 238
Hulse, W.F., 173(*571*)
Hummel, D.O., 120
Hunt, M., 135(*446j*)
Hunter, E., 243(*980*)
Hunter, W.H., 183
Husemann, E., 123, 127, 128, 130, 131, 134, 170, 174(*576*)
Huxley, R.E., 235
Hyatt, I.J., 57
Hyatt, J.W., 57

Immergut, E.H., 176(*592*)
Ingram, D.J.E., 220(*844*)
Ingram, V.M., 199, 202(*746*)
Inhoffen, E., 179
Ipatiev, V.N., 33(*99*), 132
Isihara, A., 233
Itano, H.A., 199(*731*)
Ivin, K.J., 180, 217(*825*)
Izumi, M., 238(*944*)

Name Index   293

Jackson, D.R., 186(*667*)
Jakob, L., 135(*450, 451*), 136, 182(*449, 451*)
James, H.M., 249
Jancke, W., 71(*236, 238*), 84
Jeans, J.H., 28
Jenckel, E., 242, 249
Jenkins, A.D., 220(*844*)
Jeu, K.K., 129
Johner, H., 78(*254*), 91(*252, 254*), 92(*254, 304*), 93(*304*), 94(*252, 254*), 137(*453*)
Jones, G.D., 134(*446f*)
Jones, T.T., 171
Jones, W.H., 128
Jordan, D.O., 206(*762*), 210
Joule, J.P., 28, 36, 75, 149
Joyce, R.M., 195(*715a*)

Kagawa, I., 238(*944, 949*)
Kahler, H., 210(*784*)
Kämmerer, H., 169
Karlson, R.H., 195(*717*), 231(*717*)
Karrer, E., 150
Karrer, P., 87, 105
Katchalski, E., 195, 233
Katchalsky, A., 234(*922*), 235, 236
Kattein, A., 49
Katz, J.R., 48, 74-76, 84
Katz, M., 65, 124, 193
Kaufmann, M., 57
Kaufmann, W., 242(*973a*)
Kauzmann, W., 200, 242(*973a*), 248
Kazbekov, E.N., 172(*567*), 220(*845*)
Ke, B., 244
Keeling, C.E., 181(*625*)
Kekulé, A., 9, 10, 13, 27, 28, 42, 58
Keller, A., 246, 247(*998, 1009*)
Keller, J.B., 89
Kelley, D.J., 185(*656*)
Kelvin, Lord, 36
Kemp, A.R., 215(*810*)
Kendall, F.E., 146
Kendrew, J.C., 200-202
Kenn, E., 89(*296*)
Kennedy, J.P., 184
Kern, W., 92(*304*), 93(*304*), 98, 137(*454*), 158, 159, 169, 170, 234-236
Kharasch, M.S., 175(*585*)
Kidder, G.A., 183
Kienle, R.H., 121
Kiesig, H., 246(*1005*)
Kilham, J.K., 177(*601*)
Kimball, G.E., 237
King, A.M., 244(*985*)

Kinkaid, J.F., 185(*659*), 219(*659*)
Kipping, F.S., 122, 123
Kirby, J.E., 68(*229*), 82(*229*), 129(*229*)
Kirchhof, F., 104
Kirchhof, C., 34, 37
Kirkwood, J.G., 228, 230, 250
Kitai, R., 199(*729*)
Kitanishi, Y., 179(*611*)
Klatte, F., 125, 126
Kleeberg, W., 58
Kleiner, H., 135(*448*), 136, 182(*448*)
Klug, A., 211
Knehe, E., 89(*295*)
Kohle, H., 131(*445a*)
Kohlhaas, A., 241
Kohlschütter, H.W., 79, 124, 130
Kolbe, C., 9, 14, 15, 24, 42
Kolpak, F.J., 81
Kolthoff, I.M., 215
Kondakov, I., 23
König, J., 39
Kopp, H., 8(*18*), 17
Kornberg, A., 212
Kraemer, E.O., 102, 110-112
Krafft, F., 48
Králiček, J., 195(*175b*)
Kränzlein, G., 133, 162
Kratky, O., 224, 225
v. Krauss, E., 105
Kraut, K., 114
Krigbaum, W.R., 226, 228(*889*), 231(*901, 902*)
Krimm, S., 216, 217(*821*), 243
Kronig, A., 28
Kroning, E., 194(*715*)
Kronstein, A., 25
Krüger, D., 87
Kruyt, H.R., 12
Kuhn, H., 229, 230
Kuhn, W., 104, 107, 109, 110, 117, 150, 151, 222, 229-230, 234, 235, 249, 250
Künzle, O., 234(*922*)
Kurtz, J., 233(*912*)
Kwolek, S.L., 194

Lambert, J.M., 190(*683*)
Lampl, H., 145
Landel, R.F., 251
Landler, Y., 185
Landsteiner, K., 52, 145, 146
Langhans, A., 87
Langmuir, I., 142, 233
Lansing, W.D., 102
Laplace, P.S., 20

Larsen, L.V., 6
Lasoski, S.W., Jr., 218, 219
v. Laue, M., 70, 71
Lautenschläger, L., 90, 125, 130
Lauth, H., 177
Lavin, E., 175(*587*)
Lavoisier, A.L., 20
Lawton, E.J., 176(*594b*), 180(*622*), 181(*623*), 220
Lazurkin, Yu.S., 152, 250
Leaderman, H., 249
Lebedev, S.V., 23, 35, 65
Le Bel, J.A., 14
Le Bras, J., 176
Lederberg, J., 213
Lee, T.S., 215
Legally, P., 131(*445a*)
Leuchs, H., 137
Leutner, F.S., 219(*841*)
Levin, B.Ya., 242
Levy, M., 185(*664*)
Lewis, F.M., 133, 169(*549a*), 172, 173
Lewis, J.R., 134(*446c*)
Lewis, N.B., 140(*465*)
Lewis, W.C.M., 142
Liang, C.Y., 216, 217(*821, 823, 824*)
Liebhafsky, H.A., 123
v. Liebig, J., 7, 8, 15, 27, 42, 45, 56, 137
Lifson, S., 236
Lightbown, J.E., 69(*230*)
Linderstrøm-Lang, K., 69(*230*)
Lipkin, D., 185(*662*)
Lippert, B., 123(*415o*)
Livingston, J.W., 174(*579*)
Ljubitsch, N., 105(*354*)
Lloyd, B.J., Jr., 210(*784*)
Lloyd, W.G., 240
Loeb, J., 157
Loew, O., 43(*142*)
Loewi, O., 42
Logemann, H., 170(*552a*), 174(*581*)
Lohmann, H., 79, 92(*257*)
Longi, P., 192(*698*)
Longsworth, L.G., 141
Lother, R., 15
Lotmar, W., 81
Lourenço, A.V., 113, 114
Lühdemann, P., 154
Lundberg, R.D., 196
Lungsted, L.G., 186(*667*)
Luria, S., 205
Lüthy, M., 78(*252*), 91(*252*), 92(*304*), 93(*304*), 94(*252*), 137(*453*)
Luttringer, A., 35, 66

Lynen, F., 177(*598*)
Lysenko, T.D., 213

Mc Call, D.W., 218
Mc Call, D.W., 218
Mc Carty, M., 206, 207, 208(*766*)
Mc Curdy, J.L.M., 176(*590b*)
Mac Gillavry, D., 88
Mac Innes, D.A., 141(*471*)
Mc Intire, O.R., 176(*590b*)
Macintosh, C., 53, 54
Mc Intyre, A.D., 248
Mac Leod, C.M., 206, 207, 208(*766*)
Mc Loughlin, J.R., 251
Mc Millan, F.M., 189(*680*), 190(*688a*), 191(*689*)
Macquer, P.J., 53
Madorsky, S.L., 180
Magat, E.E., 194
Magat, M., 64(*222*)
Majury, T.G., 172(*565*)
Mandelkern, L., 231(*901*), 245(*990*), 247
Mann, J., 217(*823*)
Mantico, E., 189(*679*)
Maquenne, L., 38
Marchessault, R.H., 74, 81, 217(*823, 824*)
Marcilly, L., 116
Mark, H., 73, 74, 76(*266*), 78–86, 89, 93, 95, 97, 105, 107, 108, 110, 127, 131, 143, 149–151, 155, 162, 173(*570b*), 176(*592*), 177(*602*), 204, 217(*822*), 222, 225(*868*), 227(*881*), 228(*868*), 230(*899*), 231(*899*), 244(*822*)
Marrack, J.R., 145
Marrinan, H.J., 217(*823*)
Martin, A.E., 215
Martin, A.J.P., 198
Martin, C.J., 46
Martin, H., 186(*671*), 188, 190
Marvel, C.S., 118, 134, 135, 174, 252
Marzolph, H., 225(*873*)
Maschin, A., 174
Mastin, T.W., 134(*446f*)
Matheson, J., 172
Matheson, M.S., 169(*549a*)
Matthews, F.E., 22, 23, 175, 182(*63*)
Maxwell, J.C., 111, 249
Mayo, F.R., 133, 172, 173(*571*), 174, 175(*585*), 185(*653*)
Mazzanti, G., 189(*679*)
Mead, D.J., 225
Meadows, G.W., 183
Meads, J.A., 122(*415d*)
Mecklenburg, W., 111
Medvedev, S., 186

Meehan, E.J., 214
Meerwein, H., 183(*633*), 194
Meisenburg, K., 174(*582*)
Melville, H.W., 170(*551*), 171, 172, 180, 220(*843*)
Menasse, D., 58
Merrifield, R.B., 252
Meselson, M., 211, 212
Mesrobian, R.B., 176, 179
Messmer, E., 88(*291*), 105(*354*)
Meyer, A.W., 215(*812*)
Meyer, K.H., 20, 73(*266*), 76(*266*), 79-84, 89, 93, 95-98, 138, 143, 149-151, 154-157, 162, 195, 204
Meyer, L.H., 218
Meyerson, I., 225
Michael, A., 44
Michaelis, L., 45, 52, 162
Michels, A., 132
Mie, G., 78(*254*), 91(*254*), 92(*254*), 94(*254*)
Miekeley, A., 89(*296*)
Milkovich, R., 185(*664*), 186
Millar, J.H., 38
Miller, A.A., 181(*623*)
Miller, R.G.J., 240(*959*)
Minke, R., 81
Mirsky, A.E., 142, 143, 144(*488*), 200
Misch, L., 80, 81
Mitchell, J.K., 51
Mitscherlich, E., 8
Mnookin, N.M., 67
Moffitt, W., 232
Mohan, G.R., 238(*947*)
Mohr, E., 117
Monheim, J., 170(*552b*)
Monfort, F., 183(*633*)
Mooney, M., 249
Moore, B., 49, 50
Moore, J.C., 240(*964*)
Moore, S., 198, 199
Moorhead, E.F., 246(*1001*)
Moraglio, G., 189(*679*)
Morawetz, H., 97(*324*), 179, 219, 225(*870*), 236
Morey, D.R., 239
Morgan, L.B., 170(*556*)
Morgan, P.W., 194, 252
Morgan, T.H., 203
Morita, H., 185(*657*)
Morris, G.H., 38
Morisson-Jones, C.R., 176(*590c*)
Morschel, H., 194(*714*)
Morton, A.A., 190
Mösslein, E.M., 177(*598*)

Mostynski, B., 46(*153*)
Mote, S.C., 55
Moureu, C., 24, 129
Mrowca, B.A., 217(*829*)
Mueller-Conradi, M., 68, 184(*649*)
Muggli, R., 81
Muirhead, H., 202(*751*)
Müller, E., 193
Muller, H.J., 203, 205, 213
Müller, M., 57(*207*)
Mündler, K., 154
v. Muralt, A.L., 141
Murphree, E.V., 69(*230*)
Murray, A.G., 122(*415g*)
Murray, H.A., 158
Musculus, F., 37

Nacquet, A., 27
Nagasawa, M., 238(*944, 949*)
Nägeli, C., 48, 70, 87
Nasse, 37
Nastukoff, A., 41
Natta, G., 188-194, 217(*695*), 243, 253
Naylor, R.E., Jr., 218, 219
Nernst, W., 50, 51
Newburg, N.R., 191
Nichols, J.B., 101
Nielson, H., 90(*297*)
Niemann, C., 198
Niggli, P., 86
Nishii, M., 179(*611*)
Nishikawa, S., 72
Noether, F., 236(*934*)
Norrish, R.G.W., 131, 134, 135, 172(*442*)
North, A.C.T., 202(*751*), 220(*846*)
Northrop, J.H., 144, 208
Noyes, A.A., 50
Nozaki, K., 169(*547*)

Oakes, W.G., 243(*980*)
Ochoa, S., 213
O'Connor, A., 246, 247(*998*)
Oden, S., 99, 100
Oka, S., 222
Oken, L., 15
Okamura, S., 179
Olby, R., 204, 209
Ono, S., 72
Onsager, L., 233
Ortiz, P.J., 213(*797*)
Osborne, T.B., 141
Osgan, M., 191(*690*)
Ost, H., 40, 41

Name Index

Oster, G., 179(*607*)
Osterheld, J.E., 237
Ostromislensky, I., 21, 64, 176
Osswald, W., 81(*263a*)
Ostwald, Wi., 47, 49
Ostwald, Wo., 47, 49, 52, 91, 95, 105, 111, 148, 154
O'Sullivan, C., 37, 38
Ott, E., 84, 92
Otto, M., 68, 184(*642, 649*)
Overbeek, J.T.G., 235, 237
Overberger, C.G., 169(*549b*), 184(*646*)
Ozanam, 56

Pake, G.E., 218
Pals, D.T.F., 237
Pampus, G., 174(*581*)
Paneth, F., 172
Pape, N.R., 82
Park, G.S., 170(*555*)
Parks, G.S., 241, 242
Parrish, R.G., 202(*748*)
Parsons, J.L., 217
Pasteur, L., 10, 11
Paterno, E., 12, 49
Patrick, J.C., 67
Patterson, A.L., 83, 85
Patton, J.G., 132(*445e*)
Paul, D.E., 185(*662*)
Pauli, W., 47, 50, 51, 157
Pauling, L., 143, 146, 147, 199, 200, 202, 209, 210, 212, 231, 232
Payen, A., 38, 39
Pease, R.N., 6(*11a*)
Peat, S., 76(*250*)
v. Pechmann, H., 21, 126
Pedersen, K.O., 101
Pedone, C., 223
Peiser, H.S., 190(*681*)
Pelouze, J., 39
Pepper, D.C., 184(*643*)
Peraldo, M., 192(*695, 696*), 217(*695*)
Perkin, A.G., 40
Perrin, J., 49, 101, 103, 153
Perrin, M.W., 132
Persoz, J.F., 37
Perutz, M.F., 85, 162, 200-202
Peterlin, A., 230
Peters, H., 215(*810*)
Petersen, S., 193(*700*)
Peterson, W.R., 130, 132(*436d*)
Pfannemüller, B., 123(*415o*)
Pfeil, E., 194(*715*)

Pflüger, E., 28
Philippoff, W., 250
Phillips, D.C., 202(*748, 749*)
Pickles, S.S., 34, 35, 75, 78
Piepenbrick, H.F., 193(*700*)
Pilch, K., 174(*575*)
Pino, P., 189
Pitzer, K.S., 108, 222
Planck, M., 30
Plesch, P.H., 184
Polanyi, M., 71-73, 87, 183
Pollett, W.F.V., 251
Pontecorvo, G., 213
Porath, J., 240
Porod, G., 224, 225
Porter, A.W., 31
Posnjak, E., 153
Posnjak, G., 125
Pouyet, J., 237
Powell, R.E., 244
Powell, R.S., 176(*594b*), 220(*849*)
Price, C.C., 169, 173, 183, 191
Price, F.P., 248
Priesner, C., 93, 108
Priestley, J., 19, 53
Pringsheim, H., 87, 88
Procter, H.R., 157
Ptitsyn, O.B., 223, 231(*903*)
Pummerer, R., 87, 90, 95
Putnam, F.W., 200(*736*)
Putzeys, P., 112

Quinn, F.A., Jr., 247(*1012*)

Raff, R., 127
Raine, H.C., 243(*981*)
Raman, C.V., 111, 226
Rånby, B.G., 139, 246
Raoult, F.M., 29-31, 88, 108, 153
Rasuvayev, G., 130(*437*)
Ray, N.H., 225(*871*)
Rayleigh, Lord, 111, 112
Read, A.T., 175(*585*)
Reader, W.J., 194(*705*)
Redtenbacher, J., 8(*19*), 23
Regnault, V., 21, 28
Rehberg, C.E., 245(*993*)
Rehner, J., Jr., 249
Reid, E.W., 49
Rempp, P., 252
Reppe, W., 133
Rhian, E., 218(*832*)
Rice, F.O., 128

Name Index   297

Rice, S.A., 228
Rich, A., 214
Richards, R.B., 243(*981*), 244
Rideal, E.K., 127, 173(*574*)
Rinde, H., 100
Riseman, J., 230
Ritzenthaler, B., 135
Rivlin, R.S., 249
Roche, E., 81
Rochow, E.G., 123
Robinson, C., 233
Rodewald, H., 49
Roe, C.P., 228(*885*)
Roe, J.W., 235
Röhm, O., 126, 177 (*600*)
Roithner, E., 115(*386*)
Rona, P., 46(*154*)
Rose, G., 181(*625*)
v. Rosenberg, G., 69
Ross, M., 180(*621*)
Ross, P.D., 214, 238(*946*)
Rossmann, M.G., 202(*751*)
Rötger, H., 250
Rouse, P.E., 250
Rouvé, A., 118
Roux, E., 38
Rubin, I.D., 179(*609*)
Rudorfer, H., 174(*575*)
Rundle, R.E., 232
Rushbrook, G.S., 155
Russidis, D., 92(*304*), 93(*304*)
Ruzicka, L., 117
Ryder, H., 243(*981*)
Ryle, A.P., 199(*729*)

Sachse, H., 117
Sailors, H.R., 186(*688*)
Sakurada, I., 105(*356*)
Samelson, H., 237(*937*)
Saminskii, E.M., 172(*567*), 220(*845*)
Sanger, F., 198, 199, 234
Sarid, S., 195(*718a*)
Sarko, A., 81
Sauer, J.A., 218(*832*)
Saunders, W.M., 239(*956*)
Sauter, E., 79(*255*), 92(*255*)
Sawitsch, V., 21
Schacht, H., 67
Schelling, F.W.J., 15
Schellmann, J.A., 232
Scherrer, P., 71, 72
Schertz, G.L., 134(*446f*)
Schidknecht, C.E., 190

Schidrowitz, P., 52, 55(*194*)
Schiel, J., 8(*17*)
Schindewolf, H., 238(*950*)
Schlack, P., 120, 195
Schneider, B., 97(*325*)
Schneider, E.E., 176(*594*), 220
Schneiders, J., 134
Schön, K.G., 225(*872*)
Schönbein, C.F., 56
Schramm, C.H., 195
Schrödinger, E., 204, 205
Schröter, G., 114(*384*)
Schuller, H., 178
Schultz, J., 206
Schulz, G.V., 127, 128, 130, 131, 154, 169, 170, 174, 225(*872, 873*), 239
Schulze, E., 39
Schwarz, G., 238(*952*)
Schweitzer, O., 92(*304*), 93(*304*)
Scott, N.D., 185(*660*)
Šebenda, J., 195(*715b*)
Seemann, H., 189
Sela, M., 195, 233(*913*)
Shakespeare, N.E., 176
Shambelan, C., 191(*690*)
Sharonov, Yu.A., 223
Shashoua, V., 194
Shipman, J.J., 216(*820a*)
Shore, V.C., 202(*749*)
Sigmund, F., 137
Signer, R., 78(*254*), 91(*254*), 92, 93(*302, 304*), 94(*254, 302*), 102, 103, 107-109, 124(*418*), 128, 137(*453*), 141, 182(*302*), 239, 246
Simha, R., 104, 107, 162, 180
Simon, E., 16
Simon, F., 241
Simons, E., 141(*473*)
Simonson, J.L., 134(*446c*)
Singer, S.J., 199(*731*)
Siriani, A.F., 186
Sjögren, B., 140(*464b*)
Skavronskaya, N.A., 23(*68*)
Skraup, Z.H., 39
Slichter, W.P., 218
Smets, G., 176
Smith, F.C., 145
Smith, L.F., 199(*729*)
Smith, R.R., 131, 177
Smith, S.G., 245(*992*)
Smith, W.V., 178
v. Smoluchowski, M., 104
Snow, R.D., 180
Snyder, R., 174(*584*)

Name Index

Sobecki, W., 114
Soda, A., 238(*949*)
Solzhenitsyn, A., 204
Sørensen, S.P.L., 140
Soret, J.L., 20
Spanagel, E.W., 118
Sparkman, D.H., 199(*730*)
Sparks, W.J., 69(*230*), 184(*649*)
Spell, A., 185(*659*), 219(*659*)
Spěváček, J., 97(*325*)
Spiegel-Adolf, M., 142
Sponsler, O.L., 76–78, 80
Sprenger, L., 79
Springer, A., 174(*575*)
Squire, P.G., 239(*956*)
Srinivasan, P.R., 42(*138*)
Stahl, F.W., 211, 212
Stalin, J.V., 213
Stanley, H.E., 228(*888*)
Stanley, W.M., 85
Starck, W., 126(*419e*)
Starkweather, H.W., 128
Stas, J.S., 8
Staudinger, H., 66, 78–80, 86, 87, 90–97,
  105–110, 113, 116, 118, 119, 124–131,
  133–135, 139, 140, 149, 150, 158, 160–161,
  169, 175, 177, 182, 185, 189, 224, 229, 246
Staudinger, M., 79(*255, 256*), 92(*255*), 106, 160,
  161
Staveley, F.W., 185, 191(*655*)
Staverman, A.J., 225
Stearns, R.S., 185(*658*)
Stein, G., 220(*847*)
Stein, R.S., 227
Stein, W.H., 198, 199
Steinberg, I.Z., 233(*914*)
Steinhofer, A., 128
Stern, H.J., 55
Steward, J.M., 174(*584*)
Stewart, G.W., 108
Stipanovic, A.J., 81
Stobbe, H., 125
Stockmayer, W.H., 122, 224(*862*), 228
Stokes, G., 50
Stoll, M., 117, 118
Stoll-Comte, G., 117
Storks, K.H., 246
Strachan, D.R., 194
Strandberg, B.E., 202(*749*)
Strange, E.H., 22, 23, 182(*633*)
Straus, S., 180(*617*)
Strauss, U.P., 234(*921*), 238
Street, A., 83(*268*)

Stroupe, J.D., 185(*659*), 219(*659*)
Strutt, J.W., 111, 112
Stuart, H.A., 246(*1002*), 247(*1002*), 248
Sturtevant, J.M., 214
Suess, H., 174
Sumner, J.B., 84, 144
Sundararajan, P., 81
v. Susich, G., 150
Sutherland, G.B.B.M., 215, 216(*817, 818*),
  217(*821*)
Svedberg, T., 99–102, 108, 139, 141, 142, 145,
  197, 198
Swallow, J.C., 132
Swan, J.W., 56
Symons, N.C.R., 220(*844*)
Synge, R.L.M., 198
Szwarc, M., 185, 186, 239

Talalay, A., 64(*222*)
Tamann, G., 241
Tamblyn, J.H., 239
Tarr, H.L.A., 123(*4151*)
Tate, B.E., 169
Tatum, E.L., 212
Taylor, G.B., 128
Taylor, H.S., 128, 129, 179
Taylor, R.P., 170(*551*)
Taylor, W.J., 222
Temple, R.B., 215(*814*)
ter Meer, E., 58
Thénard, J., 36, 38
Thiele, J., 34(*149*)
Thierfelder, H., 45
Thomas, A.W., 158
Thomas, C.A., 69(*232*)
Thomas, R., 214
Thomas, R.M., 69(*230*), 184
Thompson, D., 180(*617*)
Thompson, E.O.P., 199(*728*)
Thompson, W.H., 215
Thomson, W., 36
Thovert, J., 50, 51, 102
Tiers, G.V.D., 218(*837*), 219
Tilden, W.A., 33
Till, P.H., Jr., 246
Timoféev-Ressovsky, N.W., 204
Tinoco, I., 214
Tipson, R.S., 123(*4151*)
Tiselius, A., 141, 145
Tobolsky, A.V., 179, 185(*656, 657*), 243, 250,
  251
Tollens, B., 39, 40
Tongberg, C.O., 130, 132(*436c*)

Tornqvist, E.G.M., 32, 53(*93*), 54(*93*), 65(*93*), 67(*93*)
Townsend, M.G., 220(*844*)
Trainer, E., 58
Treloar, L.G.R., 249
Trevor, J.E., 31
Trieschmann, H.G., 194(*710b*)
Trogus, C., 81(*263a*), 89, 105(*356*)
Trommsdorff, E., 93(*303*), 126, 131, 158, 177(*600*)
Tschirmer, F., 21
Tschunkur, E., 66-67(*226*)
Tsuboi, M., 217(*823*)
Tuckett, R.F., 244
Tuppy, H., 199

Udy, D.C., 236(*931*), 238(*931*)
Ueberreiter, K., 242, 249

Valkó, E., 149(*507*), 150, 162
Van Bitter, K., 244(*985*)
Van Den Berghe, J., 190(*687*)
Vanderhoff, J.W., 178, 179(*607*)
van der Waals, J.D., 28, 153
Van Kerpel, R., 194(*707*)
van Natta, F.J., 118(*399*)
van't Hoff, J.H., 12-15, 17, 30, 31, 108, 117, 153
Vaughan, W.E., 6(*11b*)
Vaughn, M.F., 240
Vendrely, R., 207
Vernon, A.A., 129
Veselý, K., 184(*648*)
Vickery, H.B., 141
de Vigenère, B., 8
Vinograd, J., 211(*792*), 212
Vischer, E., 214(*803*)
Vitkuske, J.F., 178(*606*)
Volkenstein, M.V., 222, 223
Vollmert, B., 110(*370e*)
Voorn, M.J., 240
Vorländer, D., 115
Voss, A., 95

Wada, A., 228(*892*), 231(*904*), 232(*904*), 236
Wagner, G., 17(*44*), 182
Wagner-Jauregg, T., 133, 173, 175(*585*)
Wakefield, H.F., 150(*513*)
Wald, G., 30
Walden, P., 182
Waldschmitz-Leitz, E., 141(*473*)
Walker, J.F., 185(*660*)
Wall, F.T., 135, 172, 224, 238(*945*), 249
Wall, L.A., 176(*595*), 180, 181(*624*)

Wallach, O., 22, 33
Walling, C., 173, 185, 219(*842*)
Walter, N.M., 246(*1001*)
Ward, J.C., 233(*916*)
Warth, H., 110, 175
Waters, W.A., 130(*438*)
Watson, J.D., 200, 205, 209-212
Watson, R., 29
Weaver, W., 197
Weber, C.O., 52
Weese, H., 153
Wegner, G., 79(*255a*), 179
Wehrli, S., 92(*302*), 93(*302*), 94(*302, 311*), 124(*418*), 182(*302*), 189(*311*)
Weidinger, A., 74, 243
Weigert, K., 108, 151(*367*)
Weiler, J. 128
v. Weimarn, P., 95
Weiss, J., 170
Weissberg, J., 129
Weissenberg, K., 72-74, 76, 87
Weissman, S.I., 185
Welch, C.K., 55
Wells, I.C., 199(*731*)
Weltzien, W., 88(*291*)
Wenner, G., 194(*710b*)
Wenz, A., 135(*451*), 136(*451*), 182(*451*)
Wertyporoch, E., 183-184
Wessely, F., 137
Wesslau, H., 190
Wheeler, D.J., 224(*863*)
Wheeler, O.L., 175
Whinfield, J.R., 194
Whiffen, D.H., 220
Whitby, G.S., 64(*221*), 65, 124, 149
White, D.M., 179(*610*), 216(*610*)
White, F.H., 200
Whitmore, F.C., 183
Wichterle, O., 195, 252
Wiedersheim, V., 96(*319*)
Wieland, H., 86, 130, 161
Wiggam, D.R., 56(*197*)
Wilkins, M.H.F., 207, 209
Will, G., 202(*751*)
Willer, R., 182(*627*)
Willfang, G., 194(*715*)
Williams, E.G., 132
Williams, G., 21, 22, 33
Williams, I., 68(*229*), 82(*229*), 129(*229*)
Williams, J.L.R., 190(*687*)
Williams, J.W., 102, 239
Williams, M.L., 251
Williams, R.C., 210(*784*)

Williams, R.J.P., 239
Williamson, A.W., 26
Williamson, L., 180(*617*)
Willstätter, R.M., 144
Wilms, H., 185(*654*)
Wilson, C.W., III, 218
Wilson, J.A., 157(*541*)
Windemuth, E., 193(*700*)
Winter, U., 246(*1006*), 247
Wippler, C., 228
Wislicenus, J., 9, 13, 15, 24, 182
Witbecker, E.L., 194, 711
Wittorf, V., 33(*99*)
Wohl, A., 183(*639a*)
Wöhler, F., 7, 8
Wöhlisch, E., 149
Wolf, H., 55(*192*)
Wolf, R.F., 55(*192*)
Wollgast, S., 15
Wollman, E., 153, 239
Wollthan, H., 135(*449, 451*), 136(*449, 451*), 174(*583*), 182(*449, 451*)
Wolz, H., 173
Wood, L.A., 247
Wooding, N.S., 185
Woodruff, W., 53(*187*)
Woods, H.J., 83(*269*)

Woodward, R.B., 195
Worsfeld, D.J., 186(*669*)
Wright, S., 206(*760*)
Wrinch, D.M., 142, 205
Wróblewski, A., 45
Wu, F.C., 236(*931*), 238(*931*)
Wu, H., 142, 143
Wurtz, A., 9, 14, 113, 114
Wyckoff, H., 202(*748*)
Wyckoff, R.W.G., 85

Yang, J.T., 231(*904*), 232(*904, 907*)
Yohe, R.V., 183

Zahn, H., 246(*1006*), 247
Zamenhof, S., 208
Zancker, E., 174(*582*)
Zhurkov, S.N., 242
Ziegler, K., 135, 136, 182, 185–192, 215, 253
Zimm, B.H., 224(*862*), 225, 227, 228, 232
Zimmer, K.G., 204, 205
Zinoffsky, O., 44
Zlámal, Z., 184(*648*)
Zletz, A., 190
Zoss, A.O., 190(*683*)
Zsigmundy, R., 47
Zwick, M., 235(*925*)

Subject Index

Acrolein, polymerization, 23, 24
Acrylamide, polymerization, 24, 179
Acrylic acid, 24
   esters, polymerization, 126, 136, 182
   polymerization, 124
   salts, polymerization, 179
Acrylonitrile, polymerization, 220
Albumin, 49, 50, 71
Aldehydes, polymerization, 18, 79, 132, 137
α-Amino acid N-carboxyanhydride
      polymerization, 137, 195, 196
Amylopectin, 38
Amylose, 38, 123
Antibodies, 141, 146–147
Antigens, 141, 145, 146, 195
Antioxidants, 129
Avogadro hypothesis, 3, 4, 26–28
Atomic theory, 26–28
Atomic weights, 8

Benzene, 4, 8, 14
Benzoic acid, 8
Birefringence, 70, 71, 238, 247
   of flow, 108, 141
Block polymers, 186, 228
Brownian motion, 100–101
Butadiene:
   dimerization, 6
   polymerization, 23, 135, 136, 179
   synthesis, 64, 65
Butene, 4, 5
   polymerization, 189
   structure, 18
Butyraldehyde, polymerization, 132

Carbanions, 184

Carbohydrate, origin of term, 36
Carbon black, 55, 65
Carbonium ions, 183, 184
Casein, 71, 112
Cellulose:
   cellobiose residues in, 39, 40, 77, 78, 80, 88, 89
   complex with Cu^{++}, 88
   crystal structure, 72, 73, 76–78, 87, 217
   definition, 39
   discovery, 38–39
   early formulas, 40
   elastic modulus, 81
   hydrolysis, 39
   IR spectrum, 217
   mercerization, 81
   molecular association, 74, 79–80, 96
   molecular weight, 41, 95
   regenerated, 57, 81
   starch, relation to, 39
   tensile strength, 81
Cellulose acetate, 81, 88, 102, 227, 250
Cellulose nitrate, 56, 57, 87, 102, 108, 153
Chain molecules:
   entanglements, 248, 249
   excluded volume, 109–110, 222, 224, 226, 230
   expansion, 109, 110, 222, 228, 230, 236
   expansion coefficient, 223, 224, 231, 234, 237
   free draining, 229, 238
   impermeable coils, 110, 229, 238
   internal viscosity of, 230
   partially draining, 230
   radius of gyration, 227
   random coil, 109, 110, 222
   rodlike, 106–108, 149, 150, 231, 233–234, 252
   statistical chain elements of, 225
   unperturbed dimensions, 224, 231

Chain molecules (*Continued*)
   wormlike, 224–225, 232
Chitin, 79, 81
Chloroprene:
   polymerization, 67, 68, 129
   synthesis, 68
Chromatography, 198, 199, 214
   gel permeation, 240
Collagen, 84
Colligative properties, *see* Cryoscopy;
   Ebullioscopy; Osmometry
Colloids, 28, 38, 47–52, 80, 87, 91–95, 101, 148, 157, 221
Copolymerization, 114, 133–135, 172–173, 192
Copolymers of maleic acid, 236
Crankshaft-like motion, 230
Crosslinking, 24, 28, 58, 134, 240
   by ionizing radiation, 180–181
Cryoscopy, 29–31, 38, 49, 88–90, 94, 139, 225
Crystallinity, *see* Polymers, crystallinity, degree of
Crystallography:
   of cellulose, 76–81, 217
   of cellulose acetate, 81
   of chitin, 79, 81
   of collagen, 84
   of DNA, 209–211
   of fibrous proteins, 83, 84, 199, 200
   of gelatin, 84
   of gutta percha, 82
   of hemoglobin, 85, 200–202
   of *Hevea* rubber, 80
   of insulin, 85, 202
   isomorphism, 45, 245
   of myoglobin, 200, 202
   of pepsin, 85
   of poly (1-butene), isotactic, 189
   of polychloroprene, 82
   of polyethylene, 83
   of polyisobutene, 82
   of polyoxymethylene, 77–78
   of polypropylene, isotactic, 189
   of polypropylene, syndiotactic, 243
   of poly(propylene oxide), 79
   of polystyrene, isotactic, 189
   of poly(vinyl alcohol), 189–190, 217
   of poly(vinyl chloride), 82
   of poly(vinylidene chloride), 217
   side chain crystallization, 245
   of silk fibroin, 72–73, 79
   of tobacco mosaic virus, 85, 211
Cyclopentadiene, polymerization, 182

Depolymerization, 179–181

Dextrin, 37, 47, 104
   Schardinger, 87, 88
Diacetylenes, polymerization, 179
Diazomethane, polymerization, 21
Dibromethane, 12
Diffusion, 28, 47, 48, 50–52, 101–103, 217, 221, 239, 253
Dimerization, 6
Dimethylbutadiene:
   polymerization, 23, 65
   synthesis, 65
DNA, 158, 205–212
   enzymatic synthesis, 212
Donnan effect, 46
Dynamic testing, 152, 249–250

Ebullioscopy, 41, 48, 225
Electron diffraction, 246
Electrophoresis:
   of polyelectrolytes, 238
   of proteins, 45–46, 141
Enzymes, 198
   destroyed by pepsin, 45
   nature of, 45, 143–144
   polypeptide substrates for, 195
   stereoselectivity, 45
Ethylene, 4–5
   dimerization, 6
   polymerization, 20, 21, 128, 132, 186–188
Excluded volume effect, 110, 222, 224, 226, 230

Fertilization, 52
Fibers:
   artificial, 56, 57, 119–120, 125, 195
   birefringence, 70, 71
   creep and creep recovery, 249
   modulus, 81
   tensile strength, 81
   X-ray diffraction, 70, 71, 72, 247
Forces:
   interatomic, 7
   intermolecular, 28, 91, 95–97
   intramolecular, 149, 157–158, 242
Formaldehyde:
   cost of, 58
   polymerization, 79, 137
Formulas:
   rational, 9, 10
   structural, 9
Free radicals, 127, 128, 130, 131, 169–171, 175–176, 178–179, 181, 219–220
   in irradiated polymers, 176
Frictional coefficient, 99, 102, 141

Gas theory, 28, 30
Gelatin, 49, 50, 52, 64, 84, 104, 157
Gelation, 52, 105, 121–122
Gel permeation chromatography, 240, 253
Gels, hydrophilic, 252
Gel topology, 240
Genes, 203–205
Glass transition, 217–218, 241–242, 251
α-Globulin, 141
Glucose structure, 72, 76
Glycogen, 123
Graft polymers, 175–177
Gum arabic, 158
Guttapercha, 34, 82

Hair, 83
Hapten, 145
Helical structures, 11, 82, 157–158, 195, 231–233, 243
Hemoglobin:
   molecular weight, 44, 45, 101, 154
   oxygen binding, 52
   sickle cell, 199
   structure, 85, 200–202
Hydrogen bonds, 93, 143, 199, 200, 242
Hydrophobic bond, 200, 236–238

Immunochemical reactions, 52. See also Antibodies; Antigens
Indene, polymerization, 124
Insulin, 84, 198–199, 201, 202
Isobutene, polymerization, 21, 24, 68, 183, 184
Isomers, 5, 7
   cis-trans, 13
   rotational, 12, 13, 222–223
Isoprene:
   polymerization, 22, 23, 63–65, 130, 185, 191, 230
   from rubber pyrolysis, 22, 33
   structure, 33
   synthesis, 64

Latex with uniform particle size, 178–179
Levan, 123
Life, 28, 43, 51, 203, 205
Light scattering, 110–112, 221, 226–229
Liquid crystals, see Solutions, anisotropic

Maleic acid, copolymers of, 236
Mechanochemistry, 235
Methacrylic acid:
   esters, polymerization, 24, 170, 171, 177, 220
   polymerization, 24
   structure, 18

4-Methyl-hexene-1, polymerization, 191
α-Methylstyrene, polymerization, 182
Micelles, 48, 70, 80, 96, 97, 158, 178, 235, 247
   fringed, 84
Molecular biology, origin of term, 197
Molecular weight:
   distribution, 103, 120, 123, 127, 138, 186
   by end group analysis, 38, 41, 102
   of hemoglobin from iron content, 44–45
   number and weight average, 102, 227, 228
   of ovalbumin from diffusion coefficient, 50
   of proteins from titration and conductance, 50
   See also Cryoscopy; Ebullioscopy; Light scattering; Osmometry; Ultracentrifuge; Viscosity, melt
Muscular contraction, 235
Mutation, 204–205
Myoglobin, 200, 202
Myosin, 141

Nascent state, 19, 34
Naturphilosophie, 14
Neutron scattering, 223, 252
Nylon, 120

Oil of wine, 5, 6
Oligomers, origin of term, 6
Optical activity, 10–12, 14, 17
   of biological molecules, 11
   of polypeptide solutions, 232–233
   of proteins, change during denaturation, 232
   of starch and dextrin, 37, 38
Osmometry, 30, 49, 153–154, 158–159, 221, 225–228
   Donnan effect, influence on, 46, 237
Ovalbumin, 50, 64, 141

Pentene, polymerization, 18–20
Pepsin, 44, 85, 141, 144
Petroleum:
   polymerization in formation of, 24
   source of rubber monomers, 66
Phenol-formaldehyde resins, 58–59
Photopolymerization, see Polymerization, free radical; Polymerization, photochemical
Pinene, polymerization, 18–20
Plasticization:
   of cellulose nitrate, 57
   of poly(vinyl chloride), 125
Poly(acrylic acid), 158–159, 234–236, 238
   esters, 245
Polyamides, 119–120, 194, 195, 218, 245
Polycaprolactam, 120, 195, 247

Subject Index

Polycondensation, 113–123, 193–194
  chain-ring equilibrium in, 117–118
  interfacial, 194
  kinetis, 120–121
  in living cells, 123
Poly(2,3-dimethylbutadiene), 65
Polyelectrolytes, 92, 157–159
Polyesters, 118–119, 248. See also Poly(ethylene terephthalate)
Polyethylene:
  crystallinity, degree of, 243–244
  first use of term, 20
  high density, 186–188
  IR spectrum, 215, 216
  low density, 132–133, 243, 244
  radiation crosslinking, 180–181
  single crystals, 246
  viscoelesticity of melt, 251
Poly(ethylene terephthalate), 194, 245, 247
Poly(p-fluorostyrene), 243
Polyisobutene, 68–69, 82, 241
Polymerization:
  anionic, 22–24, 64, 135–136, 184–186
  cationic, 18–21, 24, 64, 68, 126, 182–184
  coordination, 186–192
  in crystalline state, 24, 79, 179, 216
  of elements, 20
  free radical, 16–25, 64, 67, 124–135, 169–180
    azo initiators for, 169
    chain transfer in, 130, 173–174
    in emulsion, 66, 126, 177–178
    in gas phase, 171–172
    gel effect in, 131–132, 172, 220
    initiator fragment incorporation, 169–170
    kinetics, 124–133, 171–173
    peroxide initiation, 125, 128, 130, 169
    photochemical, 17, 21, 22, 95, 125, 129
    redox initiation, 170
    in suspension, 64, 177
    termination mechanism, 127, 128, 131
  See also Acrolein; Acrylamide; Acrylic acid; Acrylic acid, esters; Acrylonitrile; Aldehydes; α-Amino acid N-carboxy anhydrides; Butadiene; Butene; Butyraldehyde; Chloroprene; Cyclopentadiene; Diacetylenes; Diazomethane; Dimethylbutadiene; Ethylene; Formaldehyde; Indene; Isobutene; Isoprene; Methacrylic acid; Methacrylic acid, esters; 4-Methylhexene-1; α-Methylstyrene; Pentene; Pinene; Propylene; Propylene, CHD=CHCH$_3$; Propylene oxide; Styrene; Vinyl acetate; Vinyl bromide; Vinyl chloride; Vinylidene bromide; Vinylidene chloride; Vinyl isobutyl ether; Vinyl methyl ether
Polymers:
  association in solution, 74, 80, 95–97
  ceiling temperature, 180
  chemical heterogeneity, 228
  as colloidal aggregates, 87–93
  crystallinity, degree of, 243–244
  crystallizability, 93, 119, 189
  crystallization kinetics, 247–248
  definition of, 5, 6, 117
  dielectric dispersion, 250
  fractionation, 127, 239
  for medical use, 133, 152
  melting point, 119, 120, 191, 194, 244, 245, 247
  paracrystals, 244
  rotational isomers, 222–223
  side chain crystallization, 245
  single crystals, 79, 92, 179, 246
  spherulites, 247
  stereoregular, 79, 189–192, 217, 243
  see also Chain molecules
Poly(methacrylic acid), 236
  esters, 245
Polyoxyethylene, 79, 91, 113, 114
Polyoxymethylene, 78, 79, 91–93, 246
Polyoxypropylene, 79
Polypeptides, 44, 138, 139, 141, 158, 195–196, 199–200, 221, 231–234
Polypropylene, 189–190, 243
Polyribonucleotide synthesis, 213
Polysaccharides, 145. See also Amylopectin; Amylose; Carbohydrate, origin of term; Cellulose; Chitin; Glycogen; Levan
Polysiloxanes, 122–123
Polysoaps, 238
Polystyrene, 14, 16, 17, 91–94, 102–103, 149, 218, 227, 228, 242, 243, 249
Polyurethanes, 194–195, 247
Poly(vinyl alcohol), 125, 126, 190, 217, 242
Poly(vinyl bromide), 225
Poly(vinyl chloracetate), 125
Poly(vinyl chloride), 125, 216, 249, 251
Poly(vinylidene chloride), 216
Poly(vinylidene fluoride), 219
Poly(vinyl pyridine), 234, 238
Poly(vinyl pyrrolidone), 133
Propylene:
  CHD=CHCH$_3$, cis and trans, polymerization, 192
  oxide, polymerization, 79, 191
  polymerization, 19–21, 24, 189, 190

Proteins:
   amino acid composition, 42, 43, 198
   amino acid sequence, 147, 198–199, 213, 245
   as association colloids, 89
   "cyclol" theory, 142
   denaturation, 46, 105, 142–144
   electrophoresis, 45, 141, 199
   fibrous, 83, 197, 199
   flow birefringence, 141
   frictional coefficient, 141
   globular, conformation, 85, 142–143, 200, 202
   α-helices in, 199, 200, 231, 232
   isoelectric point, 46, 111, 112
   "living" and "dead", 42, 43
   molecular weight, 44, 45, 50, 85, 101, 112, 139, 140, 198
   osmotic pressure, 49
   renaturation, 143, 144, 200
   titration, 157, 235
   see also Albumin; Antibodies; Casein; Collagen; Enzymes; Gelatin; Pepsin; Ribonuclease; Silk fibroin; Trypsin; Urease

Ribonuclease, 200
RNA, 206, 208
Rotation:
   free, 12
   hindered, 12, 108, 151, 152, 154, 229
Rubber, Hevea, 32–34, 53, 54
   cis structure, 82
   covalently bonded chain molecule, 34, 35, 91, 95
   crystallization, 74–76, 148
   destructive distillation, 21, 22, 33
   diffusion through, 51, 217
   dynamic testing, 152, 250
   elemental composition, 33
   heating on stretching, 35, 36, 75
   molecular association, 74–76
   molecular weight, 50, 90
   ozonolysis, 33, 34
   polymer of dimethylcyclooctadiene, 34
   racking, 149
   solution viscosity, 104
   spectra, 215, 217, 218
   from starch, 34, 37
   unit cell, 75
   vulcanization, 52, 54, 55, 151
Rubber, synthetic:
   acrylonitrile-butadiene, 67
   cis-1,4-polyisoprene, 185, 191, 215
   cost, 63, 66, 67
   polybutadiene, 23
   polychloroprene, 67, 68
   poly(2,3-dimethylbutadiene), 23, 65
   polyisobutene, 68, 69, 82, 223, 241
   polyisoprene, 22, 23, 63–66, 185, 191, 215
   polysulfide, 67
   styrene-butadiene, 67, 214
Rubber elasticity, 35, 36, 74, 75, 148–152, 223, 249, 252
Rubber reinforcement, 55, 65

Second virial coefficient, 225–228
Silk, 56
   fibroin:
      association, 74
      crystal structure, 72, 73, 83, 139, 140, 245
      molecular weight, 89
Solutions:
   anisotropic, 85, 221, 233–234, 252
   non-ideality, 153–156
   see also Viscosity, of solutions
Spectroscopy:
   ESR, 219–220
   IR:
      of cellulose, 217
      dichroism, 215–216
      of diene polymers, 215
      of polyacrylonitrile during pyrolysis, 217
      of polyethylene, 188, 215, 216
      of poly(vinyl alcohol), 217, 242
      of poly(vinyl chloride), 216
      of poly(vinylidene chloride), 216
   NMR:
      of bulk polymers, 217–218, 252
      high resolution, 218–219
      indicating polymer association, 97
   UV:
      copolymer composition, 214
      DNA denaturation, 214
      polyribonucleotide complexation, 214
Starch, 5, 36–38, 47, 71, 87, 92
   amylose and amylopectin fractions, 38
   complex with iodine, 37, 49
   enzymatic degradation, 37, 38
   hydrolysis, 37
   maltose residues in, 37
   molecular weight, 38
   optical activity, 37, 38
   solution viscosity, 105
   suspension agent, 64
Stokes' law, 50
Styrene, 14, 16
   and "cinnamol", 16, 17
   polymerization, 14, 16–18, 25, 124, 126–129, 136, 182, 183, 185–186, 189

Styrene (*Continued*)
　structure, 17
　synthesis, 17
Swelling of gels, 157
Swelling pressure, 153–154

Tartaric acid, 10–11, 14
Tobacco mosaic virus, 85, 115, 233, 238
Transition state, 10
Trypsin, 144
Turbitometric titration, 240

Ultracentrifuge, 101–103, 139, 141, 197, 211, 221, 239
Unit cell:
　of cellulose, 72, 73
　of *Hevea* rubber, 72, 75, 76
　of pepsin, 85
　of silk fibroin, 72, 73
Urease, 84, 144

Vinyl acetate, polymerization, 95, 125, 128, 129, 172
Vinyl bromide, polymerization, 21
Vinyl chloride, polymerization, 21, 125, 216
Vinylidene bromide, polymerization, 21, 182
Vinylidene chloride, polymerization, 21
Vinyl isobutyl ether, polymerization, 190
Vinyl methyl ether, polymerization, 24
Viscoelasticity, 152, 249–251
Viscose, 57
Viscosity:
　intrinsic, 110, 229–231
　melt, 248, 250
　of solutions, 41, 50–52, 94, 103–110, 158–159, 229–231

Wool, 83

X-ray diffraction, 71–85, 200–202, 208–211. *See also* Crystallography